Relationship Marketing

Strategy and implementation

Helen Peck, Adrian Payne, Martin Christopher, Moira Clark

Published in association with The Chartered Institute of Marketing

OXFORD AUCKLAND BOSTON JOHANNESBURG MELBOURNE NEW DELHI

Butterworth-Heinemann
Linacre House, Jordan Hill, Oxford OX2 8DP
225 Wildwood Avenue, Woburn, MA 01801-2041
A division of Reed Educational and Professional Publishing Ltd

℞ A member of the Reed Elsevier plc group

First published 1999
Reprinted 2000

British Library Cataloguing in Publication Data
A catalogue record for this book is available from the British Library

ISBN 0 7506 3626 2

Typeset by Avocet Typeset, Brill, Aylesbury, Bucks
Printed and bound in Great Britain by Biddles Ltd
www.Biddles.co.uk

Relationship Marketing

The Chartered Institute of Marketing/Butterworth-Heinemann Marketing Series is the most comprehensive, widely used and important collection of books in marketing and sales currently available worldwide.

As the CIM's official publisher, Butterworth-Heinemann develops, produces and publishes the complete series in association with the CIM. We aim to provide definitive marketing books for students and practitioners that promote excellence in marketing education and practice.

The series titles are written by CIM senior examiners and leading marketing educators for professionals, students and those studying the CIM's Certificate, Advanced Certificate and Postgraduate Diploma courses. Now firmly established, these titles provide practical study support to CIM and other marketing students and to practitioners at all levels.

 The Chartered
Institute of Marketing

Formed in 1911, The Chartered Institute of Marketing is now the largest professional marketing management body in the world with over 60,000 members located worldwide. Its primary objectives are focused on the development of awareness and understanding of marketing throughout UK industry and commerce and in the raising of standards of professionalism in the education, training and practice of this key business discipline.

Books in the series

Contents

Foreword

This book completes a trilogy that began in 1991 with *Relationship Marketing*, which was one of the first attempts to define the newly emerging concept from which the book took its name. Following this in 1995 the current team of authors produced a companion volume of selected readings, *Relationship Marketing for Competitive Advantage*. These readings were chosen to provide a broad, multi-faceted view of the by now rapidly developing arena of Relationship Marketing.

Such has been the interest amongst teachers and students of Relationship Marketing that we felt it appropriate to augment these first two volumes with a third, this time constructed around a number of case studies and case histories.

The framework used within this book is a modified version of the 'Six Markets' model, first advanced in the original *Relationship Marketing*. This simple but practical framework suggests that there are a number of market domains as well as the final marketplace that must be addressed if long-term profitability is to be assured.

Since the development of the original Six Markets model, the authors, their colleagues and students have used this framework as the basis for assessing the extent to which true, pan-company relationship marketing strategies are applied in the world of business. It has proven to be a powerful diagnostic tool, capable of aiding the development of successful relationship marketing strategies by highlighting potential weaknesses in one or more of the market domains and the stumbling blocks that may arise as a result of such neglect.

The Six Markets model, though continually evolving, has been used by the authors and their students in over 50 real world applications. During this time the framework has demonstrated its merits through the insights it provides, establishing its worth as a practical analytical tool to explain and predict success or failure in the marketplace.

The book itself takes the reader first through an exploration of the Six Markets model and its development from inception to date. The contributions of other leading academics working towards similar frameworks are also acknowledged and discussed in Chapter 1. This introductory chapter is then followed by a further four chapters, each taking as its focus one, or in some instances two, of the six market domains. In each of these chapters, the roles and significance of the market domain(s) are explained and explored in greater detail. Key themes for the management of relationships relevant to each market domain are presented, along with in-depth discussions of established theory, recent developments and new thinking in the given field. Supporting the texts are a number of case studies or histories selected to illustrate the importance and practicalities of managing some of the relationships in question. The sixth and final chapter then draws together these market domains and the themes from the earlier chapters. Two further case studies are presented here, this time exercising the Six Markets model in its entirety, thereby presenting the reader with an holistic overview of the application of Relationship Marketing.

Acknowledgements

We would like to thank the case writers who have granted us permission to use their work in this book and indeed the practitioners who generously gave their time during the development of our own case studies. We would also like to thank those of our colleagues and our students, past and present, who have contributed to our thinking and in particular we are grateful for the patient support provided by our secretaries, Tracy Brawn and Anna Newman-Brown.

Helen Peck
Adrian Payne
Martin Christopher
Moira Clark

Biographies

Helen Peck DMS
Research Fellow, Cranfield School of Management
Helen Peck is a Research Fellow in the Marketing and Logistics Group. She joined Cranfield in 1983, from a major UK retail bank, working initially with the School's Library and Information Services and Management Development Unit, before taking up a research post within the Group. In addition to developing a research interest in collaborative approaches to supply chain management, she is currently nearing completion of a PhD in Relationship Marketing as a staff candidate. Her published work to date includes papers and journal articles, joint editor and authorship of one book and contributions to many others. She is also an award-winning writer of management case studies, whose work is used extensively on marketing and logistics programmes at Cranfield and by other teaching institutions in Europe, North America and Australasia.

Professor Adrian Payne PhD, MSc, MEd, FRMIT
Professor of Services and Relationship Marketing, Cranfield School of Management
Adrian Payne is Professor of Services and Relationship Marketing and Director of the Centre for Relationship Marketing at the Cranfield School of Management, Cranfield University. He has practical experience in marketing, market research, corporate planning and general management. His previous appointments include positions as Chief Executive for a manufacturing company and he has also held senior appointments in corporate planning and marketing. He is an authority on Customer Relationship Management and is an author of four books on this topic. His research interests are in Relationship Marketing, Customer Retention Economics, the impact of IT on CRM and Marketing Strategy and Planning in Service Businesses. Adrian is a frequent keynote speaker at public and in-company seminars and conferences around the world. He is also a

consultant and educator to many service organizations, professional service firms and manufacturing companies.

Martin G. Christopher BA, MSc, PhD, FCIM
Professor of Marketing and Logistics, Cranfield School of Management

Martin is Head of the Marketing and Logistics Group and teaches those subjects in the School of Management. He has lectured widely in Europe, North America and Australasia and has had appointments as Visiting Professor at the University of British Columbia, University of South Florida and the University of New South Wales. Professor Christopher is currently a Deputy Director of the School and Chairman of Continuing Studies in the School of Management. He has written many books and articles on marketing and logistics and is Joint Editor of the International Journal of Logistics Management. He is non-executive Director of a number of companies and is an active consultant on Marketing and Logistics. He is a Fellow of the Chartered Institute of Marketing and a member of the Council of the Institute of Logistics and Distribution Management.

Moira Clark BA(Hons), MBA, DIPMM, DIPM, MCIM, MBIM
Lecturer in Services Marketing, Cranfield School of Management

Moira Clark is a Lecturer in Marketing at Cranfield School of Management. She also serves as a consultant to a number of leading UK and European companies. Her major area of research and consulting is in Relationship Marketing, Customer Retention and International Marketing in the service sector. She has done particular work on culture and climate, its impact on retention and loyalty and the critical linkages between employee behaviour and customer retention. She has published widely on this subject and is co-author of *Relationship Marketing for Competitive Advantage, Winning and Keeping Customers*. Moira is also a judge for the prestigious Management Today/Unisys Customer Service Excellence Awards.

Prior to joining Cranfield, Moira was an international marketing consultant based in Munich, where she was involved with a wide range of industries including construction and related fields, engineering, local radio, consumer and industrial goods manufacturers and service industries. She has also worked for the toiletries subsidiary of Dunhill International and as a Marketing Co-ordinator for an international health food manufacturer. Moira is a graduate in Business Studies and Marketing and has an Executive MBA from Cranfield.

Chapter 1

Relationship marketing:

The six markets framework

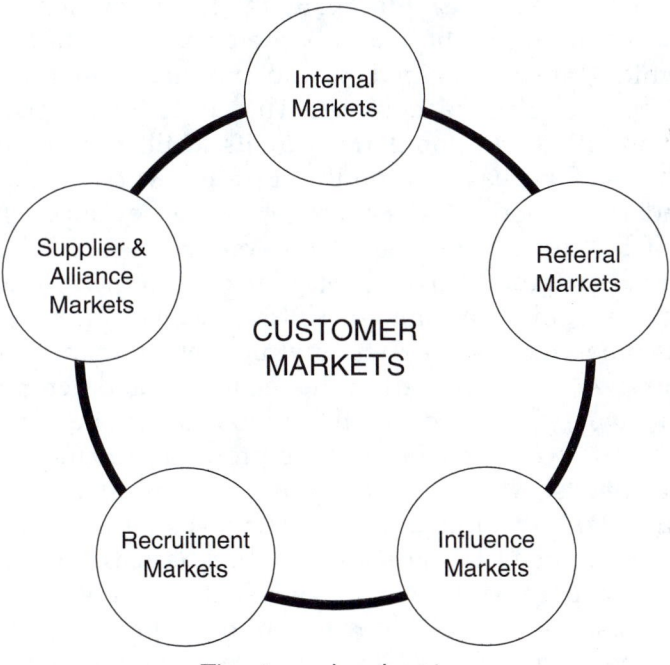

The six market domains

Introduction

'These are turbulent times in the world of organizations', observed Miles and Snow[1] more than a decade ago, and there is every indication that the bumpy ride will continue beyond the turn of the century and into the new millennium. Behind the turbulence lies a series of frequently cited environmental factors: technological advances and the deregulation of markets, creating intensified global competition. These forces have changed and continue to change the dynamics of the marketplace, raising the profile of time-based competition and causing shifts in channel power. The world is becoming a buyers' market, where increasingly discerning customers are freer than ever to select from their global marketplace – something that many corporations in the Western world were woefully slow to grasp.

As the effects of deregulation and technological change have rippled through international trade, classical models of marketing have been found to be wanting. The classical models are based on the microeconomic market model and built around the '4 Ps' framework for marketing decision making, the latter emerging from the work of Borden during the 1960s. Borden isolated 12 factors or elements which, when combined, would produce a 'marketing mix' that served to influence demand. The underlying concept was quickly simplified and popularized by its distillation into the four key elements of the teacher-friendly 4 Ps framework: product, price, place and promotion.[2, 3] These models were developed from US studies of the indigenous market for consumer goods during the post-war boom of the 1950s and 1960s, an environment where rising consumer demand gave companies little reason or incentive to consider customer relationships as anything other than brief single transactions. As such they reflect the realities of another era.

Critics have long argued that these models and the assumptions on which they were based were inappropriate for industrial and services contexts, where relationships with customers were often on-going and of pivotal importance. They were also felt to be inadequate when applied to marketing in the international arena.[4, 5, 6] Marketing management, as it was usually taught, represented neither the aspirations nor the reality of these branches of marketing. With the arrival of recession in the 1990s it became widely recognized that, even in consumer markets, this classical marketing paradigm had lost its potency.[7]

From the early 1980s an alternative approach to marketing theory and practice – Relationship Marketing – was in the ascendancy. The term itself can be traced back to the services marketing literature, though arguably it can be said to have originated in industrial marketing.[8, 9] In its earliest guises, relationship marketing focused simply on the development and cultivation of longer-term profitable and mutually beneficial relationships between an organization and a defined customer group.

The concept quickly broadened to encompass internal marketing in acknowledgement that the successful management of external relationships was largely dependent on the alignment of supporting internal relationships.[10] The proposition by writers Christopher, Payne and Ballantyne that relationship marketing represents the convergence of marketing, customer service, and the total quality movement underscores the notion of internal alignment, and stresses the cross-functional and process-dependent nature of relationship marketing.[11] Explicit in this proposition is the recognition that customer satisfaction and loyalty are built through the creation of superior value for the customer, and that value is created throughout the organization and beyond. The writers factored relationships with a range of other parties, including distributors, suppliers and public institutions, into the relationship marketing equation, bringing their broadened interpretation of relationship marketing into line with a view of marketing put forward earlier by some of the leading writers of the IMP Group and 'Nordic School'.[12] While some well-known writers in the field still seek to limit the scope of the concept to the customer–supplier dyad,[13] there is evidence that this broadened perspective is gradually gaining wider acceptance among scholars of relationship marketing.[14–17]

Several writers have pointed out that the all-pervasive philosophy of relationship marketing within the firm (and beyond) represents, in part, a revival of the marketing concept[18], which, though eclipsed by the rise and misapplication of strategic planning during the 1970s, is once more finding favour as a guiding management philosophy. The renaissance of the marketing concept is due to a growing acceptance of its potential as a strategy for dealing with market turbulence because 'at its roots, the marketing concept calls for constant change as market conditions evolve', but for most organizations change does not come easily.[19]

Competitive pressures have, however, encouraged organizations to re-examine their supply chains, reducing costs and improving quality at every stage. The search for competitive advantage through improved efficiency has led them to reconfigure their operations and consequently their organizations. New forms of organization are emerging, characterized by intraorganizational and interorganizational cooperation, as businesses reconfigure around core processes, outsourcing those activities which do not directly add value. Some organizations have gone further still, turning to partnerships and strategic alliances with customers, suppliers and competitors to further enhance and exploit their capabilities. In doing so these organizations are moving along an evolutionary continuum towards the type of network structures thought, by a growing band of authoritative writers, to be the most appropriate way to balance the rival competitive demands of greater organizational specialization and flexibility.[20-22] This combination of specialization and complexity should theoretically make them ideally suited for the creation and delivery of customer value. Furthermore, while not all networks are formed around the premise of delivering superior customer value, the most successful network designs are those that are customer-driven.[23]

The reshaping of organizations towards flatter, more responsive network forms and the rise of relationship marketing are related, not as cause and effect, but as part of the same phenomena. Both are responses to environmental turbulence and pursue a common goal – the creation of competitive advantage in a changing world. The fact that they are rarely recognized as symptomatic of a common cause means that they are not always approached in a deliberate or coordinated manner.

Whether organizational changes are overtly marketing driven or not, marketing's role within these new functionally desegregated organizations has been transformed. The long-term survival of the functionally defined marketing department remains a moot point, but the widely quoted survey by the London branch of Coopers & Lybrand concluded that 'marketing as a discipline is more vital than ever'.[24] Marketing's new remit will revolve around maximizing customer value through the boundary spanning roles of customer advocate, internal integrator, strategic director and, within network organizations, partnership broker. Against this backdrop there are calls for effective new frameworks that conceptualize the properties and scope of relationship marketing. In 1991, Christopher, Payne and Ballantyne put forward the prototype of their Six Markets model as such a framework.

The Six Markets model

The Six Markets model addresses relationship marketing at the organizational level. It presents for consideration six role-related

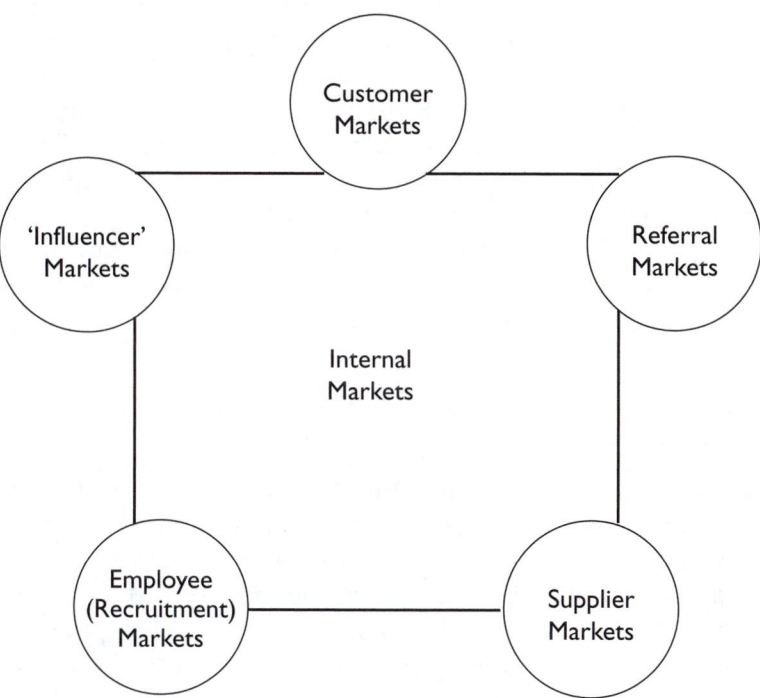

Figure 1.1   The six market domains.
Source: Christopher, Payne and Ballantyne (1991).[11]

market domains or 'markets', e,ach representing dimensions of relationship marketing and involving relationships with a number of
parties – organizations or individuals – who can potentially contribute, directly or indirectly, to an organization's marketplace effectiveness. The six market domains were initially presented as is
shown in Figure 1.1, with the focal firm, the 'internal market',
placed at the centre of the model. This configuration emphasizes
internal marketing's role as an integrator and facilitator, supporting
the management of relationships with parties within the other
'markets'.

The model has since been subtly revised on a number of occasions
as understanding of the nature of relationship marketing and the
potential contributions from the various parties deepened. Most
importantly, though, subsequent representations all place customers
rather than the focal organization at the centre of the framework, a
perspective endorsed by Cravens and Piercy in their discussion of

the cornerstones of relationship marketing.[25] Placing the customer at the centre of the Six Markets model focuses on the purpose of relationship marketing, the creation of customer value, satisfaction and loyalty, leading to improved profitability in the longer term. A brief examination of each market domain or 'market' follows. The theoretical basis for the inclusion of each domain is presented using established bodies of literature, together with a selection of empirical and descriptive examples to show how relationships with parties in each can contribute to or, if badly managed, impede overall marketplace performance and competitiveness.

Customer markets

The work of management consultants Bain & Co, directly linking customer retention to profitability in a number of mainly service situations, has done much to promote the benefits of customer retention through relationship building to the business community as a whole.[26, 27] A study of marketing in key British enterprises, commissioned by the Chartered Institute of Marketing in 1994, confirms that in the views of experienced practitioners 'relationship building is rapidly becoming the most powerful weapon in the professional marketer's armoury', and that 'this relationship building and maintenance is taking place at each level of the organisation'.[28] The issues of customer retention and relationship building will be explored in greater detail in Chapter 2 of this book.

Whether a customer is the end user of a product or service does of course depend on the position a supplier occupies in a particular value delivery sequence. Many organizations market both to trade customers (intermediaries, distributors or retailers) and consumers (end purchasers, users and consumers), but their relative power within the value system is likely to determine which relationships are cultivated most assiduously. For the manufacturers of consumer goods, the rising power of retailers has focused their attention on these relationships. Meanwhile retailers and distributors are pouring considerable effort into managing direct relationships with increasingly fickle consumers.

A point which must not be overlooked, however, is that relationship marketing is not a universal panacea. There are situations, often involving low-involvement or commodity products, when a swift and simple transaction approach is most appropriate and most valued by the customer. For businesses offering professional or financial services, regularly replaced consumer durables such as cars, and for many organizations involved in business to business marketing, the long-term investment in building relationships with individual customers is easily justified. Similarly, for manufacturers of some low-priced consumer products with high frequency purchase rates and easily identifiable target groups (e.g. some baby products), the approach can readily prove its worth. Whether the same can be proven to be an economic proposition for the marketing of mainstream fmcg (fast-moving consumer goods) products is more questionable. Nevertheless, Swiss food giant Nestlé was willing to spend millions of pounds on its Casa Buitoni Club, a database-marketing driven initiative devised to bypass the influence of the large retailers in the hope of raising customer loyalty towards what is essentially a commodity product (see Case 2.1).

Referral markets

Referrals can be a decisive element in the creation of relationships between an organization and its customers. The professional services sector has always used informal networks and reciprocal referrals to direct business towards established contacts. Word-of-mouth recommendations are certainly known to be an important part of the information search undertaken by consumers before buying high value or high risk services.[29] Recommendations may also be used by consumers as a convenient way of reducing choice between many seemingly similar products or services.[30] Similarly, in situations where the product or service may be complex or difficult to evaluate, customers will seek the advice of trusted third parties to reduce the perceived risk associated with the purchase.

Given that satisfied customers will happily endorse the products or services of the supplier if prompted, relationships with existing customers are an unrecognized or underutilized facility for many organizations. Noticeable exceptions are insurance broker Direct Line and First Direct, the world's first completely branchless cashless bank. The former used them as the theme for a high profile and highly successful advertising campaign. The latter makes much of its above average customer satisfaction ratings by publicizing the fact that around 30 per cent of would-be new customers approach the bank following a personal recommendation from an existing customer.

In manufacturing contexts too, companies sometimes create formal and informal referral agreements between themselves and suppliers of complementary products. Furthermore, in these markets, closer relationships with referral sources can provide early access to specifications and a better understanding of non-product related buying criteria. In some instances though, there may be no direct benefit to the organization making the referral, other than maintaining the goodwill of customers whose requirements fall beyond its scope or capabilities. The fact that an organization is willing and able to provide referrals can itself be seen by the customer as a benefit of the relationship.

Internal markets

In 1987 Judd observed that substantive attempts to conceptualize the employee of an organization as an element of the organization's marketing strategy were noticeably absent from the academic marketing literature.[31] There is now an abundance of 'internal marketing' literature addressing the matter, providing insights into how and why employees in all parts of an organization can contribute towards marketing effectiveness.[12, 32] In addition, Schlesinger and Heskett build on earlier research, emanating initially from organizational behaviour specialists, which clearly linked the constructs of employee satisfaction and retention to customer satisfaction and retention in service businesses.[33–35] Evidence that the links between employee retention (particularly front-line employee retention) and customer or business retention also exist in product-centred, business to business marketing situations can be found in an internal study into the cost of employee defection undertaken by the Digital Equipment Corporation in the UK (see Case 5.3).

Recruitment markets

The move away from traditional employment practices towards contract working, outsourcing and partnering allows organizations to access a wider range of specialist skills on a temporary basis. Nevertheless, there are certain categories of employees whose skills and experience create and sustain the organization's core competencies. Christopher, Payne and Ballantyne's recruitment market represents those potential employees who possess the attributes needed to sustain and enhance these core competencies. It also refers to third parties – colleges, universities, recruitment agencies or other employers – who have early access to pools of these potential employees. The logic is that if a would-be employer wants to attract the best people, it must present itself to influential third parties and to the individuals themselves as the employer of first choice. But if it also wants to *keep* these valuable employees, it must *be* the employer of first choice.

Towards the end of the 1980s there was widespread concern that a demographic shortage of young high calibre workers would reduce the competitive capabilities of businesses throughout Northwest Europe.[36] Skilled and talented employees, it was predicted, would become a scarce and valuable resource. These fears quickly faded with changes to the political landscape and widespread redundancies in the wake of prolonged recession, re-emerging with a vengeance in the UK in the late 1990s as economic recovery gained momentum.

Influence markets

Whereas Webster and Wind list influencers and gatekeepers among the members of a buying unit within the firm,[37] Christopher, Payne and Ballantyne look beyond the confines of customers' internal buying units and into the wider business environment. They apply the term 'influencer' to a range of third parties who exercise influence over the organization and its potential customers. These influencers may be governments and their agencies, press and other media, professional bodies, investors and pressure groups. In fact 'influence markets' will likely include all of the constituencies that have traditionally fallen within the domain of public relations and corporate affairs. While relationships with these parties may not directly add value to a product or service, they can directly influence the likelihood of purchase or prevent an offer from even reaching the market.

If carefully and proactively managed, these relationships can not only open doors to markets, but they can enhance or even replace some other marketing activities. The skilful management of media relationships can, in some instances, be cheaper and more effective than formal advertising, as the founders of The Body Shop and The

Virgin Group have shown. While well-managed relationships with other influencers might not be so overtly beneficial, they can be used to influence public opinion and legislators in the organization's favour. They can also mitigate the effects of potentially disastrous operational mishaps. There can, however, be few better illustrations of the consequences of mishandling these relationships than the example of Fisons Plc, a financially oriented, British-based pharmaceutical company, with horticultural and other scientific interests. The company's cavalier attitude towards key influence groups resulted, in the UK, in consumer and retailer boycotts of its horticultural products. Meanwhile, in North America, a public row with the US regulators left its most important pharmaceutical products locked out of its largest and most lucrative markets, leading eventually to the company's near collapse (see Case 4.2).

Supplier markets

Relationships with suppliers have been the focus of a great deal of interest in articles on relationship marketing in recent years, as the point where the marketing and network literature most clearly merge.[38, 39] During the 1980s, changes began to occur in purchasing behaviour of some large manufacturing companies – noticeably those threatened most seriously by Far Eastern competitors. The traditional adversarial approach to procurement that played multiple suppliers off against each other began to take on a more cooperative nature. This followed the gradual realization that, when suppliers were squeezed to the point of collapse, they were unwilling and unable to invest in the new plant and technologies required to allow them to deliver better products and services, faster and more cheaply. Instead these manufacturers were choosing to build less exploitative relationships with fewer suppliers. In doing so they are creating integrated and relatively stable supply chains, which allow quality and flexibility to be engineered into the systems while costs are reduced. Often this will involve shared infrastructural investments and the merging of some business systems. The result is improved competitiveness through the creation and delivery of a better value proposition for the end customer. Nowhere is this more obvious than in the motor industry, where collaborative Japanese management practices have been widely adopted. An early example in the UK was the Rover Group, a dismally unprofitable car manufacturer whose fortunes were revived with the help of an alliance with Honda, a Japanese competitor. Rover embraced the notion of an 'extended enterprise' and, despite its premature 'divorce' from Honda following its sale to BMW, Rover continues to work with its suppliers and dealers towards the development of a seamlessly integrated supply chain.

Further perspectives

The Six Markets model provides the basis for a simple framework to convey the complex reality of relationship marketing. While it does not attempt a detailed identification of individual relationship forms or partners (these are time and situation specific), it has the potential to provide a strategic overview of relationship marketing; its scope, nature and purpose. However, the business landscape has changed a great deal since Christopher, Payne and Ballantyne first attempted to produce a conceptual framework for relationship marketing, and early indications from on-going empirical research suggest that certain aspects of the model would benefit from further consideration.[40]

Firstly, throughout the 1990s there has been an upsurge in management interest in all aspects of process integration. This interest extends beyond the notion of the internal value chain, to entire value systems reaching upstream to suppliers and downstream through intermediaries, distributors and retailers to the end user or ultimate consumer. The importance of supplier relationships and their role in relationship marketing was already well established within the Six Markets model; more problematic is its treatment of the downstream market relationships. It does not overtly distinguish between intermediaries or distributors as customers and end users or consumers. For the many organizations involved in business to business marketing, this is an important distinction.

Secondly, there is the matter of scope. The original Six Markets model subsumed alliances and partnerships within the 'supplier' market domain and did not make explicit enough some of the more sophisticated horizontal forms of interorganizational collaborations and strategic alliances, including joint development projects between competitors or equity sharing joint ventures. From his globalist vantage point, management consultant and writer Kenichi

Ohmae has long argued that for those organizations which find themselves in a global marketplace, such strategic alliances are not an option – but a necessity.[41] Fewer and fewer markets are immune to the effects of global competition, so while it might once have been possible to believe that most businesses would be unaffected by such things, it would be unwise to do so today. Moreover, if the prophecies relating to the network or virtual corporation are more widely realized, these relationships will certainly demand greater consideration than they have so far been afforded by the Six Markets model.

In the years since the first Six Markets model was published, the ideas of some other eminent academics have moved in a similar direction to Christopher, Payne and Ballantyne, adding weight to the ascertainment that relationship marketing involves relationships with many other parties beyond the organization and its customers. Taking the work in chronological order of publication we will examine some of these more recent conceptualizations, which, when added to our own empirical observations, illuminate further this multiple stakeholder perspective of relationship marketing.

In 1992, Kotler put forward his 'Total Marketing' framework as 'a structural view of marketing performance and success', where the traditional marketing mix is not replaced, but is 'repositioned as the toolbox for understanding and responding to all the significant players in the company's environment'.[14] For the purposes of this discussion, 'Total Marketing' is taken to be relationship marketing in all but name. In the executive briefing document under that title, Kotler draws heavily on the literature of the Nordic School and IMP Group as he crisply and concisely outlines many of the same fundamental concepts identified by other writers as the foundations of relationship marketing, as his opening paragraph illustrates:

> The consensus in American business is growing: if U.S. companies are to compete successfully in domestic and global markets, they must engineer stronger bonds with their stakeholders, including customers, distributors, suppliers, employees, unions, governments, and other critical players in their environment. Common practices such as whipsawing suppliers for better prices, dictating terms to distributors, and treating employees as a cost rather than an asset, must end. Companies must move from a short-term *transaction-orientated goal* to a long-term *relationship-building goal.*

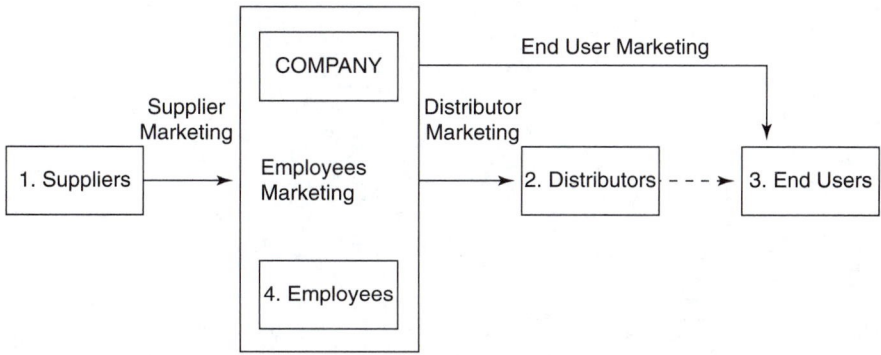

Figure 1.2 Four players in the company's immediate task environment.
Source: Kotler (1992).[14]

Kotler goes on to state that there are at least ten critical parties or
'players' in a company's marketing environment. He identifies four
of these parties – suppliers, distributors, end users and employees –
within the immediate environment of the firm. A further six
'players' – financial firms, governments, media, allies, competitors
and the general public – he places within the wider macroenvironment (see Figures 1.2 and 1.3).

Next, Morgan and Hunt presented relationship marketing as
encompassing ten discrete forms of relational exchanges, involving

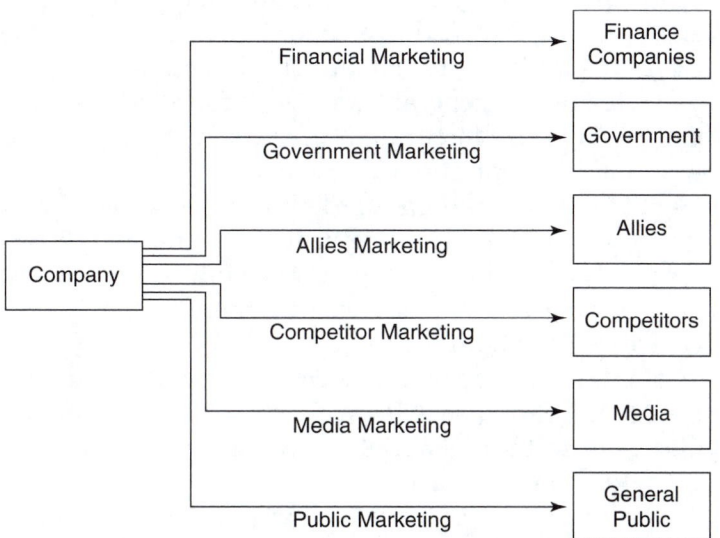

Figure 1.3 Six other key players in the company's macroenvironment.
Source: Kotler (1992).[14]

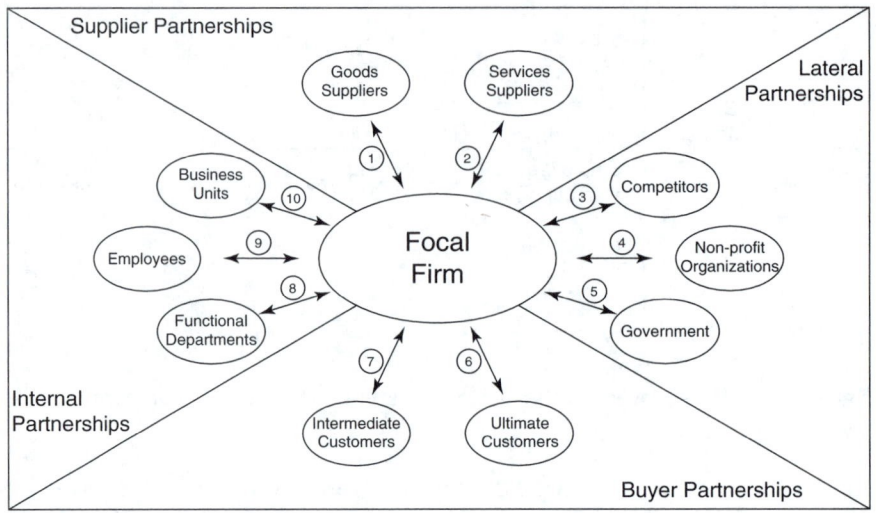

Figure 1.4 The relational exchanges in relationship marketing.
Source: Morgan and Hunt (1994).[15]

interactive relationships for various purposes, between the focal firm and an array of other parties (see Figure 1.4).[15] The parties are: goods and services suppliers; the firm's own business units, employees and functional departments; immediate and ultimate customers; and competitors, governments and non-profit organizations. Morgan and Hunt proceed to categorize these into four broad partnership types: supplier, internal, buyer and lateral.

Similarly, Doyle offers a general framework for relationship marketing 'which permits the integration of the key concepts of core capabilities, strategic intent and value creation'.[16] His framework deconstructs relationship marketing, identifying a series of dyadic relationships between the firm's central core and ten types of 'network partners'. He follows Morgan and Hunt's general taxonomy, explicitly adding strategic alliance partners to the lateral, or 'external', category (see Figure 1.5).

There are obvious similarities between these new frameworks and those postulated earlier by Christopher, Payne and Ballantyne, and by Kotler, both in terms of content and general structure, but they differ in detail and approach.

The texts supporting each of the frameworks appeared to concentrate on the identification and classification of relationships with either specific parties or compound groups, providing usually very

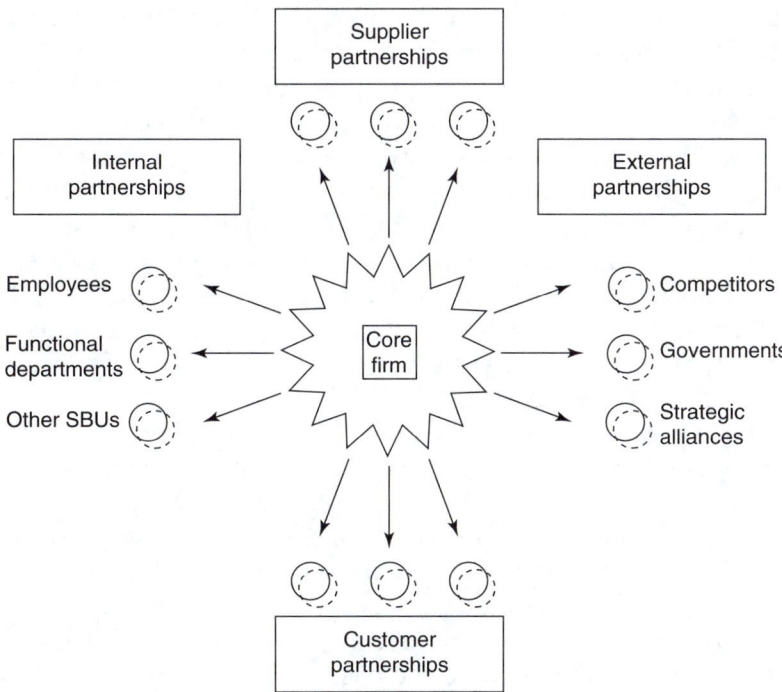

Figure 1.5 Core firm and its partnerships.
Source: Doyle (1995).[16]

specific means–end justifications for their inclusion. These parties or groups are then categorized (usually) according to the nature and/or form of the parties' relationships with the focal firm (though it should be said that most of the frameworks appeared to display some internal inconsistencies). In the majority of instances, these categories provided the superstructure for the frameworks.

The common denominators between the superstructural categories were as follows. All agree that the 'internal market' is an essential element of relationship marketing, with each framework incorporating a category for employee or internal relationships within the focal firm. In all instances they also acknowledge that the concept encompasses vertically connected relationships with supply chain parties, upstream to suppliers and downstream to customers, buyers, end users or consumers, but consensus breaks down beyond this point.

The treatment of the 'customer/buyer/end user' category differs considerably between writers. Kotler affords distributors totally separate treatment from end users, advocating very different marketing approaches for the two (trade marketing for the former and the more traditional consumer marketing approach for the latter). These two distinct 'customer' groups, along with suppliers and employees, are the four types of players he places within the company's immediate 'task environment' (i.e those most directly involved in the supply chain). Morgan and Hunt and Doyle also each subdivide their customer/buyer categories to distinguish between intermediaries or distributors as customers or buyers and ultimate consumers. Christopher, Payne and Ballantyne do discuss the subdivision of their customer category, but choose to introduce a time dimension, differentiating between new customers or prospects and existing customers.

Moving beyond the supply chain, Morgan and Hunt and Doyle use the terms 'lateral' and 'external', respectively, for their fourth and final category. Both offer 'competitors' and 'government', together with either 'non-profit organizations' or 'strategic alliances', as subcategories; qualifying their inclusion with a number of specific examples of cooperative alliances between the focal firm and one of these subcategory parties. In each instance resource sharing appears to be the basis for the relationships cited, a point which is explicitly acknowledged by Morgan and Hunt:

> Strictly speaking, in strategic alliances between competitors, partnerships between firms and government in public-purpose partnerships, and internal marketing, there are neither 'buyers' nor 'sellers', 'customers' nor 'key accounts' – only *partners* exchanging resources.

Christopher, Payne and Ballantyne also include 'government' within their framework, but categorize it according to its influential role as macromarket 'gatekeeper', placing it alongside other regulators and facilitators – such as members of the financial community, media and pressure groups within their 'Influence Market'. Interestingly, Kotler takes most of the parties identified by other writers' influence/external/lateral parties, including them for reasons of either resource provision, market access or competitive co-existence, as players within the company's macroenvironment, each warranting their own marketing considerations.

More recently, Gummesson proffered his own framework of relationship marketing from on-going research into the scope and generic properties of the phenomenon.[42] Guided by his own definition of relationship marketing as 'marketing seen as relationships, networks and interaction' he produced a more complicated classification system featuring 30 relationships, judged by the author to be of 'practical and theoretical relevance' (see Figure 1.6). Within the framework, Gummesson classifies these as either 'market relationships' or 'non-market relationships'. The market relationships are relationships between suppliers, customers, competitors and intermediaries. These are then subdivided into 'classic' market relationships – the relationships and interactions that have long been the preoccupation of traditional mainstream of marketing management – and 'special' market relationships. Special market relationships focus on certain aspects of the classic relationships, such as the interfaces between parties within the supply chain, the means through which they interact, and the status and condition of the relationships. Also listed here are examples of parties whose interactions have a direct impact on the marketplace, but whose objectives and activities (either altruistic or criminal) fall outside the usual commercial frameworks.

In addition, Gummesson identifies two categories of 'non-market relationships' which indirectly influence the efficiency of the market relationships. The first are 'mega' relationships – relationships which exist above or independent of the immediate marketplace. These appear to explore variations on the alliance or network themes (organizations as networks, markets as networks and non-market networks), including the influence of such diverse entities as supranational trading alliances, social networks and the media. The second non-market subcategory is 'nano' market relationships, involving relationships either within the focal firm or between the focal firm's functional marketing department and outside agents. Here the author raises (implicitly) central issues relating to the management of marketing as a function and a department within the firm.

Gummesson himself acknowledges that the complexity of his framework stems from his desire for completeness, but concedes that by rearranging the classifications and changing the emphasis of the earlier frameworks, they would reveal similar results.

The work by Bain & Co on the economics of customer retention and its role in establishing the credibility of relationship marketing has already been mentioned in this chapter (and will be referred to on other occasions throughout the book). However, the on-going research by the consultancy, particularly the body of work by Reichheld, one of its directors, deserves further consideration at this

CLASSIC MARKET RELATIONSHIPS

R1 The classic dyad: The relationship between the supplier and the customer

This is the parent relationship of marketing, the ultimate exchange of value which constitutes the basis of business.

R2 The classic triad: The drama of the customer–supplier–competitor triangle

Competition is a central ingredient of the market economy. In the competition there are relationships between three parties: between the customer and the current supplier, between the customer and the supplier's competitors, and between competitors.

R3 The classic multidimensional network: Physical distribution

The physical distribution consists of a network of relationships which is sometimes totally decisive for marketing success.

SPECIAL MARKET RELATIONSHIPS

R4 Relationships via full-time marketers (FTMs) and part-time marketers (PTMs)

Those who work in marketing and sales departments – the FTMs – are professional relationship-makers. All others, who perform other main functions but yet influence customer relationships directly or indirectly, are PTMs. There are also contributing FTMs and PTMs outside the organization.

R5 The service encounter: Interaction between the customer and front-line personnel

Production and delivery of services involve the customer in an interactive relationship with the service provider's personnel, often referred to as the moment of truth.

R6 The many-headed customer and the many-headed supplier

Marketing to other organizations – industrial marketing or business marketing – often means contacts between many individuals from the supplier's and the customer's organization.

R7 The relationship to the customer's customer

A condition for success is often the understanding of the customer's customer, and what suppliers can do to help their customers become successful.

R8 The mental and physical proximity to customers vs. the distant relationship

In mass marketing, the closeness to the customer is lost and the relationship becomes distant, based on surveys, statistics and written reports.

R9 The relationship to the dissatisfied customer

The dissatisfied customer perceives a special type of relationship, more intense than the normal situation, and often badly managed by the provider. The way of handling a complaint – the recovery – can determine the quality of the future relationship.

R10 The monopoly relationship: The customer or supplier as prisoners

When competition is inhibited, the customer may be at the mercy of the provider – or the other way around. One of them becomes a prisoner.

R11 The customer as 'member'

In order to create a long-term sustaining relationship, it has become increasingly frequent to enlist customers as members of various marketing programmes.

R12 IT: The electronic relationship

An important volume of marketing today takes place through networks based on IT. This volume is expected to grow in significance.

R13 Parasocial relationships: Relationships to symbols and objects

Relationships do not only exist to people and physical phenomena, but also to mental images and symbols such as brand names and corporate identities.

R14 The non-commercial relationship

This is a relationship between the public sector and citizens/customers, but it also includes voluntary organizations and other activities outside of the profit-based or monetarized economy, such as those performed in families.

R15 The green relationship

The environmental and health issues have slowly but gradually increased in importance and are creating a new type of customer relationship through legislation, the voice of opinion leading consumers, changing behaviour of consumers and an extension of the customer–supplier relationship to encompass a recycling process.

Figure 1.6 The 30 relationships of RM, the 30Rs.

Source: Gummesson (1996).[42]

R16 The law-based relationship
A relationship to a customer is sometimes founded primarily on legal contracts and the threat of litigation.

R17 The criminal network
Organized crime is built on tight and often impermeable networks guided by an illegal business mission. They exist around the world and are apparently growing but are not observed in marketing theory. These networks can disturb the functioning of a whole market or industry.

MEGA RELATIONSHIPS

R18 Personal and social networks
The personal and social networks often determine the business networks. In some cultures even, business is solely conducted between friends and friends-of-friends.

R19 Megamarketing: The real 'customer' is not always found in the marketplace
In certain instances, relationships must be sought with a 'non-market network' above the market proper – governments, legislators, influential individuals – in order to make marketing feasible on an operational level.

R20 Alliances change the market mechanisms
Alliances mean closer relationships and collaboration between companies. Thus competition is partly curbed, but collaboration is necessary to make the market economy work.

R21 The knowledge relationship
Knowledge can be the most strategic and critical resource and 'knowledge acquisition' is often the rationale for alliances.

R22 Mega-alliances
EU (The European Union) and NAFTA (The North America Free Trade Agreement) are examples of alliances above the single company and industry. They exist on government and supranational levels.

R23 The mass media relationship
The media can be supportive or damaging to the marketing. The way of handling the media relationships is often crucial for success or failure.

NANO RELATIONSHIPS

R24 Market mechanisms are brought inside the company
By introducing profit centres in an organization, a market inside the company is created and internal as well as external relationships of a new kind emerge.

R25 Interfunctional and interhierarchical dependency: The relationship between internal customers and internal suppliers
The dependency between the different tiers and departments in a company is seen as a process consisting of relationships between internal customers and internal providers.

R26 Quality providing a relationship between operations management and marketing
The modern quality concept has built a bridge between design, manufacturing and other technology-based activities and marketing. It considers the company's internal relationships as well as its relationships to the customers.

R27 Internal marketing: Relationships with the 'employee market'
Internal marketing can be seen as part of RM as it gives indirect and necessary support to the relationships with external customers.

R28 The two-dimensional matrix relationship
Organizational matrices are frequent in large corporations, above all in the relationships between product management and sales.

R29 The relationship to external providers of marketing services
External providers reinforce the marketing function by supplying a series of services, such as those offered by advertising agencies and market research institutes, but also in the area of sales and distribution.

R30 The owner and financier relationship
Owners and other financiers can sometimes determine the conditions under which marketing works. The relationship to them may influence the marketing strategy.

Figure 1.6 (*continued*)

point. The central theme of much of the work to date has been around aspects of relationship marketing, specifically customer retention and loyalty.

Reichheld's best-known work was, initially, explicitly service industry based and focused on the economics of customer retention.[26] Subsequent writings by Reichheld have taken customer loyalty as the central theme, advocating a strategic systems approach to its creation. Early iterations described a strategic business system based on loyal employees delivering superior value to increasingly loyal customers.[27] Loyal investors were soon added as a third necessary component of this symbiotic system, bringing Reichheld's thinking into line (in some respects) with ideas expressed by Bill Marriott Jnr, chairman of Marriott Hotels, who believes that customers, employees and stockholders are the three groups that the business system should aim to satisfy. The chairman of the Marriott chain is just one of several service industry specialists, working in high customer contact service industries – such as hotels, restaurants, banks and airlines – who believe that employee satisfaction should be ranked first among these, because employee satisfaction drives satisfaction for the other two.[43] Reichheld, who increasingly applies his ideas to consumer goods marketing (loyalty as repeat purchases) as well as loyalty in on-going service situations, takes the more conventional line. He applies a hierarchy of loyalty in which loyalty to the customer, through the goal of creating customer value, takes ultimate precedence, followed by loyalty to employees and investors.[44]

Importantly, the scope of Reichheld's loyalty system was broadened further in his book *The Loyalty Effect: The Hidden Force Behind Growth, Profits, and Lasting Value*.[44] Here he does acknowledge that other obvious stakeholders – such as local communities, distributors and vendors (goods and services suppliers) – are absent from the model, but maintains that this is a deliberate omission and a justifiable stance, even in the face of emerging network organizations. Reichheld concedes that other stakeholders do have a part to play in the dynamics of a loyalty system, but draws a nominal line beneath the trinity, stressing that the focus of the organization should remain on the essential three because they are 'the forces of loyalty' which create stability within the system (see Figure 1.7).

Applying the analogy of business systems as a series of linked molecules, Reichheld suggests that the trinity, of customer, employee and investor, makes up the essential 'subatomic particles'.

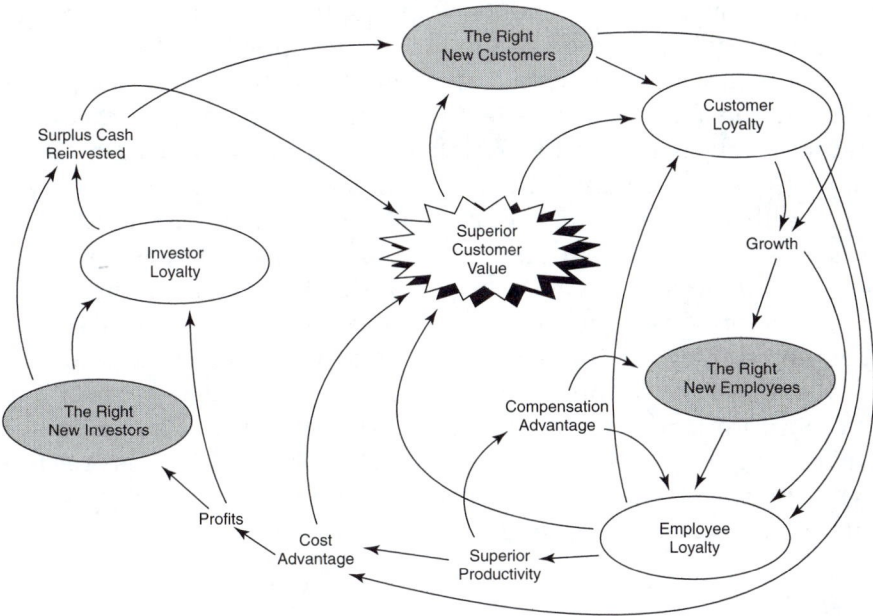

Figure 1.7 The loyalty-based cycle of growth.
Source: Reichheld (1996).[45]

Other parties involved in the value system might also be considered as linked atomic clusters of customers, employees and investors, with each element looking downstream at maximizing value for its own immediate customers. Relationships with vendors (including strategic alliances with vendors), and all other critical upstream linkages in a value delivery system, can be viewed as customer relationships, because each vendor has the responsibility to create value for their own customers. This implies that customers can behave in a passive manner and that the reduction of system costs is primarily the vendor's responsibility. Such a stance may be justified in situations where most of the end customer value is created within the organization, close to the customer interface, and where the organization is operating in a stable and predictable manner (e.g. such as a fast-food restaurant or in some financial service situations). In such circumstances a company may do very well by focusing on service-related core competencies. Elsewhere, this narrow focus could spell corporate suicide. Value creation is at the nub of the matter here, value as perceived by the customer.

In sectors where much of the value for the customer is created further away from the end customer interface, perhaps through bringing together unique technologies developed by the suppliers of subassemblies, or in instances where downstream intermediaries manage the customer interface, managers would be unwise to assume such a limited focus. Likewise in business environments fraught with uncertainty (e.g. high technology industries), relationships with suppliers, alliances and network partnerships are critical concerns. Without them an organization may be unable to create and sustain an attractive and appropriate value proposition for the customer – no value, no future. In such circumstances, the customer retention framework, with its emphasis on front-line execution, may only deliver tiny improvements to the overall value proposition. These improvements would become an irrelevance if a competitor makes a significant advance in a core value creating process or activity, or, for that matter, if the business environment or industry structure were to change suddenly.[46, 47]

The Six Markets model revised

In the light of the changes that have occurred in the business environment in recent years, a revised version of the original Six Markets model is used as the structural backbone for this book (see Figure 1.8). The market categorizations outlined in the original model did not make explicit provision for collaborative relationships between the core firm and strategic alliance, joint venture or network partners, other than those with its immediate suppliers. We have therefore added an 'Alliance' category to the framework, alongside the Supplier category, for all of those suppliers whose relationships with the focal firm continue to be conducted in a more traditional manner. Furthermore, we address in a more explicit manner the division of each market domain into different groupings or segments. For example, the 'Customer Market' domain is divided to acknowledge explicitly buyers, consumers, and channel intermediaries. In a consumer goods or services scenario, this market domain represents end customers, users and consumers. In a business-to-business marketing situation, it also embraces channel intermediaries – including all those parties, agents, retailers and distributors who are customers of the organization, but operate between them and the ultimate end users. In both

Figure 1.8 The Six Markets model.

instances it includes all members of the customer's buying unit.

More questions of priority and structure

One more point should be made, however, and this is that further empirical research may in time replace, collapse or extend these market categories again as our understanding of the field continues to crystallize. Nevertheless, there are certain pointers in the literature which shed further light on how a conceptual framework of relationship might develop from here.

Millman suggests that the Stakeholder model, from the strategic management and corporate governance literature, is worth borrowing from to provide a partial understanding of relationship marketing.[48] Indeed there are certain pointers in this literature which shed further light on the scope and structure of a conceptual framework of relationship marketing's multiple markets. The similarities of scope between relationship marketing and the Stakeholder model have also been acknowledged in recent years by many other writers including Berry, who wrote of 'using the strongest possible strategies for customer bonding, marketing to employees and other stakeholders, and building trust as a marketing tool'.[17]

The Stakeholder model itself was developed in the 1960s in response to rising influence of non-shareholder groups on corporate life and policies, but following the work of Dill it found new favour as a model for managing in turbulent times.[49] While the Stakeholder and Six Markets models differ in emphasis and purpose, the stakeholder literature can provide useful insights into the dynamics of relationship marketing. For example, Freeman and Reed note that some employees may also be shareholders, customers and influencers, to make the point that some parties or individuals will have multiple stakes or interests in an organization.[50] Webster draws a similar conclusion, this time using the example of the partner who is also simultaneously customer, competitor and vendor.[22] In short, these are multidimensional relationships.

This has implications for their management in that the management of one dimension can influence several other dimensions of the relationship and that the consequences of the management of one can manifest itself in the working of others. Expressed in the terms of the Six Markets model, this means that there are overlaps between the 'markets'.[51] This notion is supported by Gummesson following his own attempts to analyse the first three conceptualizations of relationship marketing presented in this chapter, and indeed from his own on-going efforts in this direction.[42] He concluded that:

> However desirable it would be for the sake of orderliness and simplicity, there is no single dimension along which a relationship can be organised. The relationships partly overlap. This is no surprise as a phenomenon in business — as well as in all social sciences — lacks clearly delimited definitions. The reason for this 'shortcoming' ... is that the studied phenomena are not themselves clearly delimited.

For example, Peck has since put forward a further refinement of the Six Markets model, based on on-going empirical work undertaken at Cranfield and insights from conceptual frameworks developed by other academics.[40] Earlier in this chapter we mentioned that relationship marketing's antecedents could be found in both services and industrial/international marketing literature and practice, rising simultaneously from these two fields. Peck contends that the original version of the model provided an adequate foil for the exploration of relationship marketing in service situations, where service quality was the principal means of differentiation but that the industrial and international marketing dimensions of relationship marketing were less well served. This manifestation of the model attempts to redress the balance between the two.

The old 'Customer Market' has been replaced with two separate categories – Consumer and Intermediary. The separation of the category allows students and managers to give greater consideration to the needs of these quite different customer groups. It retains the original Supplier category, but also affords the Alliance category separate treatment. The Influencer category has also been retained within this iteration of the framework. It continues to represent the public at large, and all those parties – including media, analysts and other members of the financial community, pressure groups, trade unions, industry associations, governmental bodies, legislators and regulators – who are not involved in the creation of customer value, but seek to influence or control the dynamics of the organization's macroenvironment.

Two of the other original categories, 'Referral' and 'Recruitment', have disappeared from the framework. This is not to say that the activities and parties they represent are not important – they are, and Christopher, Payne and Ballantyne make a valid case for their consideration. Reichheld also makes frequent reference to the significance of word-of-mouth referrals from satisfied customers throughout his writings on loyalty management. Referrals provided by customers and other third parties are significant promotional opportunities, but referrals are in fact *benefits* arising from successfully managed relationships with these parties, who are acting (formally or informally) as Intermediaries or Influencers.

Likewise, recruitment is an important activity, especially when it concerns the recruitment of employees whose skills and experience are most pivotal to the creation and delivery of customer value. The use of tactical marketing activities to attract the best possible people into the organization, to create a stronger and more coherent core, is a worthwhile proposition. Peck has argued that to afford the recruitment market such prominent treatment may, however, overstate its significance within the framework, possibly at the expense of other factors. The parties to whom the marketing activities would be directed – presumably schools, universities and recruitment agencies – can all be judged to be acting again as Influencers or Intermediaries. As for the would-be employees themselves, only time separates relationships with future employees from those with existing ones, who fall within the 'Internal Market', which also stands unchanged from previous iterations, to include current employees, along with other divisions and SBUs of the focal firm. Note, though, that, as Figure 1.9 indicates, it is the Consumers who remain central to the framework, as the ultimate end users of the value created and delivered through, or by virtue of, the relationship with stakeholders in all of the other 'markets'.

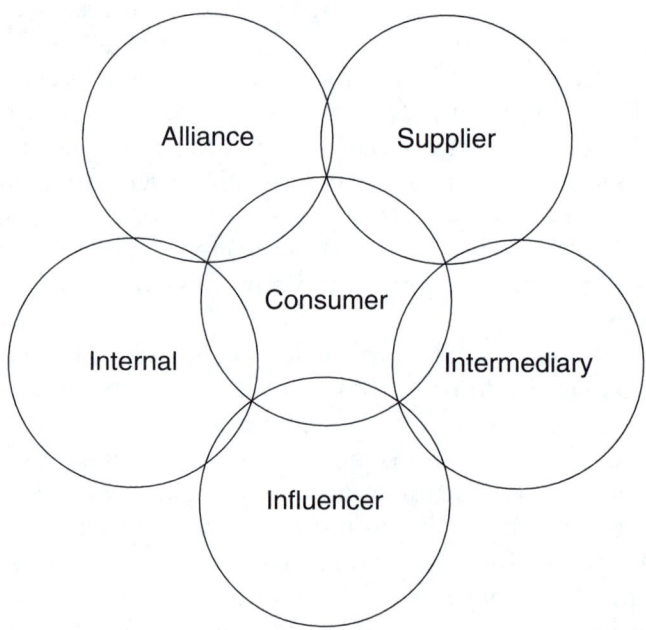

Figure 1.9   Later iteration of the Six Markets model.
Source: Peck (1997).[40]

Experience in using the Six Markets model

Since the development of the Six Markets model our work has emphasized the need to break all of the markets, or what are now called 'market domains', into relevant constituent parts. The discussion in the first publication to incorporate the Six Markets model[11] also described the downstream relationships in the marketing channel, including final consumers and intermediaries or direct customer (p. 15); it also emphasized the need to undertake segmentation based on both traditional criteria as well as more relational-based segmentation criteria such as service requirements. With the customer market, particular emphasis should be placed on understanding two major tasks – attracting new customers and retaining existing customers. These two broad pivotal activities are important regardless of what level of the customer chain is considered.

To operationalize the Six Markets model, a process was developed to address each market domain in greater detail.[52] This consisted of:

1 Identify key participants, or segments, within each of the markets.
2 Research to identify expectations and needs of key participants.
3 Review current and proposed level of emphasis in each market.
4 Formulate desired relationship strategy and determine if a formal marketing plan is necessary.

This process starts with the examination and analysis of each market domain to identify the key groups of participants and market segments. A consideration of the expectations and needs of each of these groups of segments is then undertaken.

To assess the present level of emphasis and desired level of emphasis on each of the market domains, the relationship marketing network diagram, also known as the spidergram, was developed. This configures each of the major market domains, including customer markets (which are subdivided into existing and new customers), on a series of axes and enables a group of managers within a firm to make an assessment as to the current and desired levels of emphasis on each market domain by means of a jury of executive opinion.

This represents the first level of diagnostic review of the overall process in order to make an initial judgement as to the existing and desired relevant emphasis. Once this has been completed, further network diagrams or spidergrams can then be developed for each market domain. The second level of analysis explores each market domain in much greater detail and enables the subsegment with the domains to be analysed in terms of present and desired emphasis. This approach is discussed in further detail in Chapter 6.

From our work it is clear that a contingency approach to the six markets is appropriate and the issue of what are relevant market domains needs to be considered at the firm level. The Six Market model we use in this book is not meant as a straitjacket; it may well be developed into further variants as our on-going experience in specific market sectors develops and empirical research is undertaken. What is important is that an individual organization needs to recognize that relationship marketing activities directed at customers are necessary but not sufficient. The organization also needs to identify the other relevant market domains, and then groups or segments within them. Appropriate marketing strategies can then be developed for them.

Summary

This chapter has demonstrated that relationship marketing has emerged in parallel to the new organizational forms which continue to evolve, to exploit a business environment characterized by constant change. The new organizational structures and relationship marketing are related and complementary, but not as cause and effect. Both are more likely to succeed if guided by a customer orientation, because a customer orientation is adaptive to environmental change. Relationship marketing, through internal marketing, fosters the development of this customer orientation within the company and therefore aids market-led organizational transformations. Relationship marketing is therefore a market-led, customer oriented, general management concept, based in part on a return to marketing's roots and the original marketing concept. Its wider remit is to form and sustain profitable, mutually beneficial, relationships by bringing together the necessary parties and resources to deliver the best possible value proposition for the customer.

References

1   Miles, R.E. and Snow, C.C. (1986). Organizations: New concepts for new forms. *California Management Review*, **28**, No. 3, 62–73.

2 McCarthy, E. J. (1960). *Basic Marketing: A managerial approach*, Richard D. Irwin Inc., Homewood, IL.

3 Borden, N.H. (1964). The concept of the marketing mix. *Journal of Advertising Research*, June, 2–7.

4 Jackson, B.B. (1985). Build customer relationships that last. *Harvard Business Review*, November/December, 120–128.

5 de Ferrer, R.J. (1986). A case for European management. *International Management Development Review*, **2**, 275–281.

6 Gronroos, C. (1990). Marketing redefined. *Management Decision*, **28**, No. 8, 5–9.

7 Brady, J. and Davis, I. (1993). Marketing in transition. *The McKinsey Quarterly*, No. 2, 17–28.

8 Berry, L.L. (1983). Relationship marketing. In Berry, L.L., Shostack, G.L. and Upah, G.D. (eds), *Emerging Perspective on Services Marketing*, American Marketing Association, Chicago, 25–28.

9 Payne, A., Christopher, M., Clark, M. and Peck, H. (1995). *Relationship Marketing for Competitive Advantage: Winning and keeping customers*, Butterworth-Heinemann, Oxford.

10 Flipo, J. (1986). Service firms: Interdependence of external and internal marketing strategies. *European Journal of Marketing*, **20**, No. 8.

11 Christopher, M., Payne, A. and Ballantyne, D. (1991). *Relationship Marketing*, Butterworth-Heinemann, Oxford.

12 Gummesson, E. (1987). The new marketing – developing long-term interactive relationships. *Long Range Planning*, **20**, No. 4, 10–20.

13 Sheth, J.N. (1994). *The domain of relationship marketing*, handout at the second research conference on Relationship Marketing, Centre for Relationship Marketing, Emory University, Atlanta.

14 Kotler, P. (1992). Total marketing. *Business Week Advance*, Executive Brief, **2**.

15 Morgan, R.M. and Hunt, S.D. (1994). The commitment–trust theory of relationship marketing. *Journal of Marketing*, **58**, July, 20–38.

16 Doyle, P. (1995). Marketing in the new millennium. *European Journal of Marketing*, **29**, No. 13, 23–41.

17 Berry, L.L. (1995). Relationship marketing of services – growing interest, emerging perspectives. *Journal of the Academy of Marketing Science*, **23**, No. 13, 236–245.

18 Drucker, P.F. (1954). *The Practice of Management*, Harper & Row, New York, 37.

19 Webster, F.E. (1988). The rediscovery of the marketing concept. *Business Horizons*, May–June, 29–39.

20 Johnston, R. and Lawrence, P.R. (1988). Beyond vertical integration – the rise of the value-adding partnership. *Harvard Business Review*, July/August, 94–101.

21 Achrol, R.S. (1991). Evolution of the marketing organization: New forms for turbulent environments. *Journal of Marketing*, **55**, October, 77–93.

22 Webster, F.E. (1992). The changing role of marketing in the corporation. *Journal of Marketing*, **56**, October, 1–17.

23 Powell, W.W. (1990). Neither market nor hierarchy: Network forms of organization. *Research in Organization Behaviour*, **12**, 295–336.

24 *The Economist* (1994). Death of the brand manager, 9 April, 79–80.

25 Cravens, D.W. and Piercy, N.F. (1994). Relationship marketing and collaborative networks in service organizations. *International Journal of Service Industry Management*, **5**, No. 5, 39–53.

26 Reichheld, F.F. and Sasser, W.E. (1990). Zero defections: Quality comes to services. *Harvard Business Review*, September/October, 105–111.

27 Reichheld, F.F. (1993). Loyalty-based management. *Harvard Business Review*, March/April, 64–73.

28 Chartered Institute of Marketing (1994). *Marketing – the challenge of change: A major study into the future of marketing in key British enterprises*. CIM, London.

29 Webster, C. (1988). The importance consumers place on professional services. *Journal of Services Marketing*, **2**, No. 1, 59–70.

30 File, K.M., Judd, B.B. and Russ, A.P. (1992). Interactive marketing: The influence of participation on positive word-of-mouth and referrals. *Journal of Services Marketing*, **6**, No. 4, 5–14.

31 Judd, V.C. (1987). Differentiate with the 5th P: People. *Industrial Marketing Management*, **16**, 241–247.

32 Gronroos, C. (1990). Relationship approach to marketing in services contexts: The marketing and organisational behaviour interface. *Journal of Business Research*, **20**, 3–11.

33 Schlesinger, L.A. and Heskett, J.L. (1991). Breaking the cycle of failure in services. *Sloan Management Review*, Spring, 17–28.

34 Schneider, B. (1973). The perception of organisational culture: The customer's view. *Journal of Applied Psychology*, **57**, No. 3, 248–256.

35 Schneider, B. (1980). The service organisation: Climate is crucial. *Organizational Dynamics*, Autumn, 52–65.

36 Atkinson, J. (1989). Four stages of adjustment to the demographic downturn. *Personnel Management*, August, 20–24.

37 Webster, F.E. and Wind, Y. (1972). *Organisational Buying Behaviour*, Prentice Hall, Englewood Cliffs, 78–80.

38 Hunt, S.D. and Morgan, R.M. (1994). Relationship marketing in the era of network competition. *Marketing Management*, **3**, No. 1, 18–28.

39 Christopher, M. (1995). *Networks and logistics: Managing supply chain relationships*, paper presented to the third international colloquium in Relationship Marketing, Monash University, Melbourne.

40 Peck, H.L. (1997). *Towards a framework of relationship marketing: Methodology and initial case study*, Cranfield Working Paper Series SWP 97. Cranfield School of Management, Cranfield University.

41 Ohmae, K. (1989). The global logic of strategic alliances. *Harvard Business Review*, March–April, 143–154.

42 Gummesson, E. (1996). Towards a theoretical framework of relationship marketing. In *Proceedings of the International Conference on Relationship Marketing, Berlin*, 5–18.

43 Rosenbluth, H. and Peters, D.M. (1992). *The Customer Comes Second and Other Secrets of Exceptional Service*, Marrow, New York.

44 Reichheld, F.F. (1994). Loyalty and the renaissance of marketing. *Marketing Management*, **2**, No. 4, 10–21.

45 Reichheld, F.F. (1996). *The Loyalty Effect: The hidden force behind growth, profits, and lasting value*, Harvard Business School Press, Boston.

46 Coyne, K.P. and Subramaniam, S. (1996). Bringing discipline to strategy. *The McKinsey Quarterly*, **4**, 14–25.

47 Coyne, K.P., Hall, S.J.D. and Clifford, P.G. (1997). Is your core competence a mirage? *The McKinsey Quarterly*, **1**, 40–54.

48 Millman, A.F. (1993). *The emerging concept of relationship marketing*. Paper presented at the ninth annual IMP conference, Bath, 23–25 September.

49 Dill, W.R. (1975). Public participation in corporate planning: Strategic management in a Kibitzer's world. *Long Range Planning*, February, 57–63.

50 Freeman, R.E. and Reed, D.L. (1983). Stockholders and stakeholders: A new perspective on corporate governance. *California Management Review*, **25**, No. 3, Spring, 88–106.

51 Peck, H.L. (1994). Lessons from Laura Ashley: A relationship marketing case study. In *Proceedings of the 2nd international colloquium in Relationship Marketing, 13–15 November 1994*, Cranfield School of Management, Cranfield University.

52 Payne, A. (1993). *Relationship Marketing: the Six Markets Framework*. Working Paper, Cranfield School of Management.

Chapter 2

The customer market domain:
Managing relationships with buyers, intermediaries and consumers

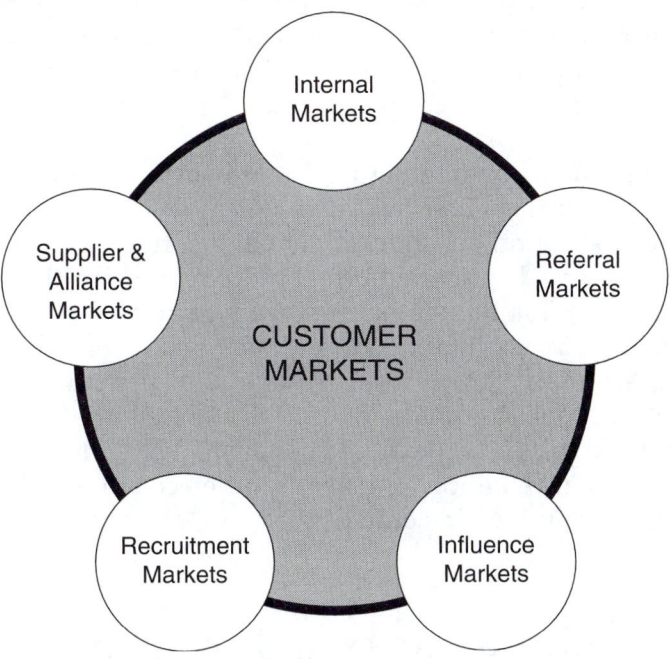

The customer market domain

Introduction

In this chapter we examine the customer market domain as the central market within the Six Markets model described in

Chapter 1. The customer market domain addresses three broad groups: direct buyers, intermediaries and final consumers.

To illustrate these groups, consider a manufacturer of domestic appliances. This manufacturer sells to a number of approved wholesalers, who in turn sell these products to retail outlets. Finally, these retail outlets sell the appliances to individual consumers. In this example the term Buyer refers to the wholesaler. The wholesaler who sells to the retailer is termed the Intermediary. The individual who purchases the appliance from the retailer is termed the Consumer. The term 'customer' will be used generally in this chapter to apply to all these groups.

For this example, the three groups in the customer market domain are:

- Buyer – the direct customer of the manufacturer, i.e. the wholesaler.
- Intermediary – the retailer to whom the wholesaler sells the appliances.
- Consumer – the individual at the end of the channel who purchases the appliance from the retailer.

These groups are shown in Figure 2.1. However, in some industries there may be further intermediaries which create additional steps within the distribution channel shown in this figure.

This chapter addresses five broad areas relating to the customer market domain. Firstly, the nature of the major subgroups within the domain – buyers, intermediaries and consumers – is discussed. Secondly, segmentation of these groups is examined. Thirdly, the nature of the decision-making unit (DMU) is briefly reviewed. Fourthly, the topic of customer acquisition and retention is explored. Fifthly, customer retention strategy and the economics of customer retention are examined in some detail.

Buyers, intermediaries and consumers

The terms Buyers, Intermediaries and Consumers are used in a specific sense in this chapter. They represent commonly used terminology in describing the customer market domain. However, it should be recognized that some organizations may use these terms in other ways or use different terms. For example, professional service firms and organizations in the healthcare sector typically refer to clients

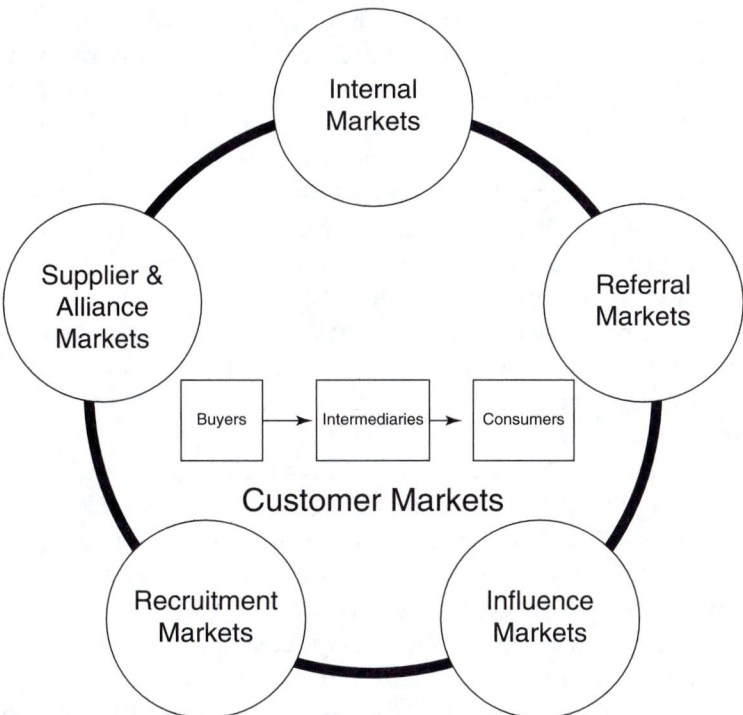

Figure 2.1 The customer market domain.
Source: Payne (1998).[1]

(rather than customers). Therefore, when we are considering a specific industry we must first understand the terms that are commonly used, develop our own relevant definitions, and then ensure that these definitions are used within our own business in a clear and unambiguous manner.

For example, in the insurance industry brokers are a typical means of distribution of the insurance products developed by an insurance underwriting company. The channel consists of the underwriting company who appoints insurance brokers who in turn sell the insurance products to final consumers. These consumers may be organizations or individuals.[*] Here there are only two levels in the distribution chain. In this case the insurance broker is usually termed

---

[*]In this example the term 'consumer' is used to refer to both organizational and individual purchasers of the end of the distribution channel. It should be noted that other terms may be used for the final business-to-business organizational purchaser in particular industries.

the intermediary and the final customer is termed the client. It should be noted that terminology used on a day-to-day basis within a given industry sector may also vary from one firm to another.

Many organizations adopt multiple channels in seeking to serve the final consumer, whilst others use only one channel. For example, some insurance companies, such as Direct Line, market directly to the final consumer. Other insurance companies sell both through the traditional broker channel as well as selling direct. For example, Guardian Royal Insurance, a large UK composite insurer, markets indirectly to final consumers through a large network of insurance brokers. However, it also has a separate division, Guardian Direct, which markets directly to final consumers.

There are a wide range of distribution options by which a company may seek to serve the final consumer. Some of these are shown in Figure 2.2. The choice regarding distribution options should be made following a determination of the value proposition relevant to the final consumers in the desired segments that a company wishes to serve. The distribution options should also be put under regular scrutiny as circumstances change and new opportunities present themselves. There is now an increasing recognition that for a firm to be successful it needs to create a supply chain that is more effective than that of its competitors. Therefore it is supply chains or market networks that compete, rather than just companies. Thus the task that needs to be addressed is how to create a superior value delivery network.[2]

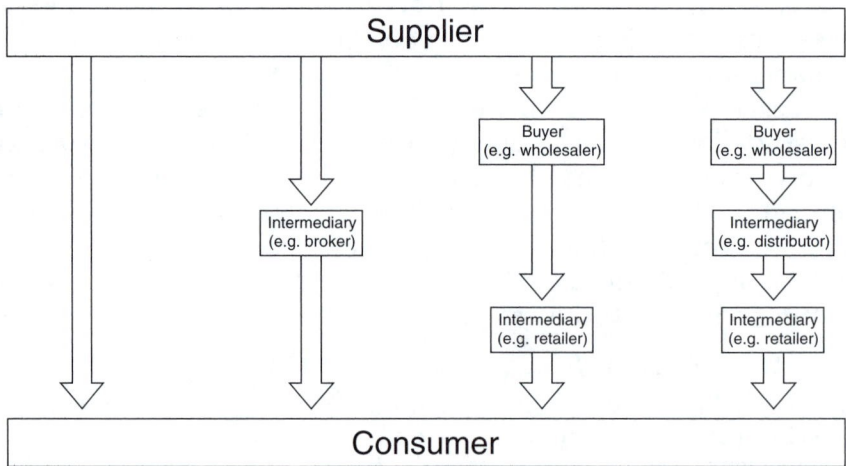

Figure 2.2   Alternative channel options in the value delivery network.

We are now in an era of 'electronic commerce'. Many new opportunities have arisen in recent years following advances in information technology and computing and, as a result, new channels to market have emerged. The internet is the most obvious example of this. For example, by 1997 Dell Computers was selling $2 million worth of computers to final customers every day via the internet. This distribution channel is expected to grow dramatically over the next five years.

Ultimately the use of new and existing channels will be driven by the desire to build channel relationships such that the lifetime value of desired consumers will be maximized. This task will be considered later in this chapter.

Segmentation and analysis of the customer market domain

Market segmentation

Market segmentation is a process of dividing up a broad generic market into a number of smaller groups, or market segments, based on characteristics or responses of customers in those segments. In the past much market segmentation has been done using very broad segments; in some cases organizations have undertaken no real market segmentation at all.

Segmentation needs to be performed at all levels of the customer market domain, not just with the organization's direct buyer. For example, a large international manufacturer of consumer durables, serving many international markets, may segment its buyers (the wholesalers to which it sells directly) in a number of ways, including by:

- country
- size
- volume
- level of sophistication
- ownership, etc.

The manufacturer may also seek to segment the retailers (to whom the wholesalers sell) – its intermediaries – according to relevant segmentation criteria, which may include:

- location
- size
- type of merchandise which it specializes in
- number of brands stocked
- socio-economic catchment area
- ethnic catchment area
- hours of opening
- number of branches, etc.

Finally, the manufacturer may wish to segment and gain an understanding of its final consumer. The final consumer (to whom the retailer sells) may be segmented on the basis of:

- family income
- socio-economic status
- size and status of family
- type of occupation
- type of residence
- need for special features, etc.

Once the relevant segmentation base (or bases) has been determined the market segments or subgroups within the buyer, intermediary and consumer levels in the distribution chain can be identified. We can then examine the opportunities offered by these segments. This leads to the identification of the most attractive segments and the development of appropriate strategies for winning and retaining customers within them.

Many companies adopting relationship marketing are now seeking to undertake much more specific and targeted market segmentation. Increasingly organizations are moving to a second level of submarket segmentation by practising micro-segmentation. In some cases this leads to marketing to a 'segment of one'[3] at the consumer level. A good example of this is Tesco Plc, the leading UK supermarket chain. Tesco has introduced a loyalty card called the Tesco Clubcard which enables it to mass-customize[4] individualized offerings to final consumers. (Case 2.3 explores this initiative in greater detail.)

A detailed review of market segmentation is beyond the scope of this book. Those readers wishing to explore this area in further detail should see McDonald and Dunbar,[5] Bonoma and Shapiro[6] and Weinstein.[7]

Reviewing marketing emphasis within the customer market domain

A key issue for a company considering a relationship marketing programme is to undertake a detailed market analysis at each level in its value delivery network and identify the type of marketing activity that needs to be directed at each of the various channel members, including direct buyers, intermediaries and final consumers. Then further analysis, in terms of the segmentation and understanding the decision-making units of different levels, needs to be completed before determining the appropriate amount of marketing expenditure and effort that is relevant at each level. The analysis of the value delivery network can be assisted by the construction of a market map that defines the distribution activities and identifies the volumes of product and services sold and the sales values associated with them. An example of a market map is provided in Figure 2.3.

The relative amount of marketing effort aimed at different channel members needs to be regularly evaluated and changed when appropriate. Whilst in some industries intermediaries may be a valuable channel member, in others the value of intermediaries is

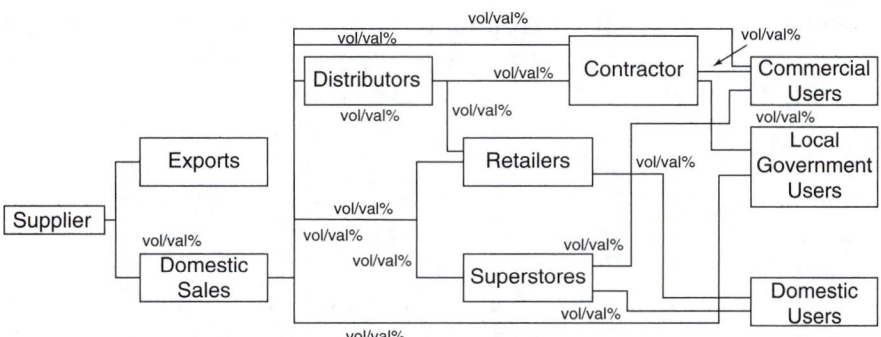

Figure 2.3 Market map with volumes and/or values at each stage.
Source: Based on McDonald and Dunbar.[5]

being challenged. Unless the intermediary is adding value to the customer relationship, it may prove to be an unnecessary cost and may be bypassed. Many organizations are now finding that in order to build stronger relationships with final consumers they need to change the emphasis and expenditure at different channel levels or, alternatively, refocus the existing expenditure in ways that build deeper and more sustained relationships. Some examples will serve to illustrate this point.

A manufacturer of domestic dishwashing machines may have traditionally spent a large proportion of its marketing efforts and marketing budget on trade marketing aimed at getting the dishwashers into large retail outlets such as department stores. Much of the marketing expenditure may have been directed at developing strong key account management; providing appropriate promotional activity; undertaking in-store point-of-sale merchandising; creating a discount structure based on volume; and establishing training programmes for sales staff in the retailers. This may have been supplemented by a considerable amount of trade advertising and trade promotion, whilst only a limited amount of advertising may have been aimed at final individual consumers. The manufacturer may decide to review its marketing approach and implement an alternative marketing strategy that focuses more closely on the needs of the consumer. It may seek to identify the means of final consumers through warranty cards or some form of direct promotional activity; send them a questionnaire to help identify their needs for a range of products and services; set up a major telephone call centre; create a customer club, etc. These and other options can be considered as a means of building relationships with the final consumers.

General Electric's (GE) Appliance Division in the USA is a good example of an organization which has built a closer relationship with its final consumers through the establishment of a major call centre. GE's Answer Centre is widely regarded as one of the best in the world. In setting up this call centre in 1981 GE sought to 'personalise GE to the consumer and to personalise the consumer to GE'. Unlike most manufacturers, who avoided any contact with the final consumer, GE did an unusual thing and gave its phone number to customers. The Answer Centre has now evolved over a 16-year period into an increasingly important relationship marketing capability where the current network of five call centres receives several million calls each year. Wayland and Cole[8] have outlined how GE's

Answer Centre has contributed to increased customer relationship value in three key areas:

> First, resolving immediate problems results in a probability-of-repurchase rate of 80 per cent for the previously dissatisfied customer, as compared to 10 per cent for the dissatisfied but uncomplaining customer and 27 per cent for an average customer. In other words, by making it easier to reach the company and by responding effectively, GE gets more opportunities to convert dissatisfied customers and to strengthen relationships. Second, contact with the centre significantly increases customers' awareness of the GE appliance line and their consideration level. Finally, the knowledge that is generated through customer interactions provides valuable input to the sales, marketing, and new product development processes.

The failure of many manufacturers to develop relationships with their final consumers is fairly common. Many readers will have had disappointing experiences in terms of purchasing a range of consumer durable products like motor cars. They might have been motivated to purchase a motor car as a result of promotional activity by a motor car manufacturer. However, they may have been very disappointed by the subsequent lack of interest the dealer who sold them the car has in maintaining the car on an on-going basis and satisfactorily rectifying faults that occurred within the warranty period. Consumers may be further upset when they seek to obtain redress directly from the manufacturer of the motor vehicle and find that the manufacturer is totally uninterested in having any dialogue with them. Within the motor car sector, radical changes in both distribution and other marketing practices, such as the approach being adopted by Daewoo, are now causing other motor car manufacturers to question what can be done to develop closer relationships with the final consumer.

Organizations in other industries are now seeking to re-evaluate how the marketing budget is spent with final consumers. In the UK, Heinz have recently undertaken a radical departure from normal marketing practices to final consumers. In this experiment Heinz have dramatically reduced their product-based television advertising and are seeking to develop one-to-one relationships with final consumers. This has been achieved by building detailed lists of Heinz consumers and then creating a customized direct mail dialogue with them.

Other companies are also rethinking their marketing strategy and are developing direct relationships with the consumer. Procter and Gamble are now focusing their attention on developing direct relationships with the consumer through direct response promotion; with their Pampers brand of nappies, consumers are offered the opportunity of obtaining discounts by completing a coupon which provides valuable data including name, address, telephone number, number of children and their age. This allows them to track consumer needs more closely and make appropriate and timely offers to them. Many other manufacturers within the retail sector are now looking at these activities with enormous interest.

The decision-making unit

Central to the understanding of customer behaviour is the decision-making unit. The concept of the decision-making unit (DMU) is important at all levels within the distribution chain. The DMU model suggests that the 'customer' comprises a number of individuals who have different roles in the purchase decision process. These include users, influencers, deciders, buyers and gatekeepers.[9] The roles of these buying unit members in the DMU are shown in Figure 2.4. Whilst this model was proposed in the context of the pur-

---

- *Users*   Users are the members of the organization who will use the product or service. In many cases, the users initiate the buying proposal and help define the product specifications.
- *Influencers*   Influencers are persons who influence the buying decision. They often help define specifications and also provide information for evaluating alternatives. Technical personnel are particularly important as influencers.
- *Deciders*   Deciders are persons who have the power to decide on product requirements and/or on suppliers.
- *Approvers*   Approvers are persons who must authorize the proposed actions of deciders or buyers.
- *Buyers*   Buyers are persons with formal authority for selecting the supplier and arranging the terms of purchase. Buyers may help shape product specifications, but they play their major role in selecting vendors and negotiating. In more complex purchases, the buyers might include high-level officers participating in the negotiations.
- *Gatekeepers*   Gatekeepers are persons who have the power to prevent sellers or information from reaching members of the buying centre. For example, purchasing agents, receptionists, and telephone operators may prevent salespersons from talking to users or deciders.

Figure 2.4   The decision-making unit (DMU) concept.
Source: Based on Webster and Wind.[9]

---

chase of industrial products, the same concept has applicability to the purchase of services and consumer goods.

The complexity of the buying process may vary: for example, with the value of the purchase and the familiarity of the buyer with the product. However, within any one customer organization there is typically a number of individuals who can make an impact on the buying decision. Selling organizations can fail in their endeavours to fully develop a customer, unless appropriate relationships are developed throughout the decision-making unit.

Often an organization's interest, in terms of the DMU, is in its most direct customer – the buyer. For example, the manufacturer of automotive components will be interested in developing a deep understanding of the DMUs at the major motor car manufacturers such as Ford and General Motors. They will seek to understand who are the various members of the DMU and what each of them requires. However, they may neglect other areas, such as the trade 'after market' and retail consumers.

Within certain industry sectors, an enormous amount of emphasis is placed on understanding the decision-making unit. Some of the best examples of this are within the aeroplane manufacturing sector. Companies such as Boeing are renowned as 'best-in-class' exponents of intimately understanding the DMU and developing sophisticated marketing programmes based on this understanding.

Understanding the DMU is also equally applicable to an organization with a longer distribution chain. For example, in the case of a manufacturer of expensive stereo equipment, in addition to understanding the DMU for their direct buyers – the large department stores – they may well also be interested in understanding the DMU at the individual consumer level. The marketing activities of expensive stereo manufacturers such as Bang & Olufsen suggest they know a lot about the purchasing behaviour at the consumer level.

Customer acquisition and customer retention

Once the tasks of determining the appropriate emphasis of marketing effect at each level within the customer market domain, and market segmentation and analysis of the DMU have been completed,

Characteristics	Transactions focus	Relationships focus
Focus	Obtaining new customers	Customer retention
Orientation	Service features	Customer value
Timescale	Short	Long
Customer service	Little emphasis	High emphasis
Customer commitment	Limited	High
Customer contact	Limited	High
Quality	An operations concern	The concern of all

Figure 2.5 Transaction focus and relationship focus.

attention then needs to be directed at customer acquisition and customer retention strategies.

The traditional focus of marketing has been on winning customers, with an emphasis on the value of an individual sale. This
transactional approach has increasingly been replaced by a relationship marketing approach, emphasizing the value of long-term relationships and repeat purchases. Figure 2.5 describes the shift from
transactional to relationship marketing. In the former, the approach
is impersonal, rule oriented and directed to short-term customer satisfaction, whilst the latter emphasizes long-term relationship building.

Despite an increasing managerial awareness of the benefits of
relationship marketing and retention, few companies claim to have
achieved the appropriate balance between acquiring and retaining
customers. There is a real danger in placing too much focus on marketing activities directed at new customers. Too much by way of
financial resources may be used in acquisition; whilst these valuable
customers are lost because too little subsequent effort is given to
retaining them. If customer service does not meet customer expectations, it is unlikely that a customer will be retained and the reputation of a company may be damaged by adverse word-of-mouth
publicity.

New and existing customers (including buyers, intermediaries
and consumers) require different strategies and should be allocated
differing portions of the marketing budget. The appropriate allocation of funds between the two activities will depend on a number of
industry and company specific factors. A company in a start-up situation, in a growing market or in a fledgling industry will need to
spend considerable resources in developing new customers, whilst

a well-established company in a mature market will primarily need to focus on retaining existing customers and intermediaries. However, the markets in most developed economies are mature. Despite this, many companies still spend too much of their resources in terms of money and time on customer acquisition and too little on customer retention.

Relationship marketing acknowledges that different marketing strategies are needed for customer acquisition and retention. The relationship marketing ladder of loyalty, illustrated in Figure 2.6, identifies the different stages of relationship development. This ladder has relevance for all groups – buyers, intermediaries and consumers – within the customer market domain.

The first task is to move a 'prospect' up to the first rung to a 'purchaser'.* The next objective is to turn the new purchaser into a 'client' who purchases regularly; and then develop a 'supporter' of the company and its products. The next advancement up the loyalty

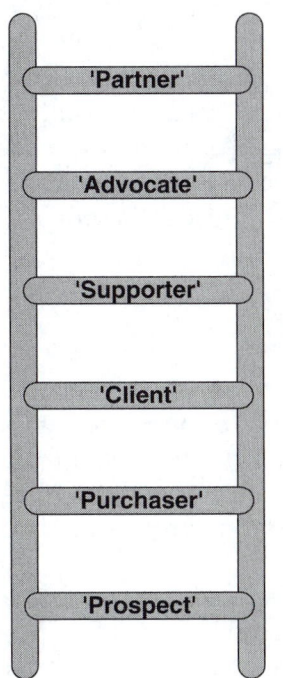

'Partner': someone who has the relationship of a partner with you

'Advocate': someone who actively recommends you to others, who does your marketing for you

'Supporter': someone who likes your organization, but only supports you passively

'Client': someone who has done business with you on a repeat basis but may be negative, or at best neutral, towards your organization

'Purchaser': someone who has done business just once with your organization

'Prospect': someone whom you believe may be persuaded to do business with you

Figure 2.6 The relationship marketing ladder of loyalty.

---

*In earlier versions of the ladder this step was called 'customer'.[10]

ladder is to an 'advocate'. This provides powerful word-of-mouth endorsement for a company. In a business to business context this may ultimately develop into a 'partner', where they are closely linked in a trusting and mutually sharing relationship with their supplier.

An important implication of the loyalty ladder is that it is not necessarily desirable to progress a relationship with every customer; some customers or customer segments may not be suitable for the investment needed in developing a relationship to 'supporter' or 'advocate', as it may prove to be too costly. Frequently there may be a need to change an organization's marketing activities and increase the marketing expenditure on the relationship building elements of the marketing mix. Managers therefore need to consider the potential life time value of a customer and determine whether it is appropriate to make this commitment.

There are a number of examples of organizations that have invested too heavily in unselective customer acquisition and have then found that they have attracted a customer base which may be unprofitable and is inappropriate for further development. For example, in the competitive mobile telephony market some suppliers have gained many low usage customers through undiscriminating advertising and poor customer targeting. As a consequence, there have been high levels of customer defection (up to nearly 50 per cent annual defection in some segments) and decreased profitability.

The value of retaining customers

Relationship marketing has acted as a catalyst to understanding the value of customer retention. Much of the early work written on relationship marketing[11-13] recognized the importance of retaining customers, but it did not explore the measurement or profit impact of keeping customers.

In Chapter 1, the work of Reichheld, a management consultant at the firm Bain & Co, and Sasser, an academic at the Harvard Business School, was briefly referred to. Their work[14] suggested that there is a high correlation between customer retention and company profitability. They researched a number of organizations and identified that even a small improvement in retention rates could make a dramatic impact on profitability. They found that a five percentage points increase in customer retention for a number of businesses yielded an improvement in profitability, in net present value terms, of between 20 per cent and 125 per cent. Some of their findings on the profit impact of retention in a range of industries are shown in Figure 2.7.

They concluded that such a dramatic effect on profitability was due to a number of factors, including:

- retained business
- sales, marketing and set-up costs are amortized over a longer customer lifetime
- increased expenditure by the customer over time
- repeat customers often cost less to service as there is a mutual familiarity with systems and processes
- satisfied customers can be an important source of referrals

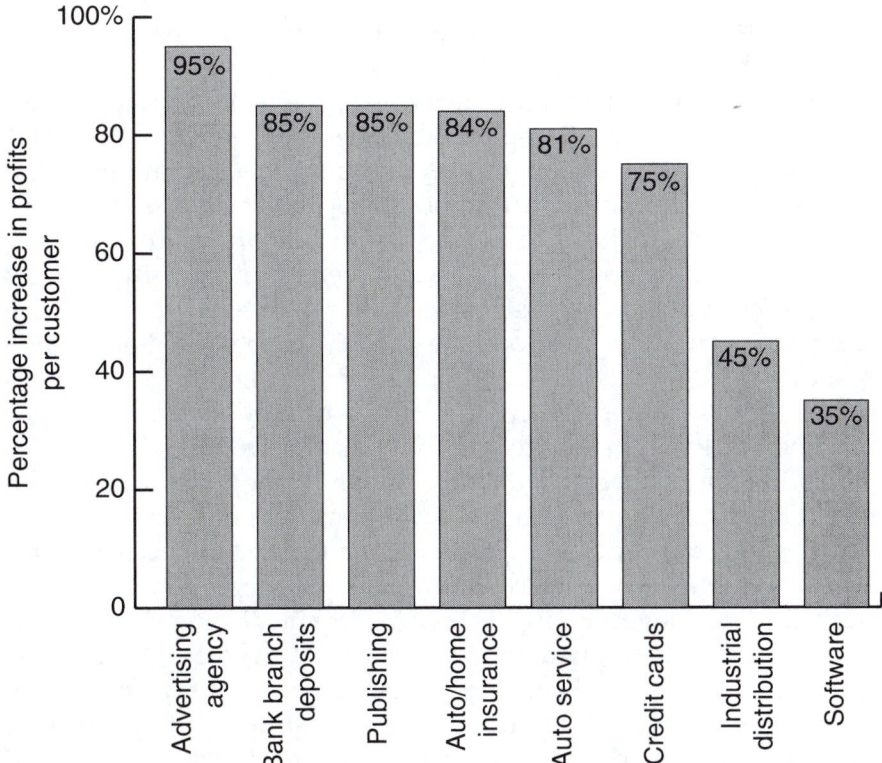

Figure 2.7 Profit impact of 5 per cent increase in retention rate.
Source: Based on Reichheld.[20]

● satisfied customers tend to be less price sensitive and may be willing to pay a price premium.

This early work on customer retention was of great interest to both academics and managers. In particular, managers in individual businesses are keen to know the value of retention for their company in general, for individual products and in different customer segments. Managers need to have information to allow them to make choices in allocation of resources between customer acquisition and customer retention.

Payne and Rickard[15] have pointed out that up to this time most of the research and results on retention shared a number of characteristics: they were based on company-specific examples from consulting work; they were descriptive and based on confidential work assignments; there was a limited specification of variables; and not all the factors identified were supported by strong empirical evidence.

Payne and Rickard have developed a mathematical model[16] of customer retention with the objective of enabling a trade-off to be made in the allocation of scarce marketing resources between strategies concerned with retaining existing customers and attracting new customers. Their retention model permits the calculation of the impact on profitability of a number of factors related to customer retention and acquisition. These factors include: the customer retention rate, the number of existing customers, the acquisition target for new customers, the cost of acquiring each customer and the profit per customer per period. This model has been used on a range of industries to understand how changes in the above variables impact on customer segment and company profitability. The basic building blocks of this model are shown in Figure 2.8.

This model has been used by Payne and Frow[17] to examine the impact of marketing programmes aimed at retaining existing customers and acquiring new customers for a major UK electricity supplier. This sector is one which is presented with special relationship marketing challenges. From 1998 competition will enter the UK residential electricity supply market and suppliers will face, for the first time, the challenge of keeping their existing customers, who are now open to competitive attack from electricity suppliers beyond their geographic boundaries, as well as the task of acquiring new customers in the geographic territories of their new competitors. Based on the results of a segmentation study of 2000 residential customers,

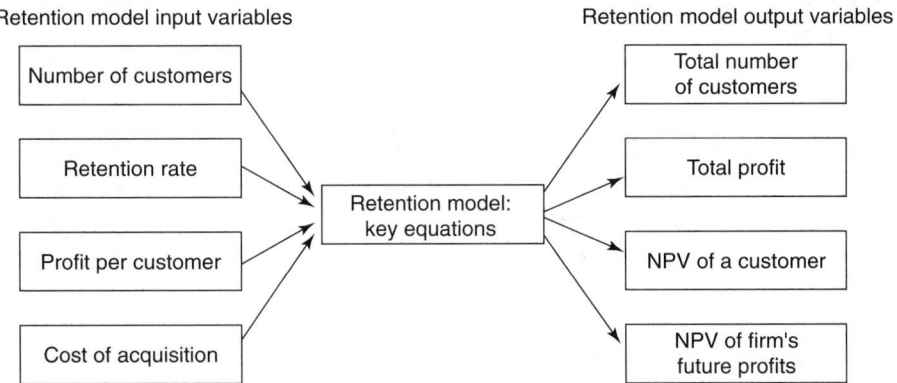

Figure 2.8 The customer retention model.
Source: Based on Payne and Rickard.[15]

they modelled long-term profitability within four key market segments. They suggest that retention strategies need to be based on understanding the relative profitability of different segments and micro-segments and that existing mass-marketing strategies will need to be replaced by those based on identifiable value propositions and retention management programmes aimed at specific segments.

Other researchers are also working on aspects of customer retention. These include Ennew and Binks,[18] who have examined the links between customer retention/defection and service quality in the context of the UK banking sector and banks' relationships with small business customers. They develop a framework for examining satisfaction and retention and present the results of some empirical research. They find support for the hypothesis that loyalty/retention is influenced by service quality and customer relationships; and find that trust in the banking relationships has the largest impact on potential defection.

Other work on customer retention has been undertaken by Page, Bethon and Money,[19] who developed a quantitative approach to analysing defections and their impact using a case study approach. They explore the role of customer defection probability and customer contribution as a function of age and then use sensitivity analysis to investigate the impact on contribution and market share sensitivity of two different strategies – reduction in defection rate and improvement in new customer acquisition. They then address different potential marketing mix strategies for new and existing customers.

Developing relationships with the customer market domain

The discussion in this chapter has suggested three key groups in the customer market domain with whom we need to build relationships: direct buyers, intermediaries and final consumers. Some of the key steps which need to be considered in addressing this domain include the following.

- Review the current distribution framework.
- Determine what represents a superior value delivery network, including the role of new distribution opportunities such as the internet and electronic commerce.
- Undertake a market segmentation of buyers, intermediaries and consumers.
- Extend broad-based market segmentation into micro-segmentation and, where relevant, one-to-one marketing strategies.
- Create a market map to define distribution activities and identify volumes and sales values of products and services sold.
- Determine the appropriate level of marketing emphasis at each level of the value delivery network.
- Identify the decision-making unit. Where there is a distribution chain, this should be undertaken for buyers, intermediaries and consumers.
- Undertake an analysis of lifetime value of customer segments and develop a detailed understanding of customer acquisition and customer retention economics for each major segment and micro-segment.
- Conduct market research to understand reasons for customer defection and drivers of customer loyalty.
- Determine appropriate retention and acquisition marketing strategies, including a consideration of whether loyalty schemes have a role to play.
- Develop a marketing plan for the customer market domain. This should take account of relationships within each of the other six market domains considered in the following chapters as part of the overall relationship marketing plan.

Summary

In this chapter we have explored the role of the customer market domain, which comprises the direct buyers of an organization's products or services and, where there is a distribution channel, the intermediaries in that channel and the final consumers. These customer groups form the key domain within the six markets framework.

Traditional marketing activity has focused on direct buyers – how to win them, and what kind of offer to make to them. Relationship marketing broadens the perspective of where marketing should focus, recognizing that success with the direct buyer is also dependent on managing relationships with the intermediaries and consumers in the distribution chain.

Increasingly organizations will need to make regular reviews of their value delivery network and determine the appropriate use of channels and channel members in order to maximize the lifetime value of attractive segments within the final consumer group. Organizations seeking to improve their relationship marketing will seek to develop appropriate segmentation strategies for direct buyers, intermediaries and final consumers. They will also seek to develop more specific and targeted marketing segmentation by moving towards micro-segmentation and, in some cases, to one-to-one marketing. A regular and informed review of marketing emphasis and marketing spend on both customer attraction and customer retention needs to be conducted. Market maps are a useful tool in helping to understand and define volumes and values which are sold through different elements of the value delivery network. Central to the successful relationship marketing process will be an understanding of the decision-making unit at all levels of the value delivery network.

In the future, organizations will seek to determine which customers or customer groups it wishes to progress up the relationship marketing ladder and which ones it does not wish to progress. In doing this organizations will move from a broad understanding that customer retention is important to a focus on customer retention economics and determining which segments and micro-segments will yield the greatest increase in profitability from improvement in retention activity.

Strategies aimed at retaining customers can be expensive, as this often involves increasing customer service levels and tailoring the product or service offer to suit individual customers or customer groups. Successful retention programmes will segment customers according to their potential lifetime profitability and then determine the type and frequency of marketing activity relevant for each group to increase this potential.

Focusing on customers is, however, necessary but not sufficient. It has been argued that with increasing competitive activity in virtually every industrial, consumer and service sector we are approaching an era of hyper-competition where increasingly it is being recognized that it is supply chains rather than firms which compete. The next chapter focuses on the supplier and alliance market domain and its role in relationship marketing.

References

1   Payne, A.F. (1998). *Relationship marketing – the six markets framework: A review and extension.* Draft Working Paper, Cranfield School of Management, Cranfield University.
2   Kotler, P. (1997). *Marketing Management: Analysing, Planning, Implementation and Control,* Prentice Hall, Upper Saddle River, NJ, 45.
3   Peppers, D. and Rogers, M. (1997). *Enterprise One-to-one,* Piatkus Publishers, London.
4   Pine, B.J. (1993). *Mass Customisation: The new frontier in business competition,* Harvard Business School Press, Boston.
5   McDonald, M. and Dunbar, I. (1995). *Market Segmentation: A step by step approach to creating profitable market segments,* Macmillan, Basingstoke.
6   Bonoma, T.V. and Shapiro, B. (1983). *Industrial Market Segmentation,* Lexington Books, Lexington, Mass.
7   Weinstein, A. (1987). *Market Segmentation,* Probus Publishing Co.
8   Wayland, R.E. and Cole, P.M. (1997). *Customer Connections: New strategic for growth,* Harvard Business School Press, Boston.
9   Webster, F.E. and Wind, Y. (1972). *Organisational Buying Behaviour,* Prentice Hall, Englewood Cliffs, NJ, 78–80.
10 Christopher, M.G., Payne, A.F. and Ballantyne, D. (1991). *Relationship Marketing: Bringing quality, customer service and marketing together,* Butterworth-Heinemann, Oxford.

11 Berry, L.L. (1983). Relationship marketing. In Berry, L.L., Shostack, G.L. and Upah, G.D. (eds), *Emerging Perspectives on Services Marketing*, American Marketing Association, Chicago, 25–28.

12 Levitt, T. (1983). After the sale is over. *Harvard Business Review*, September–October, 87–93.

13 Jackson, B. (1985). Building customer relationships that last. *Harvard Business Review*, November–December, 120–128.

14 Reichheld, F.F. and Sasser, W.E. (1990). Zero defections: Quality comes to services. *Harvard Business Review*, September–October, 105–111.

15 Payne, A. and Rickard, J. (1996). *Relationship marketing, customer retention and firm profitability*. Working Paper, Cranfield School of Management, Cranfield University (revised version of earlier working paper).

16 *op. cit.*

17 Payne, A.F. and Frow, P.E. (1997). Relationship marketing: Key issues for the utilities sector. *Journal of Marketing Management*, **13**, 463–477.

18 Ennew, C.T. and Binks, M.R. (1996). The impact of service quality and service characteristics on customer retention: The small businesses and their banks in the UK. *British Journal of Management*, **7**, 219–230.

19 Page, M., Pitt, L., Berthon, P. and Money, A. (1996). Analysing customer defections and their effects on corporate performance: The case of Indco. *Journal of Marketing Management*, **121**, 617–627.

20 Reichheld, F.F. (1994). Loyalty and the renaissance of marketing. *Marketing Management*, **2**, No. 4, 10–21.

Chapter 2 case studies

This chapter focuses on the customer market domain. It addresses marketing in the context of both the direct customer and the final consumer as well as other intermediaries who may play a role as a channel member. This is the most important of the six markets, and so six cases are included in this chapter. They focus on a range of different industry environments, including fast-moving consumer goods, retail, industrial, services and the not-for-profit sector.

Case 2.1: Nestlé Buitoni: The house that mamma built

This case was written by Edward Hickman, a Management Consultant at Robson Rhodes, and Professor Erich Joachimsthaler Visiting Professor at the Darden Graduate School of Business Administration, University of Virginia.

Abstract
The Nestlé Buitoni case study examines an early example of the introduction of a club concept within Europe. The case traces the history of the company's origins and growth until it was taken over by the Nestlé group in 1988. By this stage it had established itself as the third largest food manufacturer in Italy. The case focuses on the UK subsidiary of Buitoni and in particular its launch of the Casa Buitoni Club, aimed at British consumers who shared a love of Italian food. The foundation of the club and its objectives are covered, including how they built a database of customers and how they sought to educate British consumers, who are fairly unsophisticated about Italian food.

Learning points
This is an interesting case which gives insight into a club-type approach to promoting a food product range. It discusses creating a relationship by involving the customer in brand-building activities. In particular, Buitoni sought to strengthen brand awareness and create a core number of customers interested in getting involved in Italian cooking. Their integrated communications campaign based on 'share the Italian love of food' resulted in a database of more than 200 000 consumers.

Broad marketing and strategic issues:

- brand building as a defence against commoditization and retailers' own brands
- growing a food brand in different geographic market segments
- strategies to build a brand with potential, but having a modest position in the overall product portfolio of the group
- how to educate a new marketplace
- addressing alternative ways to serve the customer.

Relationship marketing issues:

- the concept of a consumer club
- building a customer database
- developing a value proposition for the club
- approaches to developing relationships with customers.

Since the club's inception membership has grown rapidly through marketing efforts aimed at alternative low cost options, including public relations events, promotions and invitations on packages of Buitoni pasta. As a result, both the use of Buitoni products and customers' loyalty have increased. Lessons learnt from the Buitoni programme in the UK have influenced the marketing of the brand in other countries, such as Japan, and it has also impacted on strategies adopted for other Nestlé brands.

Case 2.2: Carlton Electronics

This case was written by Professor Martin Christopher, who is currently Head of the Marketing and Logistics Group and Deputy Director of Cranfield School of Management at Cranfield University.

Abstract

Carlton Electronics is a producer of telephone answering machines and associated peripheral equipment. Its customers can be divided into three broad categories, including large chain stores, British Telecom and discount stores. The case study focuses on the need to address these different types of trade buyers and the differences between them and provides details of a customer service survey that Carlton Electronics commissioned to gain a better understanding of customer service issues. The case addresses a number of specific service quality issues, including delivery time, inventory availability, sales representation, advertising and promotion and packaging. Philip Millard, the Marketing Director of Carlton Electronics, needs to consider what is an appropriate new approach to strengthen relationships with his existing customers and the internal implications if Carlton becomes more customer oriented.

Learning points

This short case focuses on how a company selling consumer electronic products should address three separate trade markets, each of which has different demands for service. Issues of how to translate research findings into a service strategy are raised.

Broad marketing and strategic issues:

- evaluating the strengths and weaknesses of Carlton's market position
- understanding the requirements of different market segments in the retail trade
- evaluating the role of customer service in the overall marketing mix.

Relationship marketing issues:

- creating relationships with a trade
- the role of customer service in building relationships
- how to use information technology to improve physical distribution
- key account management.

This mini-case provides an opportunity for a case discussion on trade marketing in the consumer electronics industry. In particular, it enables a discussion to be made on marketing to the trade and marketing to the final consumer. Although the case does not explicitly examine marketing to the final consumer, participants in the case discussion may wish to explore what could be done to create final consumer appeal for the product and, if this is viable, what percentage of the marketing budget should be spent on doing this and which specific tools might be used to achieve this. A range of issues can be considered, including freephone numbers on the pack, customized packaging, warranty forms, etc. This can lead to a general discussion about who owns the final consumer and what manufacturers can do to develop closer relationships with the final consumer.

Case 2.3: Does Tesco hold all the cards?

This case was written by Helen Mitchell and Helen Peck, who are researchers at Cranfield School of Management, Cranfield University.

Abstract

This case study focuses on Tesco Plc, a large UK retailer. The case covers the period 1992 to 1997. In February 1995 Tesco launched the 'Clubcard' – the first major national supermarket loyalty card in the UK. The case study charts the success of the Clubcard and a range of other customer-focused initiatives and enhancements. Central to this success is the use of database marketing, which was used to underpin other elements of its marketing strategy. The subsequent introduction of a debit card – 'Clubcard Plus' – is also discussed. This case examines one of the most successful examples of a loyalty card in the UK, one which has resulted in Tesco displacing J. Sainsbury Plc from their former pre-eminent position as the number one food retailer. Competitors' reactions to Tesco's initiatives are also addressed.

Learning points

The purpose of the Tesco case is to demonstrate the implementation of a highly successful customer retention programme in the UK food retailing sector. It highlights the success that can be achieved by having a clear market focus and using all the organization's resources to achieve their strategy. In particular, it highlights how an organization such as a food retailer needs to develop an integrated marketing approach to achieve customer retention. Tesco developed an impressive marketing campaign which ensured the right goods were stocked, at the right price points, with the right store environment, in the right location and with high levels of customer service. Tesco first focused on delivery of a superior value proposition to the customer segments it sought to serve and then used the loyalty card as a method of customer bonding. This strategy is in sharp contrast to the strategy of some other retailers, who have sought to 'buy' loyalty by simply issuing a loyalty card.

Broad marketing and strategic issues:

- repositioning a major food retailer
- developing and implementing a new business strategy in retailing
- creation of a superior value proposition
- creating a new product development strategy through the launch of financial services
- first mover advantage.

Relationship marketing issues:

- the introduction of a successful loyalty card
- database marketing
- the use of micro marketing
- the use of loyalty cards as a mechanism for customer bonding
- the profit impact of customer retention on profitability
- strategic and tactical issues in the use of loyalty card schemes.

This case shows how a market leadership position can be achieved by a market follower through a highly innovative customer-focused marketing strategy. It demonstrates the tremendous value that can be generated by understanding customers at the individual level in an industry where, five years before, retailers had virtually no knowledge of customers at the individual level. It shows how a continuous dialogue can be initiated with customers, how this communication can be mass-customized at the individual customer level and how relationships can be built through the leverage of continuous transactional data. Issues from other of the six market domains are also addressed.

Case 2.4: Rover Cars: The Catalyst and Conquest '91 direct marketing programmes

This case was written by Professor Adrian Payne, who is Director of the Centre for Relationship Marketing at Cranfield School of Management, Cranfield University.

Abstract

From the late 1970s Rover became increasingly interested in making better use of the wealth of data that was potentially available from customer purchases, subsequent servicing and claims on warranty. This case study describes how, from a modest beginning, Rover developed a sophisticated approach to direct marketing. Rover benefited in a very tangible way from this approach, which was widely recognized at the time through a number of prestigious awards that it received both in the UK and the USA in recognition of the quality of its innovation in direct marketing. How direct marketing programmes evolved at Rover between 1978 and 1991 is described in detail, especially the circumstances and thinking which led to each new initiative being introduced.

Learning points

This case study shows how a major motor car manufacturer has sought to use direct marketing approaches to build a relationship with final consumers. Many motor car manufacturers have traditionally viewed marketing as a set of activities directed towards their direct customers – the dealerships. They have relied on advertising special deals and other promotional events to deal with the final consumer. Rover have progressively refined an approach to development of a relationship with the final consumer.

Broad marketing and strategic issues:

- explore strengths and weaknesses of direct mail as a marketing medium
- identify an approach that can be used to create differentiation from other motor car manufacturers aiming at similar segments
- evaluate direct response advertising, direct mail and promotions as alternative ways of acquiring purchases
- appreciate the value of being able to quantify the impact of direct marketing
- understand the concept of 'total cost' when exploring a direct marketing programme.

Relationship marketing issues:

- the problems of dealing with the final consumer when a car manufacturer is separated from them by an intermediary – in this case the motor car distributor
- the development of a lifestyle magazine aimed at different lifestyle segments and establishing a relationship with them
- developing an approach to warm up potential customers and maintain a dialogue with existing customers.

The Rover Group has probably had one of the most chequered histories of any major car producer currently still operating in the UK. It has been innovative in its direct marketing programmes and one of the first companies to use a lifestyle magazine with different tailored editions aimed at different lifestyle segments. The case demonstrates how good direct marketing programmes are ones of continuous improvement through factoring in relevant lessons that are learnt. Participants in the case discussion may wish to compare and contrast Rover's approach in the early 1990s with that of other motor car manufacturers, aiming at both the general market (e.g. Daewoo's direct marketing programme) or at specialist market niches (e.g. Porsche Club).

Case 2.5: Direct Line Insurance Plc

This case was written by Professor Derek Channon, who is a Professor in the Management School, Imperial College, University of London.

Abstract
This case describes the development of Direct Line Insurance, from its beginnings in 1985 until 1992. Over this period the company has grown to become the largest supplier of private motor insurance in the UK and a major supplier of household insurance. This was achieved by creating an insurance company which sold directly to the public and bypassed the traditional intermediary – the broker. It achieved this by transforming the cost structure of the industry, by using direct telephone selling, offering significantly lower prices based on lower costs, and extensively using information technology to gain strategic advantage. The case explains the competitive environment and outlines how Direct Line's success was at the expense of the UK composite insurance underwriters who traditionally sold their policies through the independent broker. By bypassing the insurance brokers, Direct Line saved between 25 and 40 per cent of the premiums written. The success has been imitated by a number of companies who have adopted a very similar approach.

Learning points
The case illustrates how a traditional sleepy and inefficient industry, private motor insurance, can be transformed through leveraging information technology and excluding the intermediary. It shows how an organization that has only been established for a short time can achieve a market leadership position through creating a superior value proposition and breaking the traditional 'rules' of the industry.

Broad marketing and strategic issues:

- the use of IT to create competitive advantage
- creation of a quality product through differentiation and low cost strategies
- development of an organization almost without middle management
- creating a unique advertising message to a specific target audience
- issues of target market selection, segmentation and positioning.

Relationship marketing issues:

- building relationships through service quality and information technology
- creating a responsive, process-oriented organization
- creation of a distinctive value proposition
- transformed traditional channels of distribution
- creation of relationships with final customers in the insurance market
- building an organization through referrals from advocates.

Direct Line represents one of the best recent examples in the UK of achieving both differentiation and low cost strategies at the same time and leveraging information technology to create strong relationships with the customer base. It combines a combination of high quality advertising and the use of referrals from satisfied customers to build a superior market position. Since the writing of the case the market has become even more competitive and profits at Direct Line have fallen. Participants in a case study discussion will wish to discuss what should be done by Direct Line to sustain and build on its market position in the face of competitive imitation.

Case 2.6: Relationship marketing: The RSPB – a bird in the hand

This case was written by Helen Mitchell, a researcher at Cranfield School of Management, and based on an original project written by Karen Farquharson, Stephen Barlow, Aiden Cotter, Matthew Meredith, Malcolm Stainforth and David White.

Abstract
The Royal Society for the Protection of Birds (RSPB) is the largest

wildlife conservation charity in Europe, with over 827 000 members. The case explores the development of the Society between 1991 and 1996, a stage when many changes were occurring in the UK charities market. In particular, the introduction of the National Lottery in 1994 created considerable fears as to its likely effect on reducing revenues to the charitable sector.

The case shows how the Society sought to improve its income to enable it to finance on-going expansion in a number of areas. The case addresses the need to manage the membership base and develop relationship marketing initiatives to improve customer retention and increase levels of donation.

Learning points

This case study focuses on the application of relationship marketing to the not-for-profit sector. It demonstrates how marketing activities directed at members of a charity cannot rely solely on market aggregation and how specific marketing activities need to be developed to build a long-term relationship and increasing involvement with the charity. It highlights the need to understand customer segments and how these segments have different customer retention characteristics. The case provides a good illustration of the need to manage both customer acquisition and customer retention in a not-for-profit business and how other relationship markets impact on the customer market.

Broad marketing and strategic issues:

- the challenges of marketing and management in a large charity
- development of a segmentation strategy
- issues regarding scope of mission – looking after birds versus protecting the environment.

Relationship marketing issues:

- analysing acquisition costs and retention costs
- managing customer retention and customer acquisition activities concurrently
- recognizing how emphasis needs to be placed on all markets to achieve success in the customer market
- the need to have adequate information sources.

This case illustrates many of the issues involved in developing a relationship marketing strategy. It addresses, in addition to the customer market, the need to manage relationships with relevant influence markets such as the Government as well as internal market considerations.

Case 2.1   Nestlé Buitoni: The house that mamma built

This case was written by Edward Hickman, a Management Consultant at Robson Rhodes, and Professor Erich Joachimsthaler Visiting Professor at the Darden Graduate School of Business Administration, University of Virginia. The case is reproduced with kind permission.

The Italian house of Buitoni was the third largest food manufacturer in Italy when taken over by Nestlé in 1988. Buitoni is the central plank of Nestlé's pasta and related products group and is the company which Nestlé was using to experiment with new marketing ideas during the early 1990s. Through a number of different marketing programmes, Buitoni, the UK subsidiary of Nestlé, built a large database of consumers, which became the basis for the launch of the Casa Buitoni Club. The club consisted of a collection of British consumers who share the Italian love for food. The club was the centrepiece of Nestlé's effort of building Buitoni into a global brand.

History*

The company was founded in 1827 in the small Tuscany village of Sansepolcro, Italy, by Guilia Buitoni, spouse of Giovan Battista Buitoni. Faced with a sick husband and hungry children, the young Guilia Buitoni pawned her wedding ring to pay for a pasta-making machine large enough to produce on a commercial basis. From the beginning, Guilia Buitoni insisted that her pasta should be of the finest quality, made only from 100 per cent durum wheat semolina.

From these beginnings, Guilia Buitoni founded the world's commercial dry pasta industry. The one small pasta factory at Sansepolcro was added to in 1856 and later in 1878, when further factories were opened in Città del Castello and Pérouse.

By the turn of the century, the name of Buitoni had become renowned all over Italy and had won numerous prizes at international food competitions. The company reins were passed down to the direct descendants of the founders.

In 1907, one of the founders' grandsons, Francesco, founded La Perugina, a chocolate and confectionery maker.

The company continued its steady expansion. In 1922, a new plant was opened in Rome. First steps abroad were taken in 1935, with the establishment of Buitoni SA in France, and a factory at Saint-Maur-des-Fossés near

---

*The section draws from Jean Heer (1991), *Nestlé: Cent Vingt-Cinq Ans De 1866 A 1991*, Nestlé, SA, Vevey.

Paris. By this stage, the company had also branched into packaging, with Buitoni Poligrafico.

In 1941, just before hostilities were declared between Italy and the United States, Buitoni Food Corporation was founded in the USA. It comprised a pasta plant in New Jersey, a factory making sauces in Brooklyn and a restaurant in Times Square, New York as well as a luxury Perugina chocolate shop in Fifth Avenue. Happily, the breaking of contact caused by the war was only an interruption.

The post-war period saw a phase of rapid growth for Buitoni. The plant at Sansepolcro was remodelled in 1946. New factories were opened in Aprilia and Foggia in 1961 and at Camaret, in France, in 1966.

In the meantime, Buitoni's European operations had been consolidated into one organization, Internazionale Buitoni Organizzazione (IBO) in order to compete more effectively. The US operation was consolidated into IBO in 1966 and in 1969 Buitoni and Perugina regrouped under the name Industrie Buitoni Perugina SpA (IBP). IBP went public in 1972 and was quoted on the Rome and Milan Stock Exchanges.

In the early 1970s, Buitoni extended operations into the UK with the acquisition of Bibby's, as well as into Holland and Sweden. The company was also present in Brazil. The crisis in Italy during the early 1970s caused financial disruption to the company and it was forced to float a part of Perugina to raise cash.

The company did not sparkle after this time. In 1985, IBP was taken over by Carlo de Benedetti's holding company, CIR SpA. Buitoni became Buitoni SpA. The senior management was replaced and the marketing functions in particular were reorganized. Buitoni's sales doubled between 1985 and 1987, and results improved.

During this period, Buitoni launched a premium pasta brand, Rasagnole, and two new lines, Preziose and Bella Napoli. There were also a series of acquisitions of companies in Italy producing rice, olive oil, charcuterie and condiments.

Food products were not central to de Benedetti's ambitions and in 1988 the company was sold to Nestlé for $1.4 billion, the third largest takeover deal in Europe for that year. The acquired company made losses in 1988 of $33.1 million and 1989 of $53.8 million while its activities were rationalized.

Nestlé and Buitoni

When Nestlé took over Buitoni, Buitoni products sold in more than 50 different countries. There were hundreds of manufacturers that produced over 400 different types of pasta. Barilla was the European market leader with 20 per cent, followed by BSN with 11 per cent, Birkel with 4.5 per cent, Lustucru with 4.5 per cent and Buitoni with 3.5 per cent. Private labels accounted for a huge share of the market. Stark differences existed with regard to competition and consumer preferences across the European countries.

Buitoni's share in Europe's largest market, Italy, was 7 per cent in 1988. The market leader by some margin was Barilla with, including subsidiaries Voiello and Braibanti, a 31 per cent share, more than four times its nearest rivals, Buitoni and BSN (6 per cent). The Italian market had nearly 200 producers and over 600 different types of pasta shapes. Annual per capita consumption of pasta was 25 kilograms. Other rivals of note were Agnesi (4 per cent), De Cecco (3.5 per cent), Amato (3.5 per cent), Divella (2.5 per cent), Federici (2.5 per cent) and Corticella (2.5 per cent).

One of Nestlé's first actions was to invest in the expansion of the plant capacity in Sansepolcro. Nestlé planned to increase its market share. BSN also had ambitions to increase its market shares, with a target of 10 to 15 per cent over the coming years. BSN focused on a continued strategy of acquisitions that included the purchase of the Tomadini, Spiga, Ponte, Ghigi and Mantovano brands. It also owned at that time a 30 per cent stake in Agnesi.

Buitoni was market leader in France of precooked foods (ravioli, cannelloni, couscous, paella). Following the acquisition of the brand Davigel, the company also had a strong presence supplying frozen food to the restaurant and hotel trades in France. French consumers ate about six kilograms of pasta per year.

In Germany, imports mainly from Italy accounted for around 33 per cent of the market. The leading German companies were BSB Nahrungsmittel GmbH (Birkel), Drei Glocken GmbH, and Aldi (the private brand of Aldi, one of the leading discount retailers). BSB and Drei Glocken together accounted for 45 per cent of the share of the pasta market. In total, there were 19 German pasta manufacturers. There were also certain types of pastas that only sold in Germany, one of which was Spätzle, sold mostly in the southeastern part. Noodles accounted for over a third of sales in the pasta category. On the average, Germans consumed five kilograms of pasta per year.

Since the late 1980s, the Italian market leader Barilla invested extensively in a pan-European market development effort. The company had realized that their dominant market position in Italy was threatened by cheap and lower-quality pasta imports that did not meet the stringent quality standards of the Italian manufacturers. This was the consequence of the progress towards the single European market. Barilla's response was to globalize and to build the brand across Europe and the US. The company ventured on an extensive testimonial-type advertising campaign with local adaptation. In Germany, Barilla advertised pasta featuring Steffi Graf as a spokesperson; in France, the actor Gerard Depardieu gave a helping hand; and in Sweden, the commercials starred tennis pro Stefan Edberg. In Spain, Antonio Banderas and Sharon Stone (a big hit there since the movie *Basic Instinct*) touted the pleasures and tastes of Barilla pasta shapes.

When Nestlé took over Buitoni, it was with a view to adding a significant pasta brand to its food interests. However, the role of pasta and Italian foods as part of Nestlé's overall food strategy quickly changed and the aim became to make Buitoni the leading worldwide brand of authentic Italian food. Nestlé developed a simple but encompassing concept for Buitoni: Buitoni represents all good Italian cooking. In a few years, Nestlé invested $400 million in Buitoni to build the company's competitive position with pasta, sauces and cooked dishes in a variety of ways such as dry, frozen, fresh and cooked-chilled.

In early 1992, after several earlier reorganizations, Nestlé refocused its activities away from a narrow division by product and process technology, such as frozen or canned foods. The new organization consisted of market-focused divisions with four key strategic brands plus two product areas. Buitoni was placed in this reformed structure alongside the Friskies, Nestlé and Nescafé brands. Along with the reorganization came a new strategic focus that took advantage of Nestlé's reach by picking out 'hit' or lead brands such as Buitoni from a country and sowing them across borders.

Buitoni got its own formal strategic business unit structure (SBU) and a dedicated research facility. It had been operating as a strategic business unit informally since 1990. The SBU has overall responsibility for the development of the brand and provides advice to each marketing unit on marketing, advertising and technical developments of the range.

Focus: United Kingdom

Nestlé chose the Buitoni brand and particularly the UK for an experiment of new ways of building a brand. At the time, Buitoni was the brand leader in the £72 million pasta market, with an 18 per cent value share and a 16 per cent volume share. Buitoni's total pasta business in the UK, including add-on products, amounted to between £22 million and £25 million. Its product range included a wide range of authentic Italian products, including dry pasta, tomato purée, bakery products, parmesan cheese and speciality pasta sauces.

Pasta was a small but rapidly growing market in the UK. It grew more than 10 per cent to £85 million in 1991 and £102 million in 1992, according to the Pasta Information Centre, an industry association.* Buitoni held the market share position followed by Napolina, a CPC International division, with 6 per cent. Napolina was one of the few branded pasta makers that had started to build a product portfolio around the theme of Italian-based

---

*Market sizes and share estimates varied widely from research company to research company, including Euromonitor, the Pasta Information Centre and Nielsen, due to the differences in the year-to-year reporting intervals, the rapid growth but small size of the market and the differences in definitions of the pasta market.

foods. Other brands had a 20 per cent market share in total and private label brands between them accounted for 58 per cent of the market. The leading brands were losing sales to the private labels.

Despite a steady growth of pasta as a regular household staple, UK consumption of over 75 000 tons and about two kilograms per person per year was still far below consumption in other countries. In comparison, per capita consumption in the US was four times that of the British consumers. Pasta had only a 65 per cent household penetration in the UK but had higher penetration in households with children. This boded well for the future.

Pasta was seen as healthy, wholesome, easy to prepare, cheap and a dish for all seasons of the year. Pasta was likely to be of growing appeal as consumers became more health and time conscious. There was also an overall trend towards more experiential cooking as Italian as well as Asian cooking ingredients became more available. The general awareness of these ethnic foods and their varieties and tastes increased consumption steadily and caused a substitution from other UK diets.

Nevertheless, there was a feeling that the UK consumer was still not well educated about pasta, the many varieties and the value and qualities of Italian food in general. There was a feeling that the arrival of branded manufacturers would be necessary to really develop the market. Several years ago, there was only private label pasta and branded pasta was few and far between. The arrival of Buitoni in the late 1980s was welcomed because the primary market development was to the benefit of all manufacturers.

Consumers were also confused about the several dozens of pasta availabilities partly because of the proliferation of varieties and types of Italian food in recent years. There was fresh, dry, frozen, and canned pasta. Fresh pasta grew from almost nothing to about £12 million in 1992, a 300 per cent increase from 1986. Some research houses estimated the market for fresh pasta at £18 million in 1992. The two main varieties were filled or unfilled fresh pasta. In the 1990s, the market was consistently growing at between 10 and 15 per cent a year.

The lion's share of the over £100 million market went to dry pasta. Dry pasta divided into different product forms such as spaghetti (42 per cent), shapes, including spirals, twists, shells and bows (35.7 per cent), macaroni (8.1 per cent), tagliatelle (5.8 per cent) and lasagne (6.6 per cent). The fastest growing variety was shapes, which overtook spaghetti in terms of volume in 1992. Nestlé served the frozen entries through the Findus brand.

Casa Buitoni Club

Around the years of 1991 and 1992, Buitoni announced a series of marketing activities to the press and retailers that would eventually lead to the formation of a consumer club. It had been dubbed the Casa Buitoni Club experiment since Nestlé was experimenting with the building of the Buitoni brand on the basis of a one-to-one relationship with the consumer. It was likely to have a major impact on the whole of Nestlé's marketing strategy, if successful. Because this new initiative was launched by the subsidiary of one of the largest and most well-respected worldwide companies in the marketing of food products, developments around the Casa Buitoni Club experiment had been closely monitored.

At the centre of the initiative surrounding the Casa Buitoni Club was the realization that Nestlé, like many other companies, was finding itself increasingly cut off from the final consumer. One factor that caused this separation was the usual go-between, the retailer. There was the growing concentration and much-talked-about power of retailers which, particularly in the UK, had taken on enormous proportions. There was the booming private label business that made branded manufacturers unnecessary in the retailer–consumer relationship. There was the growing sophistication of retailers. For one thing, retailers had slowly become smart about their advantage of close contact and natural proximity to consumers. Retailers had become good users of market research information given to them through Direct Product Profitability (DPP) accounting, and Electronic Point of Sale (EPOS) systems, including bar-code scanners. These data were available to retailers ever cheaper and faster. Retailers increasingly used the information to fine-tune their stores' product offerings, to test market new products, and to make quick product listing and elimination decisions.

Moreover, retailers, whether multiples or independents, figured that if consumers could be turned into regulars, the implementation of so-called customer loyalty systems would suit them just fine. It would also be an effective weapon against the store competition from, perhaps newer, retailing formats as well.

Another factor was the quality of communication of the retailers with consumers. In Europe, it was not the branded manufacturers that spent most on communicating with consumers. It was, in fact, the retailers: in the UK, Sainsbury, Tesco and others had advertising budgets exceeding those of even some of the strongest brands.* In Germany, C&A, the discount department store chain, claimed the biggest advertising spending. Retailers like Tesco spent money communicating with consumers at two levels: about the quality and value of all their products in their stores, as in a classic umbrella advertising campaign, and also in particular about the quality and value of their private label business. Because Tesco's individual stores could place the tailored messages in local media, they effectively circumvented expensive mass media that had become a pain from a cost point of view, anyhow.

---

*Figures from the 1994 Register-Meal showed that Tesco spent £2.5 million on television, press, cinema and poster advertising. This compared to the spending of Nescafé of £1.6 million, which was the top spending individual brand in the UK.

Then there was an important factor that related to some basic economics of building brands. Buitoni was the market leader but Buitoni was also a small-fry brand for Nestlé in the UK. Nestlé would not be able nor willing to spend the kind of money that is needed on only one brand. This was the age of rising media costs, proliferation of channels and increasingly sophisticated consumers who question the value of advertising as a credible and truthful source of information about new brands and products.

For the large retailers that advertise entire product ranges, or for Nestlé as a whole, the economics still may work out, but not for one strategic business unit and a £10 to £15 million brand in a company with total revenues in excess of $36.5 billion. Moreover, pasta was an undifferentiated product. There was little brand loyalty since consumers saw one brand as essentially the same as another brand.

Lastly, there was also the particular situation of the pasta market in the UK. Per capita pasta consumption in the UK was one of the lowest in the European Union. The British consumer was unaware of the health benefits of this type of Mediterranean diet. He/she did not know about the ease of preparation of delicious pasta dishes. British consumers did not understand the advantages and benefits of Italian foods, and more specifically the value and quality of branded pasta over private label pasta. Not that British consumers did not know Buitoni. Awareness was already reasonably high (over 50 per cent), which was not surprising given Buitoni's leading share of the market.

The situation called for a development of the market through customer education, so-called primary demand stimulation. It was not just a matter of picking up some market share points from the competition. For this objective, mass-marketing methods of communication with consumers would be ineffective.

The solution for Nestlé and Buitoni was to bypass, in some way, both the retailer and the traditional forms of mass-media communication and to talk directly with the consumer.

The Casa Buitoni brand identity

Nestlé started with a clear definition of the brand identity of Buitoni and the role of the Casa Buitoni, the original villa of the Buitoni family in Tuscany, Italy. The identity was defined as 'the embodiment of Italian authenticity and genuine origins, representing 165 years of tradition, dedication and above all, a continuous commitment to providing the finest that Italian cuisine has to offer'. Buitoni represents the roots and history of Italian pasta. It was hoped that this would differentiate Buitoni from competitors, and would allow the company to charge a 10 per cent premium. The objective of the Buitoni brand was to become a helpful authority and an expert about Italian food, a brand, place and company where the consumer can turn to for advice.

Central to the development of this brand identity was the purchase and renovation of the original villa in Sansepolcro, Tuscany where the company was founded. The interior of the villa was remodelled so that it could serve as a research centre focused solely on the Buitoni brand (a Nestlé first). The villa also housed a public relations office, kitchens and hotel-style accommodation.

Activities in the research centre included research studies on new frozen, refrigerated, dehydrated and sterilized products. There was extensive experimentation with new Italian recipes and products. A group of experts on Italian foods that consisted of food writers, chefs and home economists formed the Buitoni appraisal panel, which gave advice, tasted and tested the new products.

A picture of the Casa Buitoni was taken and became the logo of the brand. The picture's headline became 'Dalla Casa Buitoni – 1827' on all products. Packaging was upgraded and the new brand logo was added. It was hoped that the modernized brand logo encapsulated well the heritage of the brand and provided a statement of the company's Italian roots. In addition, the logo served as a seal of quality for all Buitoni products.

Building the brand

As far as Buitoni's marketing strategy was concerned, the company took two directions that led eventually to the launch of the Casa Buitoni Club. First, the brand was extended into different categories and product forms, and second, a database was built via different forms of communication media.

The year of 1992 saw the launch of Buitoni Risotto, the first branded risotto in the £96 million UK rice market. Il Risotto included three rice dishes. Rice happened to be a very popular dish in the Northern part of Italy.

There was also the launch of Buitoni Fresco in autumn 1992, a fresh pasta range that had been in a three-month test market. The Fresco brand came to the UK via Switzerland. It was owned by Hirz Frischprodukte, a Swiss dairy products manufacturer which had been acquired by Nestlé a few years earlier. Buitoni was considered the first major brand in the fresh pasta market, which had been mainly served by small producers such as Rossi and Pasta Reale and private labels. Because of the need of frequent store delivery of the six pastas and three ready-to-serve sauces, Nestlé distributed the Fresco range via its Chambourcy chilled dessert division.

In addition to these major launches, there were numerous additions of pasta types and varieties in every sector of the market. In the fresh pasta market, there were four additions of shapes of pasta as well as sauces in 1993, which expanded the Buitoni range to a total of ten. The dry pasta market had seen the introduction of a range of short, high-egg content pasta shapes with the brand name Le Preziose (the precious) and a huge number of other shape innovations. By the end of 1992, there were a total of 25 dry pasta products with an additional ten products in the ambient line.

Since Nestlé acquired Buitoni, the brand and these product innovations had been well supported through advertising and promotions. In 1989, Nestlé spent £1 million on television in a sort of test market of the effectiveness of TV. The same amount was spent in 1990. The 30-second commercials had the strap line 'Share the Italian Love of Foods'. The experiment lasted only for a few months. In 1992, Buitoni spent another £1.5 million in brand support, with a huge proportion spent on the relaunched 'Share the Italian Love of Food' campaign, this time in the press. The marketing budgets increased in 1993 and 1994 to £2.5 million and £3.5 million, respectively. Buitoni was by far the biggest spender in the pasta category in the UK. Important was the shift of spending from above the line marketing support to below the line marketing support. While in 1992 the spending above the line was 60 per cent of the budget, the 1993 budget allocation was reversed; 40 per cent went to above the line support while 60 per cent went to below the line support. The 1994 budget split between above the line support of 45 per cent, a third on Casa Buitoni Club related activities, and the rest went to below the line support.

Between 1992 and 1993, Buitoni's marketing strategy consisted of four main programmes: a direct-response television and print campaign, in-store promotion, sponsorship and public relations.

Direct-response print advertisement. The direct-response print campaign broke in the spring of 1992, featuring a set of quick and easy authentic recipes. These ads were placed in women's magazines with a view to developing and strengthening consumers' perceptions of Buitoni's Italian heritage and origin, the attractions of the Italian lifestyle, and Buitoni the approachable authority on fine Italian cooking. Consumers were invited to call a London telephone number for money-off vouchers, new product information and more recipe ideas. There were 50 000 consumers who called to obtain the mailed information pack. When Buitoni launched the next round of the campaign in the spring of 1993 (a £1.5 million press campaign), the on-page recipe ads invited consumers to call or write for a step-by-step recipe book, and an offer for a free packet of spaghetti. The press included mainstream women's magazines such as *Cosmopolitan* and *Good Housekeeping*. In June of 1993, another campaign was added to emphasize the Buitoni Fresco range with the strap line 'When we say Buitoni Fresco is freshly made in Italy, we mean freshly made in Italy'. This three-month campaign was run within the 'Share the Italian Love of Food' umbrella campaign.

Direct-response TV advertisement. Buitoni also launched a small £0.5 million TV campaign in 1993 after earlier test runs. Like the continuing print campaign, the direct-response ads invited consumers to call in for the recipe booklet and the like. The ad copy, as did the print campaign, emphasized the Buitoni colours, and the conviviality of Italian food. A musical theme had been added to the TV ad copy. *In-store promotions.* The press and TV campaign were supported by sampling and taste demonstrations from the pasta range of Buitoni. These sampling activities were carried out inside large multiples such as Tesco. Consumers could also pick up money-off coupons that encouraged trial purchasing. At the sampling stands, recipe booklets and other leaflets were also distributed.

Sponsorship. Buitoni also signed a two-year sponsorship agreement to become the official sponsor of the UK's largest road race, the BUPA Great North Run. Every year, the race starts in Newcastle, the location of Buitoni's factory. As part of the sponsorship, Buitoni invited 30 000 athletes the night before the race to a pasta feast. The cost of the sponsorship was in the neighbourhood of £50 000. In 1993, Buitoni was also testing the grounds on the sponsorship of certain television programmes such as the *De Medici Kitchen*, which was a cookery programme aired in the UK.

Public relations. In order to get the full benefits of the company's market development activities, Buitoni allocated part of the marketing budget to public relations activities. The PR agency signed up the Great North Run sponsorship, for example.

The Casa Buitoni Club

The objectives of the Casa Buitoni Club were not just to have a database of consumers who were responsive to Buitoni communications such as sales promotions and other consumer offers. The idea behind the club was to develop a deep and strong sense of brand loyalty through the linking of the Buitoni brand values with a large audience of Italian food lovers in the UK. Management used the buzz words of warmth, empathy and the sense of belonging and lifetime value of a customer when talking about the club. It was hoped that the club will allow Buitoni to communicate intimately and passionately with interested consumers, thus building the Buitoni brand into the trustworthy authority and expert in things about Italy and pasta.

Nestlé eventually launched the Casa Buitoni Club in November of 1993. The company mailed invitations of membership to those who had registered and had expressed an interest in Italian food during the campaigns since 1992. There were over 150 000 names and addresses. The time had arrived for the experiment in brand building to begin in earnest.

To those who responded to the invitation, Buitoni sent an Italian lifestyle information pack. The company also provided a care line for members who wanted to call and ask for cooking advice or suggestions. This created a situation where the consumer actually talked back to Buitoni, a step towards the much-hyped consumer dialogue over advertising monologue. On a regular basis, members received a full-colour newsletter with editorials about Tuscany and Italy, information about the lifestyle of the Italians, numerous suggestions for cooking pasta, and money-off vouchers.

Buitoni further offered sweepstakes to members. The winners were invited to visit the original Casa Buitoni villa. Club members were also offered travel arrangements to the villa for cookery weeks, and invited to provide input into the new product development efforts at the research centre. Even those members who could not visit the Casa Buitoni villa had the opportunity to voice their opinions. They could participate in pre-market tests of new products from the convenience of their home. Members could also suggest their own events and services as part of the club. In this way, Buitoni hoped to develop a sense of ownership of the club among the members.

Buitoni in Europe

While the Casa Buitoni Club experiment was off to a good start in the UK, Buitoni was running two steps behind Barilla and local competitors in Europe's individual country markets. Barilla was spending significantly more across Europe than Buitoni. For example, in the year ending in June of 1992, total television ad spending of Barilla was $237 million versus $19 million for Buitoni in Europe.*

The marketing strategy of Buitoni in Europe seemed to be shaped by the strategy of Barilla and other local competitors in Europe. Buitoni's proportion of above the line spending was significantly higher in the rest of Europe than in the UK. But even though the spending pattern differed, Buitoni had harmonized many of the other key features of its European strategy. There was no adaptation of the product to local tastes. All marketing communications carried the essential elements that expressed and translated the Buitoni brand identity to the consumer: the Buitoni colours and brand logo, the musical theme, the emphasis on conviviality and the Italian love for food and authenticity.

There were also Casa Buitoni Club members from other parts of Europe, but these consumers had called and written to Buitoni in the UK for the membership kit. Management from Germany and France closely monitored the developments around the Casa Buitoni Club for possible adoption in the future.

Meanwhile Nestlé chose the one-to-one relationship marketing route as the best approach to stimulate the primary demand for pasta among consumers, competition had begun to form some sort of advertising alliances with the same objectives. In Spain, for example, the manufacturers Gallo, La Familia, Barilla, Rivoire Carret, Grupo Fló (El Pavo) and Oromás pooled some of their advertising resources for a generic ad campaign. They hoped to increase consumption in the Spanish market from 3.9 kilograms to 5.0 kilograms of pasta per person per year. In Germany, three leading Spätzle manufacturers formed the Spätzle venture to develop further the market for this German pasta speciality and, more importantly, to stem off competition from Barilla and Buitoni.

---

*These advertising budgets were reported in *Campaign*, 27 November 1992.

Case 2.2 Carlton Electronics

This case was written by Professor Martin Christopher, Cranfield School of Management as a basis for class discussion rather than to illustrate effective or ineffective handling of an administrative situation.

Philip Millard, the marketing director of Carlton Electronics, was examining the results of a customer survey concerning the quality of Carlton's customer service. As he looked through the results, he began to consider possible courses of action that he might take. However, he was not clear in his own mind which direction the company should take.

Carlton was a producer of a range of telephone answering machines and related peripherals. It was not a large company, but its reputation as a quality supplier was well established in a fast-changing market. Marketing practices at Carlton had followed a traditional approach in the past, selling mainly through British Telecom. As the telecommunications market changed, Carlton had been forced to keep pace, although they had not taken a position of marketing leadership. First, there had been a wave of new competitors from continental Europe, in which competitive products were offered to customers at slight discounts from Carlton's quoted prices. Carlton had responded, not by matching prices but by investing in a marketing programme emphasizing the quality which had identified Carlton products. The next wave of competition from Japan was more difficult to deal with. There were some Japanese lines that had become established as equal quality but at a significantly lower price. In addition, there had been a flood of low-priced imports from South-east Asia which looked as if they were going to take over the consumer market completely.

Carlton's immediate customers fell into three categories: large chain stores specializing in consumer electronics, British Telecom and discount operators. British Telecom had been the basis for Carlton's original markets. However, with the growth of competition through deregulation, Carlton had to rethink its strategy. Within the last few years, new customers had emerged in the form of large specialist chain stores and discount stores. Where British Telecom had previously tended to take the full product line and maintain the established price structure, chain stores tended to be more selective and also to press Carlton for promotional discounts. The discounters were far more aggressive in their dealings with Carlton. They only took high-volume lines, but moved these products in significant volumes. They asked for substantial discounts on cumulative quantity purchases, but only passed part of these discounts on to the consumer.

The volume of business with British Telecom had held steady, while that of the chain stores had grown. The promise of future growth, however, lay with the discounters because they were getting a larger share of the consumer business.

Carlton dealt directly with retail buyers, who as a group were extremely conscious of service from suppliers. There was an attitude which Carlton's sales force sensed of 'What have you done for us lately?' This was obviously most pronounced amongst the discounters. They tended to operate on weekly order cycles, and wanted to manage inventory on a just-in-time basis.

The chain stores followed similar patterns but perhaps not so intensively. Most were on a two-week order cycle and maintained adequate stock levels in their own distribution centres as back-up to their stores. They were beginning to enquire about direct ordering from their computers to Carlton's computers, which Carlton was totally unprepared for.

British Telecom were the least aggressive, being the most traditional member of the industry. They ordered in a variety of ways: sometimes using standard order quantities, sometimes on monthly order cycles. Carlton valued British Telecom as customers because they presented the full product line, which the other customers did not do. Therefore they could not be pushed aside.

The customer service survey

Carlton had commissioned the customer service survey to gain a better understanding of customer service issues. No one in the organization had looked at customer service in any detail before. Customer complaints had usually come through the sales force which had passed them on to the sales manager who in turn referred them to the distribution manager. These complaints referred to a variety of different problems: delivery, credit, representative call frequency. They were hard to pin down in a specific pattern. The marketing director wanted several answers from this survey: were there specific areas of complaint that stood out? How was Carlton's performance compared to competition and what areas should Carlton try to improve?

The survey was conducted primarily through direct interviews with a selected number of key customers amongst chain and discount store buyers. In addition to general comments, questions were asked seeking comparative ratings of Carlton versus major competitors on specific aspects of service. The survey covered several specific aspects of customer service: delivery time, delivery reliability, the quality and knowledge of sales representatives, inventory availability, communications and order procedures, advertising and promotion, and packaging. The consultants were to report the results of the survey in a form that Carlton could use to take action.

The general results of the survey were reported as follows:

Delivery time. The largest suppliers were about equal in terms of delivery performance. However, as a group they were not equal to the performance of smaller competitors. Carlton's average times were considered to be acceptable, but their performance had slipped in terms of reliability. Customers could not be sure when their orders would arrive. On a local basis, this was a problem for Carlton's local delivery despatching. At a national level, it reflected on the performance of the carriers that Carlton was using.

Inventory availability. Carlton was again comparable in performance to its major competitors, fulfilling about 80 per cent of orders from stock on hand. Smaller competitors with their restricted product lines tended to do better. Carlton's back-order performance was better than average. Customers recognized that back-orders from Carlton would always be filled, even if it took a couple of months. When orders were picked there would be occasional errors, which meant that items, once shipped, had to be returned. This seemed to lead to considerable confusion on Carlton's part.

Sales representation. Carlton had very low turnover among its sales representatives, which was well received in the trade. They knew their products and market policies in general. Where they fell down was in dealing with distribution problems, because they had little direct contact with distribution operations. When there was a distribution problem, it seemed to take a long time to get an answer, and then it was not always satisfactory. When orders could not be filled, Carlton did not notify customers, assuming that they understood that products would be delivered when available. The chain and discount buyers, however, noted that they were getting precise indications when their orders would be served and shipped from competitors and found it useful for inventory planning for their own operations.

Orders. Customers expressed considerable annoyance about Carlton's ordering system. There were frequent errors in recording and filling orders, and there was no confirmation when the order was received. Orders were normally entered from written purchase requests from customers, sales representatives' order forms or by telephone requests by customers. When they were received, orders were taken in the sales department and entered in the computer. In the case of documentation, customers expected confirmation, which some suppliers furnished automatically but others did not.

Advertising and promotion. This was not a strong point with Carlton. They relied on their retail customers to promote specific products. Most ads for Carlton were corporate, emphasizing product quality in general, rather than specific items in the line. This was in direct contrast to the most aggressive competitors, who ran campaigns to push specific products and to support major selling events.

Packaging. This was an important area but involved conflict. Carlton's policy had always been to pack well. Items were always well-labelled. However, some of the chain and discount buyers had begun to use laser bar-code scanners and wanted suppliers to adapt their packages to the new system of item identification, requiring preprinted bar-codes on a non-reflective surface. Neither Carlton nor the competition had done this yet.

Customer issues

Further analysis of the survey results suggested that the three major customer groups (British Telecom, the chain stores and discount operators) had different service priorities. British Telecom placed greater emphasis upon the importance of sales representation and general marketing support whereas the chain stores were more concerned with on-time delivery and order-fill levels. Not surprisingly, the discount operations, primary concern was price and the terms of trade, but inventory availability during limited period in-store promotions was critical.

Underlying all of these issues was a shared concern with shelf-space profitability. Retailers were generally much more aware of the importance of using profit per square metre as the measure of store performance and as a result were beginning to examine more critically the real drivers of shelf-space profitability.

Philip Millard recognized that in order to strengthen relationships with these major customers a new approach would be required. He also recognized that to be successful a total business focus was necessary. In other words, it would not be enough just to strengthen the existing marketing mix but rather other key business functions such as distribution, materials management and procurement had to be involved in creating this new approach to customers.

Millard was also aware of the implications for organizational change that would arise if Carlton was to become more customer service oriented.

Case 2.3   Does Tesco hold all the cards?

This case was prepared by Helen Mitchell and Helen Peck, Cranfield School of Management, as a basis for class discussion, rather than to illustrate effective or ineffective handling of an administrative situation.
© Copyright Cranfield School of Management, December 1997. All rights reserved.

Introduction

On 13 February 1995, grocery multiple Tesco launched Clubcard, the UK's first national supermarket loyalty card scheme. A spokesman for Tesco explained that the principal objective of the scheme was not to lure shoppers away from competitors' stores. Clubcard was, he claimed, 'a way of saying thank you to existing customers'. He went on to add that Tesco was aiming to recreate the kind of relationship that had existed between local shops and their customers 50 years ago.

In the weeks following the Clubcard launch Tesco's leading competitors were dismissive of the whole idea of loyalty cards. None more so than David Sainsbury, who regarded the Clubcard as 'the Green Shield Stamp way of offering value'. In his opinion the scheme would 'cost at least £10 million just to administer. That's wasted money which brings no benefits at all to customers. We have no plans at all to go down that route.' The 1 per cent discount offered by Clubcard was, he believed, of insignificant value to the customer. Observers in the City were equally sceptical and the anticipated additional marketing overheads were cited by financial analysts as being partially responsible for the two pence mark-down in Tesco's share price following Clubcard's launch.

Nevertheless, speculation was rife that within a matter of weeks other leading supermarket chains would be forced to follow with their own national loyalty schemes. Market leader J. Sainsbury already had its own Saver Card scheme in place in a limited number of stores, as did third-placed rival, Safeway, with its Added Bonus Card (ABC). Both schemes were devised as promotional devices, offering small percentage discounts to shoppers in a bid to boost sales at poorly performing stores.

Industry background

Tesco's initiative had served to heighten the competition in the already fiercely competitive supermarket sector. The total value of the UK grocery sector in 1994 was £54.9 billion[1] and was dominated by larger businesses (defined as those with 1990 turnover of £4.5 million or more) who generated £49.7 billion of total sector turnover, whereas small businesses had only a 10 per cent share with £5.2 billion. (Several major multiples dominated the sector: Sainsbury, Tesco, Safeway [Argyll group], Asda, Somerfield [Gateway, and Somerfield stores], Kwik Save and Iceland, together with the co-operative societies). Since 1990 large grocery retailers had advanced their turnover by an average of 40 per cent, whilst smaller ones had grown by only 8 per cent. By 1992 chains with 100 or more stores accounted for 68.5 per cent of sector sales.

The market was commonly beset by price skirmishes, with each store trying to wrest the price initiative from their rivals. Growth had not come from traditional product categories and consumer expenditure on food as a percentage of total consumer expenditure had actually fallen over this period but it was the larger supermarkets that were better equipped to sell products in other expenditure categories to achieve growth. The market was considered to be saturated apart from the discounters, and the multiples were selling more or less the same things for the same prices. However, one area where there was seen to be a difference was in the demographic profile of the shoppers for each chain: J. Sainsbury was seen to have an older, more up-market clientele, Tesco's profile was much younger, with a larger percentage of 25 to 45 year-olds and Safeway's was very similar to Tesco. Kwik Save, with most stores situated in poorer city centres in the North, had an older, less affluent customer profile, whilst Asda catered for the younger family on a tight budget. Table 2.3.1, showing sector performance for 1993, shows Sainsbury with a clear lead, Tesco in its

Table 2.3.1   UK grocery sector, performance figures, year ending 1992/93

1992/1993	Market share[1] (% of total market)	Turnover (£ million)	Pre-tax profits (£ million)	Profits as % of turnover
Sainsbury*	17.3	9685	732.8	7.6
Tesco	14.9	7581	580.9	7.6
Safeway**	8.6	5196	417.3	8.0
Asda***	8.4	4614	142.1	3.0
Kwik Save	5.8	2651	126.1	4.7

Source: Taylor Nelson AGB: Retailer ShareTrack and Reuters.

Notes: * Sainsbury + Savacentre, ** Safeway + Presto + other Argyll stores, *** Asda + Dales.

[1] Assessed across 179 selected markets (packaged goods, fresh foods and toiletries) by Taylor Nelson AGB, which is recognized by the sector as the most comprehensive measurement. It is not related to annual turnover but represents each retailer's share of these specific markets. Annual turnover is made up of non-grocery goods, e.g. petrol, fashion clothing and in Sainsbury's case also includes Homebase sales, and as a result market share cannot be easily or fairly judged by turnover figures.

familiar second place and the remaining players fighting it out below them.

The way forward

In an industry where margins were slim, increasing market share was vital. Due to its position in the league table Tesco had been battling to improve its performance. 1991 and 1992 had been bleak trading years for them and gloomy early figures for 1993 had shown that they had lost 3 per cent of customers in the previous year. In response to this decline, research was undertaken in order to throw some light on the reasons for the fall-off in customer numbers. Results showed that customers wanted better value, service and responsiveness as well as a supermarket that was innovative. Under the charge of Terry Leahy, then Marketing Director, a strategy plan was devised on the basis of these findings which concluded that Tesco 'should aim to be positively classless, the best value, offering the best shopping trip. This will be achieved by having a contemporary business and therefore one that remains relevant by responding to changing needs. We should aim to be the natural choice of the middle market by being relevant to their current needs and serving them better, i.e. customer focused.'[2]

In the early 1990s the corporate strategy had been margins-focused and efforts had been directed at extracting higher returns in a nationwide cost-efficiency drive. Under the new plan, strategy was changed to the aim of delivering the best shopping trip for the customer. This covered a variety of attributes so, starting in 1993, Tesco embarked on a series of initiatives designed to offer better value, improve the stores, and give a higher level of service to their customers.

Better value

The Tesco shopper was found to be a value-oriented customer and, in response to requests for cheaper prices, Tesco launched a commodity range across 110 core products. In their distinctive blue and white packaging, they were designed to communicate value for money for everyday goods and negate the need for cross-shopping at discount stores. Customer response was very positive and trading improved. This was then followed by a reduction in the price of popular, key branded items, to give the Tesco shopper the cheapest prices in the marketplace. In 1995 this was extended to key lines of fruit and vegetables, the most notable of these being bananas. This sparked a mini banana war in the high street in which the price fell from 49p to 19p per pound in the space of a few weeks, one of the lowest across the world. In-store bakeries were next to offer cheaper prices (a category which had only previously been seen in the northern super-market chains, Morrisons and Asda, offer value lines), resulting in a significant increase in customer volume. Pricing initiatives culminated in the launch in September 1996 of the 'Unbeatable Value' scheme in which the prices of over 600 core products were reduced, together with a promise that if the price can be beaten then Tesco will refund twice the difference.

Improved stores

The research had shown that shoppers found it difficult to find products in existing stores and that they perceived the shops as sterile, grey and industrial in style. As a response to that, and also to forthcoming government curbs on superstore development, new, more flexible, store formats were introduced. The first were called 'Compacts', slightly smaller than superstores and more suitable for market towns, while 'Metro' marked a return to the high street and served the sophisticated urban shopper. Finally, development of the 'Express' format saw the Tesco brand enter the petrol station/convenience store sector. Existing stores were also improved and many stores were upgraded and extended with the customer in mind. The next three years saw the introduction of many new product ranges: meat counters, pharmacies, textiles, hot chicken counters, delicatessens and fresh fish counters, all of which were very popular and well received by staff and customers alike.

Customer service

Research had also shown that service was an important component of the shopping experience. In 1993 Tesco began a series of Customer Panels in which they recruited shoppers for half-day sessions with local managers and head office staff, to ask them exactly what they wanted from their shopping trip and encourage them to suggest improvements. Issues and tasks which arose from the panels were then allocated to either corporate or local personnel, whichever was more appropriate.

In order to manage the changes, Customer Service strategy was then broken down into three areas: Facilities, Standards and Culture. Facilities were upgraded with the launch of the 'New Look' initiative in 1993, which introduced tangible innovations to improve the shopping experience. For example, hand towels were placed in the meat aisle for wiping hands that were sticky after handling meat packs, barriers were removed at the entrance to the store as a result of customer requests, despite their obvious security advantages. More checkouts and customer service desks were installed and all signage was changed to blue and white and made much easier to read. The aim was that customers would reach any point in the shopping trip and think that Tesco understood shopping from *their* viewpoint. Overall, there was a range of over 100 different improvements, for example, meat counter towels, fish display units, specialized trolleys and so on from which a regional director and local store manager would choose, costing from £20 000 to £500 000, to achieve the 'New Look' for each individual store.

The next element of service was about standards, i.e. setting, measuring and managing them in each store. Queuing was known to be one of the things customers disliked most about shopping. However, further research showed that if there was only one person in front of them and they were busy unloading the trolley, they didn't consider that they were queuing. In response, autumn 1994 saw the introduction of 'One in Front', in which £15 million was spent in ensuring that Tesco could pledge to immediately open more tills if there was more than one person in front of a customer at any time. Launched with a major advertising campaign, it was regarded by management as one of the most successful initiatives to date.

Culture was considered as the most difficult and intangible element of the service strategy. In 1992 Tesco had begun to address this issue with the launch of 'First Class Service'. It was a radical change and very different to the centralized 'command and control' style that previously operated. Each member of the 130 000 staff was given responsibility to look after customers in the way they thought best and managers were then encouraged to recognize staff achievements and to treat them as individuals, so that they in turn treated customers as individuals too. After four years the scheme is considered by management as a remarkable success. Staff now perceive each and every customer as valuable and are aware that keeping them loyal to Tesco means that, over a lifetime, each individual has the potential to spend an average of £90 000 with the company. The initiative was developed further with the introduction of Customer Assistants in every store; £20 million was spent in hiring and developing staff to be responsible only for helping customers in each store, the idea behind this being that ultimately they would come to know them by name, seek them out and develop a one-to-one relationship with them. The culture and climate created by these initiatives is seen as 'pure alchemy' by the management, which they believe has allowed each individual store to develop stronger alliances with their own customers.

The Clubcard launch

The Clubcard launch in 1995 was also an integral part of the move to a customer-centred strategy. Launched as a 'Thank You Card', it allowed Tesco to develop one-to-one relationships on a corporate level. The cost of running the scheme is considerable. In addition to the 1 per cent discount on sales, it was estimated that start-up costs alone were £10 million but the company was convinced that this was money well spent. During pre-launch trials at 14 stores, over 250 000 Clubcards were issued, representing an uptake at the sites involved of between 70 and 80 per cent. During the trials, high spending customers were identified and given special treatment, including invitations to 'meet the staff' cheese and wine evenings at their local stores. The customers seemed to appreciate the events and responded favourably to the Clubcard.

This success paved the way for the UK national launch. Membership of the scheme was open to all Tesco customers through any of its (then 519) stores. Cards were issued on application and were able to be used to accumulate points on every shopping trip. The customer presented the magnetic stripe card at the checkout, where it was swiped through existing credit card reading equipment. Details of the customer's purchases were recorded, with Clubcard points automatically awarded for every £5 spent in the store, over a minimum of £10 per visit. Points are added up quarterly and (provided that the customer has accumulated a minimum of 50 points) money-off vouchers are posted directly to customers' homes to be redeemed against future spending. In addition to the vouchers, customers also received money-off promotional vouchers for specific branded goods.

The costs of the mailings were high. By October 1996 these were estimated to be £11 million alone, but Tesco was seeking a return from this investment. The card was an important tool for gathering individual customer data: what they purchase, how much they spend, when and how often they shop, with the information revealing a great deal about the lifestyles of the shoppers themselves. It allowed Tesco to segment its customer base according to real purchase behaviour, rather than a version of purchase behaviour based on demographic or socio-economic stereotypes. The intention was to use the data to build loyalty through tailored, value-based offers, mailed to the homes of specific groups of customers. Early indications were encouraging, with 300 people turning up one morning at a store for breakfast in response to a Clubcard offer.

By the end of March 1995, one month after the launch, over five million people had joined the Clubcard scheme and Tesco recorded a like-for-like increase in sales. Clubcard's impact on sales was confirmed by independent researcher AGB who announced that, according to their Retailer Share Track monitoring system, Tesco had surged ahead of Sainsbury (for the first time) to become Britain's leading retailer of packaged goods. Tesco now had 18.1 per cent of the market, with a 2.1 per cent increase in market share since Clubcard's launch. Household penetration had increased by almost 1 per cent, meaning that an additional 200 000 households had come to shop at Tesco's stores. Over a third of the gains were reported to be at Sainsbury's expense. Tesco endeavoured to play down reports of market share gains, saying it was the fact that it was pleasing its customers that really counted.

Reaction from competitors was mixed. Safeway increased the number of stores participating in its ABC scheme from 25 to 106, to fend off direct competition from Tesco at vulnerable sites. They also began to develop the ABC card data capture capabilities so that, although points were redeemed in-store, they could still accumulate customer information. Sainsbury, although widening availability of the Saver card, initially resisted any further reaction. It still saw the Clubcard as a margin-slashing initiative and as having little effect on its own performance. Their results for the year ending May 1995 appeared to justify this stance: pre-tax profits were up by 10 per cent to £808 million and yearly sales of £12 billion were the highest for any UK-based retailer. David Sainsbury, when asked about Clubcard, said that it had been launched at substantial cost, 'We are still not convinced of the benefits to customers but we are not against cards *per se*. Indeed, our own scheme rewards customers at the point of sale, which is far less costly to operate than theirs.'

Early days for Clubcard

After the enormous success of the launch and as customers became used to using the card, Tesco began to realize the full implications of running a nationwide loyalty scheme. Communication was proving to be a two-way process, hundreds of letters and up to 30 000 calls being received every week from customers keen to find out more or comment on the Clubcard service. It became clear that refinements would be needed, many in direct response to customer requests. One of the first of these was the introduction of a second card to enable another family member in the same household to have their own card, which proved to be more convenient for the two shoppers in the household. Another was to lower the minimum spend required for pensioners to £5, since many had complained that they were unlikely to achieve the £10 amount to qualify for points, as they were living alone and did not buy large volumes. It emerged from early Clubcard data that pensioners shopped very frequently but filled smaller baskets. This was also later applied to students through the introduction of a Student Clubcard with a lower spending threshold.

The scheme also allowed Tesco to form alliances with other brands. One of the earliest was with Thomson Holidays, who allowed points to be redeemed for holiday discounts, and also B&Q, where Clubcard holders could earn points by spending on any product within their stores and accumulate Clubcard points to be redeemed in Tesco stores.

However, one of the most successful refinements was the introduction of points for petrol purchases. At the time of its introduction many of the major petrol retailers were involved in competitive schemes to increase market share. Esso had recently introduced the 'Price Watch' promotion promising the lowest prices, whilst Shell was developing the 'Smart Card' scheme, in which they were recruiting a number of high profile organizations and brands to join forces behind their loyalty card scheme. At one point Sainsbury was rumoured to be committed to joining the Smart Card retailers. For Tesco, the introduction of points for petrol, combined with a price pledge, resulted in a large increase in petrol sales for the group.

However, not everything was plain sailing. One of the earlier problems to emerge with Clubcard was in the sheer amount of data that was being collected. As well as having one of the largest customer databases in the UK, this was also continually updated with purchase information every time an individual customer used their card. Initially there were rumours of data overload as the system struggled to cope with the volume and complexity of the data being collected.

Another issue to tackle was exactly how to use the information collected. The database management was initially outsourced. Meanwhile the consultants worked together with an in-house team, with the intention of Tesco developing their own expertise over time. A timely reminder of how difficult it was going to be was evident in the first mailing of money-back coupons to over five million Clubcard holders in May 1995. The first mailout, worth over £12 million in money-back vouchers, also offered money-off promotional vouchers for specific goods. Initially these were for an own-label product, PG Tips tea and Coca Cola. The response was enthusiastically received by members but many also wrote to say that they were mystified by receiving offers for Coca Cola, one typical complainant remarking that he had never drunk the product in his life and, as he was 85 years old, was very unlikely to start now!

This aspect of the scheme also attracted criticism by the Direct Marketing Association, for giving all customers the same reward. Tesco admitted that in the first mailing there were only slight regional variations in the offers. One commentator said, 'it makes no sense to spend the same amount on all customers and the more personalized and specialized the communication with customers, the better'. This proved to be an early lesson for the database team and as a result the next mailing to pensioners was an offer for biscuits, which had a more successful 25 per cent redemption.

Success of Clubcard

Six months after the launch of Clubcard, the company was able to report concrete evidence of their success. Figures showed a 16 per cent rise in half-yearly pre-tax profits to £290 million, together with a 2.1 per cent increase in market share. More importantly, the group also reported an increase in like-for-like sales of 10 per cent – its best for over five years. Whilst stressing that the Clubcard is part of its long-term customer service plan, Sir Ian McLaurin, Tesco chairman, attributed a large part of the company's success to the scheme, 'We were determined to be the first and now Tesco Clubcard is *the* loyalty card in retailing. There is no doubt that it has helped boost our profits.' He commented that the card now had almost seven million members who had received £25 million worth of vouchers so far, of which £24.5 million had been redeemed.

In contrast, Sainsbury's figures were now beginning to show the possible knock-on effect of Clubcard, with like-for-like sales down by 2.2 per cent over the same six months. David Sainsbury conceded that the Clubcard effect had eroded individual store sales by 0.6 per cent, where a Sainsbury store was in direct competition with a nearby Tesco, which in his eyes 'was not a very great one'. David Simons, CEO of Somerfield (no. 5 in the industry league), openly disputed this claim and calculated that it was much more likely that Tesco would be taking nearly 3 per cent in sales from any nearby rival. Despite the insistence that the effect was of little consequence, Sainsbury widened trials of their Saver Card to 170 stores in June (although it still did not have the data capture facilities of Clubcard). Meanwhile the Marketing Director, Antony Rees, stressed that they would not be rushing into a national scheme, 'Why come first and get it wrong? We would rather come second and get it right. We will continue to listen to our customers to see what they want and react to *them*, rather than our competitors.'

However, they did extend the Homebase Spend and Save Card (which was one of the UK's first loyalty cards), to include all purchases at the recently acquired 214 Texas Homecare stores. The vouchers earned from expenditure on DIY and gardening products in these stores could be redeemed in all of Sainsbury's UK outlets, including supermarkets. Analysts saw this as a sensible stopgap measure to counter the impact of the Clubcard. Sainsbury did confirm that the company was looking at ideas for a loyalty card with better benefits than Tesco. David Sainsbury was quoted as saying that, 'we want to find something which gives better value to the customer, which really rewards loyalty. I remain to be convinced that a card which gives just a 1 per cent discount on all your purchases gives the customer much advantage.' Another Sainsbury's spokesman said that the company's experience with Saver Card had shown that interest had waned after six months, but Spend and Save was different because spending on DIY products and the margins involved meant higher discounts could be given and it was much easier for customers to accrue points. He added that Spend and Save would never become the standard loyalty card for the company's food stores.

Coinciding with the widening of the Homebase card scheme was the launch of the Sainsbury's 'Customer First' campaign in June 1995, in which they pledged to offer a variety of customer services, including extra packers, freephone customer hotline, baby changing facilities, parent and child parking and 10 different types of trolley. A company spokesman said that this was a direct response to customers, 'We talked to them and a demand for old-fashioned shopping came through loud and clear.'

Safeway entered the arena a few months later by offering the ABC card nationally in the autumn of 1995. Many commentators saw the offer of one point for every pound spent as better value than the points system offered by Tesco. ABC cardholders could also claim money off particular products and take up special offers, such as a free car wash, dry cleaning or cinema tickets. One advantage which Safeway had over Tesco was that their points were redeemable at point of sale, so they had a greater opportunity for promotional contact with their customers than Tesco's quarterly mailouts, as well as the fact that mailing costs were not incurred. By the end of December 1995, Safeway were able to announce that they had successfully enlisted 3.5 million members to their ABC scheme (approximately half its customer base).

Other retailers were by now also rolling out their own loyalty schemes. For instance, Somerfield partnered with Argos in their Premier Points scheme and Budgen, a smaller retailer with a significant presence in the South of England, launched a Visa card. Their card gave back 5p in the pound to shoppers using it in their own stores and £10 for every £400 spent on the card elsewhere. Asda, whilst not launching a card of their own, insisted they would stick to price pledge initiatives, offering customers value for money. They also announced that they would meet Tesco head on by honouring any Clubcard voucher in their own stores, which caused Tesco to complain to the Advertising Standards Association. This was to little avail as the Association could only meet to review the case after Christmas and so Asda reaped the benefits of its offer during the busy pre-Xmas rush. Shortly after this time, Sainsbury quietly dropped their Saver Card scheme, claiming they could not justify the £30 million cost to extend it on a nationwide basis.

In February 1996, one year after the launch of Clubcard, Tesco's market share stood at 19 per cent.[3] Although a part of this rise was due to the takeover of William Low supermarkets in Scotland, the card was credited with raising the average weekly spend to £75 and the average spend of each shopping trip from £23 to £26. Over the first year the cost of Tesco's 'thank you' was estimated to be at £60 million and operating costs £8–10 million. However, it was against a background of customer-led initiatives that the full value of the Clubcard could be assessed. Tim Mason, Tesco's recently promoted Marketing Director, saw the real impact working at the local level, 'I believe that the key to the future is going to be an individual response to individual customers. It is no good talking about which is the "best shopping trip"; you have got to deliver the best shopping trip for each and every customer.' He predicted that each of the 524 stores would become a marketing unit where staff would get to know their own customers individually and also confirmed that 'Micro-Marketing will be at the forefront of our future strategy.'

Clubcard Magazine

By now, due to the database facilities that had been developed, segmentation of Clubcard members was also possible at the corporate level. In March 1996 the store announced the planned launch of the first Clubcard Magazine. Produced in five different versions to suit five different lifestyle bands (students, younger adults without children, younger people with families, older people with children and pensioners), each version of the magazine was tailored to suit its audience. Students would read features about overseas travel, parents would read about family issues and so on; offers in the magazine were also designed to be appropriate to each group. Each publication carried promotions, competitions and advertising from many suppliers as well as information on new product launches on branded and own-label goods. The magazine proved to be highly popular with high redemption of money-off tokens, which Tesco attributed to their value and relevance for each of their customers.

The launch of Clubcard Plus

By late spring, it was clear that Tesco had achieved dominance in the UK food retailing industry. This was confirmed by the subsequent news that Sainsbury had reported a dramatic fall in pre-tax profits of around £100 million, its first fall in 22 years, from £802 million to £712 million. This profit shortfall was followed by an announcement that the company was set to launch a national loyalty card in June of that year. The turnaround was seen as an attempt to pacify furious shareholders and to redeem the tarnished company image. Industry analysts felt that the move merely reinforced the already-held notion in the City, that Sainsbury's marketing strategy was lacking in ideas.

Whilst not announcing a launch date (City rumours were predicting late summer), Kevin McCarten, the newly appointed marketing director, insisted that the 'Reward Card' would be better value than Tesco's Clubcard and that they would 'be looking at things that our competitors cannot copy; we don't want to do anything they can replicate'. Brian Woolf, an international customer loyalty consultant, predicted that Sainsbury 'would go for the jugular and leap-frog Tesco, not just catch them up'.

However, first impressions of the offer suggested that it was similar to the Tesco Clubcard. Customers were offered one point for every £1 spent above £5 and this could be redeemed for a voucher for £2.50, once they had accumulated 250 points. The card also enabled customers to collect Air Miles in a tie-up with British Airways. Each voucher entitled a customer to 40 Air Miles, which could also be exchanged for ferry tickets, holidays or cinema tickets. This meant that an average family spending £60 a week could earn a free individual flight to Paris every year.

With devastating timing, Tesco announced that the launch of Clubcard Plus was to be 17 June 1996, causing Sainsbury to bring forward the launch date of the Reward Card to the same week. Modelled on a scheme run by Carrefour in France, Clubcard Plus allowed shoppers to deposit money into an account by a monthly standing order, so that when customers shopped at Tesco, the cost of groceries or petrol could be deducted from the account. However, the real value to customers was the 5 per cent gross interest paid on the account (if in credit). This was up to 20 times the interest paid by some high street banks and building societies. The scheme also allowed for an overdraft facility which charged 9 per cent APR (the cheapest agreed overdraft rate in the UK at the time of the launch) and a cash withdrawal facility was available through the ATMs of the National Westminster Bank, the retailer's partner in the scheme, as well as at Tesco checkouts. Clubcard points were accumulated in the same way as before, building up points based on the amount of money spent and converted into money-off coupons every three months.

Describing the arrangement as 'own-label banking', Chairman Sir Ian McLaurin explained at the press conference that 'since we launched Clubcard last February we have been trying to find ways to make it more user friendly; many of our customers wanted to be able to pay with Clubcard. This system allows them to do so.' A NatWest spokesman saw the deal as a 'co-branding' opportunity and said it would quickly be followed by a number of similar deals. 'We do what we are good at; they do what they are good at. What we do best is processing, administering accounts and credit management.'

Some commentators saw this not only as an assault on other food retailers but on financial institutions as well. Gill South, writing in the *Daily Telegraph*, described the move as 'the first time that customers are being offered accounts, overdrafts and cash back at extremely competitive rates, by a company that they trust more than their bank'. The move was reported to have sent tremors through the banking industry as well as other competitors in the grocery retailing sector. Conversely, some commentators failed to see how Tesco could afford to pay 5 per cent on small accounts, they also predicted some cannibalization of accounts for National Westminster and were uneasy about the clash between what they saw as two intrinsically different cultures. Despite these reservations, the

City reacted favourably and the Tesco share price edged ahead to 303p, up from its low in 1994 of 227p.

From beans to banking

This time reaction from competitors was swift. In October 1996 Sainsbury announced that they were to launch a bank in partnership with the Bank of Scotland. Going further than the Clubcard Plus debit scheme, the move would make Sainsbury the first UK retailer to offer comprehensive banking services. Sainsbury's Bank, scheduled to open in the first quarter of 1997, would offer customers fully fledged cheque accounts via a telephone banking service, combined with in-store, multi-functional teller machines. The service would not be tied in any way to Sainsbury stores, by restricting the offer to loyalty card holders, but customers would be able to use a Sainsbury Classic or Gold Visa card in the same way as any other Visa card. One analyst saw this as an extension of Sainsbury's own-label strategy, 'Bank of Scotland will manufacture the product, but it will be packaged with the Sainsbury Bank label'. The joint venture, of which Sainsbury will own 55 per cent, was seen to make sense for the Bank of Scotland as it has weak market presence in England and Wales.

As a result of this move, Dino Adriano, the new Deputy CEO of Sainsbury, signalled that the arrangement for the big four banks' ATMs, currently sited at Sainsbury stores, would be likely to be reconsidered when the contracts, usually lasting three to five years, came up for renewal. Any move to scrap the ATMs in favour of Sainsbury's Bank machines would be likely to be fiercely opposed by the banks. NatWest commented that they were likely to be a target because of their relationship with Tesco, 'Sainsbury's will probably get rid of us which will leave us with fewer supermarket ATMs.'

In the same week Sainsbury also announced an increase in the number of points that it would award Reward Card holders. The scheme, based on bulk purchases of around 200 selected products, meant that a customer purchasing three 200g jars of Nescafé could earn 300 points, which previously would have required a spend of £300. This meant that, with selective buying, a family spending £75 a week could earn enough points to get four free flights to Paris in one month. Costs were split between Sainsbury and their suppliers. At the launch, Kevin McCarten said that the scheme would make all other loyalty cards seem pointless and that they were rewarding their customers 'better than any other retailer in the UK'.

This activity was accompanied by the announcement that the retailer would also target one million 'high value, promiscuous' Reward Card customers in Sainsbury's first large-scale direct marketing campaign. A catalogue offering extra Reward benefits would be sent to families spending

£120 per week or more in order to tempt them to the store. It was hoped that this would act as a boost to the rather flat sales growth of 3 per cent, which was just in line with inflation. McCarten predicted that observers 'will see a much greater degree of targeting within our customer base and a lot more experimentation in order to target our specific customers'. He denied, however, that Sainsbury was to follow Tesco's route, taken in the previous Christmas period, when it switched a large proportion of its annual £30 million advertising spend into direct marketing. However, these announcements did little to soften the blow of less than encouraging interim results released that same week, showing half-year profits of £393 million, below the previous year's figure of £456 million and falling far short of City expectations.

Safeway was quick to follow in Sainsbury's wake, announcing that it would be joining forces with Abbey National to offer a range of financial services to its customers. The first product of the new partnership was to be a new ABC bonus card, a similar scheme to Tesco's debit account scheme, launched on 17 June 1996. However, it took the debit account scheme even further by allowing card holders to use the card to pay for goods at any of the 70 000 other retail outlets that accepted the Visa Electron scheme, whose participants included W.H. Smith, C&A and Our Price. The card would also allow cash withdrawal from more than 1500 Abbey Link machines and give customers a free 24-hour telephone banking service although, unlike Tesco, there would be no overdraft facility.

Colin Smith, CEO of Safeway, described the card as a logical extension of Safeway's customer proposition 'aimed at making shopping easier'. He hinted that in the future they would hope to offer a full range of financial services, through a planned Abbey National presence in Safeway stores. Meanwhile, Ambrose McGinn, Abbey's Marketing Director, recognized mutual benefits, 'There's a large degree of customer overlap; we can use behavioural data from each database to make timely and relevant offers to our customers.' Safeway shares subsequently climbed to a new high for the year.

Other food retailers appeared to maintain a respectful distance from these developments. As yet Asda had only tested different types of loyalty schemes, one being a 'Style card' which gave vouchers for its range of clothing and leisure goods. It had also linked up as a redemption member of the British Gas Goldfish credit card, in which holders of the card could earn discounts for gas bills and also spend Goldfish points in Asda and other redemption member outlets. Asda's Chairman, and his new CEO Alan Leighton, preferred to stress that 'simplicity is divinity', in contrast to their competitors' foray into financial services. However, rumours of discussions between Asda, Kwik Save and Somerfield with various banking and insurance companies were common. A survey by ICL Financial

Services Group gave credence to the notion that supermarkets had a strong position from which to enter this market. They found that 32 per cent of people would be happy to buy financial services from a supermarket and that more people were likely to have a loyalty card than a credit card.

The high level of financial services activity in this sector continued. Sainsbury announced the introduction of insurance services through Homebase stores in a partnership with insurance brokers Willis Coroon. Speculation then focused on Sainsbury's rumoured plans to enter the mortgage market. Many saw a natural synergy between the Homebase brand and a mortgage lending operation, although warning bells were sounded by some that mortgage lending is very different to selling groceries or even running current accounts. 'Retailers have no experience of the segmentation and management of customers; they are not used to turning them away.'

However, focus was to shift quickly from partnerships to performance as Sainsbury's were forced to issue a dramatic profits warning, causing shares to fall sharply to 341p, their lowest for five years. The company said that profits for 1996/97 would be around £640–650 million, well below City expectations of £715 million. One large institutional shareholder said the warning was particularly surprising as it came only a month after Sainsbury's management had visited investors with an upbeat message. Others were less kind. Frank Davidson, an analyst at James Capel, was quoted as saying, 'This is not the bottom. I see nothing here to say that this business has turned the corner.'

A spokesman for Sainsbury blamed the fall on high costs associated with building sales through its loyalty card scheme as well as conversion costs to change Texas stores to the Homebase brand. The warning came with an announcement that sales in the eight-week run-up to Christmas were up by 4.4 per cent but this offered little solace as Tesco's were said to be 7.5 per cent. The Reward Card was said to be contributing a 2 per cent uplift to sales which, in the words of the spokesman, 'were at the bottom end of their pre-launch expectations and barely enough to cover the card's cost'. Analysts believed that Tesco's performance not only eclipsed but was at the expense of Sainsbury's. When asked to comment, the Sainsbury press office would only say that they 'do not comment on competitors'.

The problems encountered by Sainsbury were likely to influence other organizations within the sector. Asda, who since 1995 had tested various loyalty schemes and had been involved most recently in rumours of a banking partnership with the Royal Bank of Scotland, had still to decide on its strategy. Having differentiated itself as the 'value-for-money store' with a low price proposition, it was considered unlikely that the management would lead the chain into banking unless it could offer a cut-price account for customers and some predicted that this could lead to a price war with the four high street banks. An Asda insider said that banking had long been shelved and that it was looking at a loyalty card close to the Clubcard Plus scheme. Commentators were divided about their wait-and-see stance, some believing that Asda had waited to learn from the mistakes of its rivals, others believing that it was too late in entering the market, whilst others pointed to the latest research commissioned by *Checkout* magazine, which questioned any justification of loyalty schemes at all. In a survey of 990 adults, two-thirds professed to have a loyalty card but a further two-thirds expressed a strong preference for lower prices rather than card schemes. One of the research team pointed out that 'you have high levels of holding and using, especially amongst the prime target of heavy grocery spenders, but it appears cards are not influencing people's behaviour'. Replying to its critics, an Asda source remarked that, 'you can pull out of a loyalty scheme and replace it with home delivery, catalogue schemes and other devices but you can't pull out of a bank that easily. I think retailers will get out of financial services and loyalty cards in three years, by when they will have the lists.'

Tesco took the opportunity of its rivals' troubles and deliberations to call a press conference in late January 1997 to announce strong Christmas trading figures which showed like-for-like sales up by 7.5 per cent. Total sales growth in the period was even greater, at 13.1 per cent. Only two years after analysts had so readily marked down Tesco shares after the launch of Clubcard, they were having to eat their words. Tables 2.3.2 and 2.3.3 show

Table 2.3.2   Food retailers: Turnover and pre-tax profits, 1992–97 (£ million)

Year ending	1992	1993	1994	1995	1996	1997
Tesco						
Turnover	7097	7581	8600	10 101	12 094	13 887
Pre-tax profits	545	558	435.5	551	675	750
J. Sainsbury						
Turnover	8695	9685	10 583	11 357	13 499	14 312
Pre-tax profits	628	732.8	368.8	809.2	712.2	609
Safeway						
Turnover	4729	5196	5608	5815	6069	6589
Pre-tax profits	364.5	417.3	361.8	175.6	429.4	420
Asda						
Turnover	4529	4614	4882	5285	6042	not
Pre-tax profits	−364.8	142.1	−125.9	257.2	311.5	available
Kwik-Save						
Turnover	2391	2651	2800	2992	3254	not
Pre-tax profits	74.3	126.1	135.6	125.5	2.8	available

Source: Reuters Business Briefing.

Table 2.3.3   Food retailers: Market share, 1993–97 (% of total market)

	1993	1994	1995	1996	1997
Tesco	14.9	15.7	18.1	19	20.5
Sainsbury*	17.3	17.9	17.5	17.4	17.8
Safeway**	8.6	8.5	8.4	9	9.4
Asda***	8.4	8.9	9.7	10.7	11.9
Kwik-Save	5.8	5.6	5.5	5.5	4.6

Source: Taylor Nelson AGB: Retailer Share Track.
Notes: *Sainsbury + Savacentre, **Safeway + Presto + other Argyll stores, ***Asda + Dales.

that all performance indicators had improved dramatically, charting a scenario previously unthought of. Tesco's share price was at its highest, market share had risen sharply and operating profits were predicted to overtake Sainsbury's in spring 1997 for the first time in their history.

Tesco's success in satisfying customers had not gone unnoticed in other quarters: *Wine* magazine voted the store 'Wine Merchant of the Year', readers of *Woman* magazine voted Tesco 'Supermarket of the Year' and directors of Britain's 10 largest companies across 26 sectors had unanimously voted it their 'Most Admired Company'. These accolades culminated in the award of 'Retail Brand of the Year' by *Marketing Week*.

Speculation was now mounting as to the response Tesco would make to Sainsbury's banking launch. The company were rumoured to be planning enhancements to the Clubcard Plus scheme with the introduction of a credit card and allowing customers to pay from home computers. Estimates put membership of Clubcard Plus at the 150–180 000 mark and Tesco was thought to want to boost this to the 500 000 level but research had shown that many customers would like more facilities on the scheme before they joined it. NatWest, their partner in the Clubcard Plus scheme, was quoted as saying that working with Tesco to create a new bank 'was not high on the priority list'. However, at the press conference management preferred to focus attention on their recent performance.

The recently elevated peer, Lord McLaurin, said, 'the figures have revealed another excellent trading period, our sales have continued to outperform the industry average in what remains a highly competitive climate'. He attributed the improved performance to the 'Unbeatable Value' campaign of low prices and the growing popularity of Clubcard, which now had 9.5 million members. Reaction in the City was enthusiastic, the figures were far better than analysts expected and, as a result, profit forecasts for 1996/97 were raised to £750 million and Tesco's share price rose to 369p.

References

1 *The Economist Intelligence Unit Report* (1995). No. 36, December.
2 Mason, T. (1996). *The best shopping trip ...? How Tesco keeps the customer satisfied.* Transcript of a speech delivered to The Marketing Society on 11 September.
3 AGB Taylor Nelson Superpanel. Retailer Share Track – 179 markets covering packaged groceries, fresh foods and toiletries.

Case 2.4 Rover Cars: The Catalyst and Conquest '91 direct marketing programmes

This case was written by Professor Adrian Payne, who is Director of the Centre for Relationship Marketing at Cranfield School of Management, Cranfield University.

The UK motor-car market is subject to considerable cyclical activity. Typically there is a high surge in car sales in August each year when a new letter on number plates becomes current, and a surge at the start of a new year.

Table 2.4.1 provides volumes and percentages of car sales for 1990.

The motor-car market is volatile and highly competitive. Around 14 car makers currently sell over 20 000 units each year and these account for approximately 92 per cent of all sales. With the recessionary conditions over this period all motor-car manufacturers have been under pressure to achieve sales volume and maintain profitability. Annual statistics for the period 1987–1991 are given in Table 2.4.2. In order to try and achieve similar levels of sales, some companies have sought to increase market share by significant reductions in price.

For a number of years the lion's share of the car market has gone to meet the demand of fleet users. For some manufacturers, up to 70 per cent of their output found its way on to the road as the company car. Attractive though the fleet market is, competition has always been fierce and buyers have typically been able to negotiate sizeable discounts. More recently, government policy has tended to weigh against the company car, certainly in the form

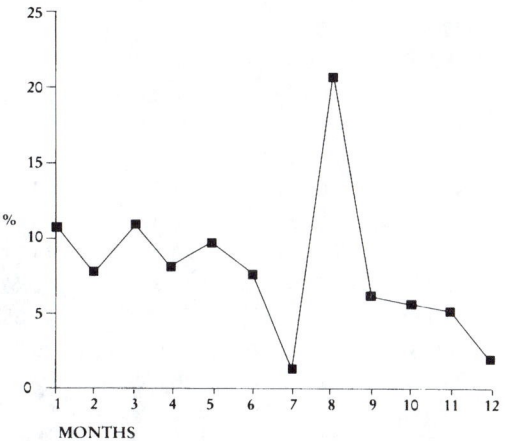

MONTHS

Figure 2.4.1 Typical calendarised sales pattern
Source: SMMT

Table 2.4.1   New motor vehicle sales (1990)

Month	Volume	%
January	206 393	10.27
February	167 420	8.33
March	208 733	10.39
April	166 913	8.31
May	173 896	8.66
June	142 697	7.10
July	45 707	2.28
August	432 867	21.55
September	153 595	7.65
October	130 377	6.48
November	117 499	5.85
December	62 837	3.13
Total	2 008 934	100.00

of a management perquisite. In particular, taxation charges to the individual have made company cars less attractive. Also, many companies are now responding to pressure from within and allowing managers to choose their own cars, as opposed to being issued with a standard vehicle.

Both of these trends have been advantageous to Rover which was,

Table 2.4.2   UK sales by manufacturer

	1991	1990	1989	1988	1987
Ford	385 944	507 260	608 617	583 790	580 119
GM	248 704	323 054	349 901	303 565	270 778
ROVER	229 291	281 385	312 306	332 607	301 811
Peugeot	115 557	123 671	138 958	126 825	101 264
Nissan	64 213	106 783	138 437	134 724	114 243
VW	74 263	95 018	108 778	102 210	91 294
Renault	63 547	67 578	88 111	85 592	78 694
Volvo	46 765	66 017	81 706	80 427	70 880
Citroën	53 424	60 899	66 409	66 930	46 014
Fiat	34 661	54 945	70 173	75 113	68 576
BMW	38 707	43 004	48 910	42 761	37 525
Toyota	41 239	42 662	42 408	39 853	38 269
Honda	28 248	31 750	27 016	26 714	24 743
Mercedes	20 672	26 559	28 335	23 896	21 752
Others	147 091	178 349	190 879	190 567	167 731
Total	1 592 326	2 008 934	2 300 944	2 215 574	2 013 693

Source: SMMT

already, less reliant on fleet sales than many of its contemporaries (the split being approximately 50% fleet and 50% retail sales). The loosening of the central buyer's grip made available a new market segment – **user choosers** – who requested particular makes and models from their employers.

Having identified this market situation, Rover designed direct marketing initiatives – the Catalyst programme and later the Conquest '91 programme – aimed at providing maximum impact on the individual retail private purchaser and also the user chooser.

Interlink and Superlink

Catalyst and Conquest '91 built upon the Company's earlier experience of direct marketing gained through its Interlink and Superlink follow-up initiatives.

These used direct marketing to exploit the sizeable database which the Company possessed as a result of registering the warranties of all cars sold. By today's standards this information was not particularly comprehensive, but the Interlink programme was innovative at the time and broke new ground.

The database provided names and addresses of customers, details of the vehicles purchased, and records of claims made on warranty. Using it, the Company mailed customers enquiring if it was time to change their car, and enclosing information about the current product range.

No evaluation exists to measure the impact of Interlink and Superlink. Even so, Interlink was seen to be a worthwhile approach which at least maintained contact between the Company and its customers.

Superlink managed to generate a slightly higher level of dealer involvement, but it was still dogged by the same inherent criticisms which could be levelled at its forerunner.

The basic problem was that both programmes were assumptive. That is, the Company had to make assumptions of what the customers would do in the future, based on their behaviour in the past. It did not take account of changed circumstances of customers or how this might affect their buying behaviour.

Clearly any new initiative, while owing a debt to Interlink and Superlink, would have to be radically different to overcome these flaws. This understanding stimulated the creativity which gave birth to the **Catalyst** programme.

The Catalyst programme

Initial considerations
To be successful the Catalyst programme had to be fuelled by higher

octane information. Much more had to be known about previous and prospective customers in terms of their lifestyle and buying intentions in order to gauge purchase timing as accurately as possible. Rover had to establish how it might induce prospects to provide this information. What could be offered in exchange and be acceptable as well as motivational? Rover was concerned that the approach should be perceived by prospects and customers as a controlled process, i.e. Rover supplied information when it was requested and required. Customers and prospects would therefore need to be prompted to initiate the approach.

As a result of these considerations, Rover decided to produce an exclusive magazine as the axis around which the new direct marketing initiative would revolve. Not surprisingly, the magazine took its name from the programme – *Catalyst*. The role of the magazine was twofold – to 'warm' potential customers to Rover and to maintain a dialogue with existing customers. From the outset, Rover was determined that the Catalyst programme would not be an ordinary direct marketing campaign.

Since the objective was to understand more about the lifestyles of customers and prospects, it was decided that *Catalyst* would feature articles relating to lifestyle issues. Moreover, it had to be of a standard and quality which reflected well on Rover, but was also comparable to, or better than, equivalent publications to be found on the newsagents' shelves. Accordingly, a top-class editorial team was assembled. Michael Parkinson was appointed Editor, and a stable of well-known writers was commissioned to write articles as and when required.

It was 'sold' to existing and prospective customers as something unique – a magazine which treated them as individuals by allowing them to choose part of the contents for themselves. Prospects could select from a range of lifestyle sections, ensuring that they received a magazine that was tailored to their interests – for example, gardening, cooking and sports. Different versions of the *Catalyst* magazine were produced, reflecting the profile of the target audiences' lifestyle and interests. Image was also a key consideration – the Catalyst programme was to play a central role in making people perceive Rover as a quality company that they would like to do business with.

How the magazine was used
The existing database included some 300 000 previous buyers who had registered warranties. The database was analysed to identify profiles of prospects. After conducting research into this group, Rover identified profiles of typical buyers. It then rented an estimated 580 000 names from lists at an approximate average cost of £95 per thousand. Lists were selected to reflect buyer profiles and after deduplication a total database of 800 000 people was established.

Rover mailed all those listed an introductory *Catalyst*, a 90-page magazine, in the autumn of 1988. This was personalised, by placing the recipient's name and address in a window on the front cover.

Industry sources have suggested an approximate total mailed cost of £1.50 for the personalised magazine. Inside the magazine was a freepost order form (for further editions), which also doubled as a questionnaire eliciting information about the reader's current car(s), whether it was bought new or second-hand, whether or not it was owned by a company or the reader, when it might be replaced and which Rover car would come into consideration.

Clearly this new information greatly enriched the original database and provided Rover with three possible next steps depending whether the readers had:

1 Chosen not to respond (industry sources estimate approximately 75 per cent of total mailed).

   After experimentation, it was found to be counter-productive to try tracking this group down in order to discover why they had not responded. It was considered better to exclude them from the mailing and build up replacements from new lists.

2 Asked for further copies of the magazine (industry sources estimate approximately 25 per cent of total mailed. Of these, an estimated 50 per cent of previous buyers and 15 per cent of 'cold' names responded positively). Here the next edition of *Catalyst* was tailored to the reader's specification, using three of six possible lifestyle sections (travel, sport, food and drink, home and garden, entertainment and female interest). The magazine was published three times a year and each issue contained a questionnaire to monitor and update the reader's details and 'purchase window'.

3 Asked for the magazine and also approaching their 'purchase window' – within six months of the time they were planning to renew their car (estimated at approximately 10 per cent of respondents – i.e. 10 per cent of the 25 per cent).

   This activated the following four stages:

   a Dealers were sent Opportunity to update contact advices, which allowed them to update the records before the next stage. About three months ahead of repurchase date, the reader received a telemarketing call by, or on behalf of, the local dealer. This checked whether they genuinely intended to renew their car or whether the situation had changed (in which case the new data was recorded). Dealers were encouraged to make the requalification calls themselves, but could ask to have them made on their behalf (for a fee) by a Rover-appointed telemarketing agency. Rover actively encouraged direct contact in order to establish a personal relationship between dealer and prospect at the earliest possible time. Participation levels by dealers were, however, very disappointing.

   b The positive telemarketing calls led to the prospective customer receiving a Vehicle Information Portfolio (VIP). This consisted of a personalised letter from the dealer and information about the relevant Rover model and the varying specifications. It was presented in brochures exclusive to the Catalyst programme.

A price list was intentionally not included, to prompt customers and prospects to request this information from the dealer thereby initiating a dialogue. However, later market research found this was not the best approach as recipients liked to receive full information 'up front'. Incentives to test drive or purchase were not usually included, unless prospects were felt to need an additional nudge, and not included at all in the later stages of the Catalyst programme.

c The dealer received a contact report listing to use to convert 'hot leads' into sales. Industry estimates suggest that the VIP pack converted about 49 per cent to test drive and 45 per cent of those people who took a test drive then went on to purchase a car.

d On the completion of a sale, the dealer fed back details about the transaction, and the communications 'loop' was completed.

The people on the database now received a regular communication from Rover. The point at which each prospect moved into the purchase window could be monitored. A tailored approach to convert the prospect into a purchaser could then be made.

The role of the dealer

Unlike the earlier programmes, the Catalyst programme put specific demands on the dealer, but it also offered the prospect of greater rewards. The programme relied upon the dealer network updating and enriching the database. To ensure that dealers and all their staff understood the Catalyst programme and their role in the total scheme of things, a special team of 15 people was set up to spearhead a dealer development programme.

A dealer pack was produced, which included a videotape explaining how to make the most of *Catalyst*. A range of material was produced for dealers, including a *Catalyst Promotions Quarterly* with promotional suggestions and sample letters for prospects and customers.

In addition to providing a videotape and other materials to explain the Catalyst programme and the processes to be followed, the specialist team encouraged every dealer to reach a 'model' standard. To this end they developed case studies and publicised success stories.

However, there were two barriers which were not always overcome and which reflected more on the dealer than the development team:

1 The level of administrative discipline required was beyond the reach of a certain number of dealers.

2 From the outset, Rover intended to cultivate a local expert at each dealership to ensure the accuracy and currency of all data and sort out any problems.

Unfortunately, the car sales side of the business is characterised by high staff turnover, even at senior levels, so the person who understood most about the Catalyst programme often disappeared, leaving colleagues floundering.

Evaluation

The campaign reflected the general increase in direct marketing, which had grown in the recession of the early '90s, in this and other markets. In many ways the Catalyst programme was successful. It overcame the drawbacks associated with earlier attempts at direct marketing. It was proactive and highly focused. It certainly achieved results but, if dealers did not always complete their part of the process, the full extent of these could not be calculated with accuracy. It also relied upon prospects updating information, to identify the purchase window opportunity.

Notwithstanding the apparent signs of success, the Catalyst programme accounted for something in the order of 20 per cent of the total marketing budget. In the hard-nosed, recessionary climate of 1990 some penetrating questions were asked. Concrete proof was sought that the Catalyst programme worked and provided good value for money.

Although cost pressures had a part to play, with management constantly seeking greater value from the marketing budget, there was also pressure to change to a new programme from within the direct marketing department at Rover. The reasons for the change were:

1 Recognition that customers and prospects are different and should be treated as such. Customers had an ongoing relationship and were interested in new information about the Company and its products. Prospects had a different level of knowledge and commitment to Rover and needed to be prompted into a relationship with the Company.

2 Need for a process that would work in August (a process that would help dealers to sell cars at even their busiest times, rather than be perceived as a hindrance).

3 Need to 'conquest' new business from competitors, capturing customers with a different profile from those on the existing database. This stimulated additional market research to establish the lifestyle and buying habits of these new target segments. The new product positioning strategy was to appeal to a more up-market, affluent, younger group of customers than before. While this did not necessarily invalidate the existing database, it certainly generated pressure to expand it.

4 Need to improve the perceived standard of leads passed to dealers, thus ensuring their commitment to the programme. Dealers complained that contacts provided through the Catalyst programme were often not ready to purchase.

5 Rover had developed a more sophisticated and competitive product range than when the Catalyst programme began. This could be marketed with greater assurance.

6 Desire for greater integration with other marketing activities. A synergy among all the elements of the marketing mix was lost with the Catalyst programme working independently of other marketing initiatives.

7 Need for greater measurability of the process ('if you can't measure it, you can't manage it'). This was also part of the desire to control the process.

8 A desire to obtain information about the purchasing behaviour of prospects in a one-stage rather than a two-stage programme – making information as relevant and recent as possible.

The stage was set for a new direct marketing initiative which would build on the success of the Catalyst programme. It would need to overcome some of its weaknesses, and meet specific objectives concerning increased showroom traffic, the number of test drives, and most important of all, increased 'conquest' sales.

Conquest '91

Initial considerations
Despite the comments above, the Catalyst approach was seen to be extremely valuable in terms of monitoring existing customers and maintaining their loyalty to Rover. That element of direct marketing therefore remained in place. The new prospecting programme – Conquest '91 – was, as the name suggests, devised to win new customers, including those with a different profile from those on the existing database, in what was becoming an increasingly competitive and depressed business environment.

Having decided to approach existing customers and new prospects differently, Rover could make its communications even more pertinent and effective. However, the new campaign had to be up and running by August in order to capitalise on this peak period of car sales. As illustrated in Table 2.4.1, if the August sales peak was missed, the cost-effectiveness of the programme would be reduced. With such a tight schedule it was imperative to start 'trawling' for prospects as early as possible.

One of the driving principles behind the new initiative was to be able to evaluate the effectiveness of individual activities and the campaign overall. However, the self-imposed time pressures under which Rover was operating meant that trials and research could not be conducted coolly in advance, but were carried out in parallel with the programme. Conquest '91 was, in reality, an 'action-research' model, particularly in its approach to finding new customers. Results from each element of the programme were fed back to further refine and improve effectiveness. For example, the VIP pack was redesigned to reflect market research findings.

Finding new prospects

It is a perennial problem for marketers to decide in advance what is going to be the most effective way of building up lists of prospects. Cost is not the sole criterion by which to judge the results. Did the most expensive approach yield the bank of prospects which had the highest conversion rate to customers? Which approach yielded the best quality leads for each model range? Often the answers to these questions could only be guesswork.

In keeping with the overriding spirit of Conquest '91, Rover set out to find the best trawling method while actually generating leads. Experience with Catalyst had shown that the best quality leads were those who intended to repurchase in the next few months. On the other hand, longer-term intentions were useful but nothing like as accurate. In this sense the best quality information had a very short shelf-life and needed to be followed up quickly while it still had value. Therefore the decision was taken to trawl actively for imminent repurchasers. Emphasis through the prospecting programme was on developing a controlled two-way relationship with suitable prospects. In order to obtain some comparative results, a three-pronged approach was adopted to generate ready-qualified leads:

1 Direct response advertising   In April–May 1991 advertising appeared in what were considered to be the most appropriate national newspapers. The message invited readers to join Rover's pre-purchase information programme, which offered the prospect of making the car buying decision easier and more convenient. Industry sources estimate that the total spend on such a campaign would be approximately £200 000. The campaign invited those who were interested to respond by a Freepost coupon or Freefone. In addition, a gift incentive was sometimes offered in advertisements, stimulating interest by phrases such as: '. . . and you could qualify for some unique offers'. Rover was anxious, however, that the incentive should not be the major motivator for a response, as serious prospective buyers were sought.

   This approach yielded a list of interested prospects estimated by industry experts to be in the order of 20 000. At the same time, because of the response mechanism, Rover was able to evaluate which advertisements and which newspapers produced the most cost-effective responses.

2 Direct mail   A variety of lists was selected by matching lifestyles and profiles of targeted prospects. Some 550 000 names were rented and, after deduplication, some 500 000 direct mail packs were sent, inviting the recipients to take part in the Rover pre-purchase campaign. These were not just letters, but also press endorsements of the product and details of the pre-purchase information programme and an application form with a response telephone number. Industry estimates suggest that this would have yielded a further 25 000 replies at a cost of around 60 pence per mailing pack including personalisation and postage.

3 Promotions   While in the past the Company had willingly participated in promotions and competitions where the prize was a Rover car, the client value of
such events was seen as mainly PR potential. For Conquest '91, Rover again
contributed a car for a competition in a major credit card magazine (selected
because of its readership profile).

This was one of a series of promotional opportunities and the most successful example. Entrants to the competition were not asked directly if they
wanted to participate in the pre-purchase programme. They were simply asked
to supply their repurchase details and to indicate if they were happy for Rover
to contact them again. Each entrant was subsequently sent a letter explaining
what would happen to their information and offering them the chance to opt
out. In this way Rover was 'controlling the process'. Entrants perceived that they
were initiating further contact with Rover, and the database was only updated
with interested prospective customers.

This campaign – which industry sources suggest would cost around £20 000
(including the cost of the car) – produced some 36 000 contacts out of a total
competition entry of some 130 000.

A total budget, suggested to be in the region of £100 000, was given to this
promotional element of the marketing mix and is thought to have yielded a
total of 85 000 prospects.

An overview of the Conquest '91 process is shown in Figure 2.4.2 together
with the key steps. The process will now be described in further detail.

What happened next?
All prospects approaching their purchase window received a redesigned
VIP pack from the Company about two months before they intended to
purchase. The whole emphasis of this was on selling the new image of
Rover as well as its products. With a covering personalised letter the pack
included:

1 A highly visual, specially designed brochure about the model said to be of most
   interest to the prospect.
2 A booklet which indicated the various specifications available for that model.
3 A price list.
4 A booklet which promoted the Company and the overall range of cars, showing
   alternatives and reassuring prospects about the 'new' Rover.
5 Incentive vouchers and a gift booklet. The strategy here was to provide a
   voucher for a test drive, a purchase, or both. Once the prospect had completed
   one of these actions, the dealer would stamp the voucher by way of confirmation and it could then be submitted by the prospect to claim a 'thank you for
   participating' gift from the booklet.

Special design features of the new VIP pack were its high-quality image
and its flexibility. Image was especially important to emphasise Rover

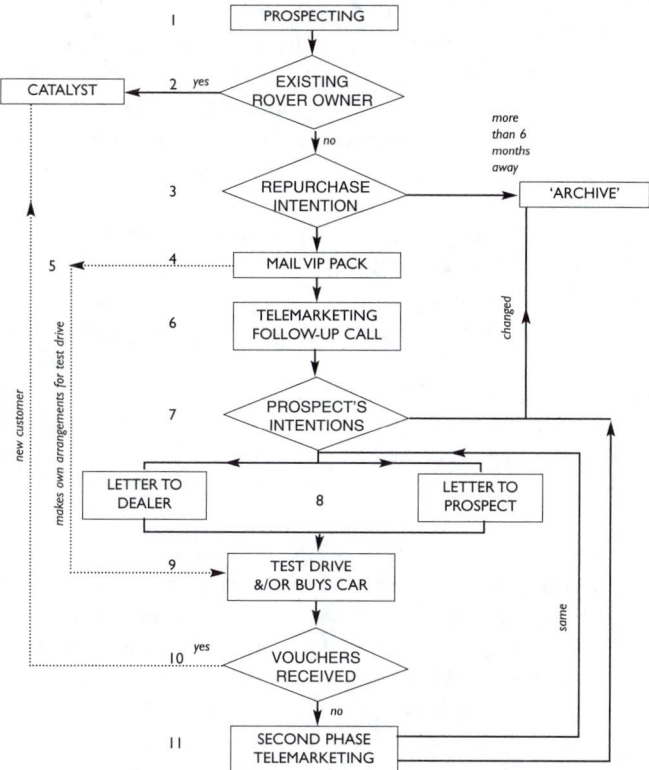

Figure 2.4.2 Overview of the Conquest '91 process

1 Prospecting using a variety of methods.
2 If the prospect is an existing Rover owner, allocate to the Catalyst programme. If not, allocate to Conquest '91 programme.
3 If the repurchase intention is more than six months away, 'archive' for requalification at a later date. If not, load on to mailing files for the appropriate month.
4 Mail the VIP pack approximately three months ahead of stated repurchase date.
5 Prospects might respond immediately by making their own arrangements for a test drive.
6 Telemarketing follow-up call after two weeks to confirm receipt of the VIP pack and arrange a test drive if necessary.
7 If circumstances have changed or prospect refuses a test drive, update database and 'archive' information as Step 3. Otherwise, confirm prospect's choice of dealer.
8 Simultaneous letters of confirmation sent to the dealer and the prospect. (At this point, Rover effectively hands over responsibility for the sale to the chosen dealer. Note that if the dealer does not fulfil the commitment, the prospect has the option to get back in touch with Rover direct via a Freefone helpline. If necessary, Rover takes over the process again at Step 11 – see flow chart.)
9 Prospect takes test drive (and/or buys car) and submits validated incentive vouchers for fulfilment.
10 If vouchers are received, Rover updates database and, if vehicle has been purchased, allocates new customer to Catalyst.
11 If vouchers are not received, Rover implements second phase of telemarketing to establish if anything went wrong (in which case attempts are made to repair the damage) or if circumstances have changed (in which case the database would be updated as Step 3).
12 The 'archive' is a database of longer-term prospects. Periodic requalification exercises keep the repurchase information up to date, and prospects are then selected for mailing at the appropriate time when they move into the purchase window.

values of quality and prestige to this 'new' audience and at the same time to distance the VIP pack from lower-quality direct mail used by competitors. By having separate documents for car models, specifications, prices and so on, it was relatively easy to update the contents without incurring the massive printing costs associated with reprinting the whole pack.

Telemarketing

In order not to lose momentum or allow warm leads to cool down, about two weeks after sending the VIP pack Rover would call the prospect. The trained caller would follow an easy-paced, conversational, but nevertheless largely scripted routine which was designed to establish:

- If the prospect had received the VIP pack.
- If the pack contained the relevant information.
- If there was still a firm intention to buy as stated.
- How Rover could help.

If the prospect's personal situation had changed and the intention to buy had been delayed, then the telemarketer would take notes about the new situation and the database would be updated accordingly.

Help would take the form of offering prospects the opportunity of a test drive at the dealership of their choice. Giving this element of choice was important because some prospects would have personal reasons for not wanting to use the dealer nearest to their home address. For example, some might prefer a test drive during their lunch break and therefore need contact with a dealer near their place of work.

For reasons of cost efficiency, up to five attempts were made to call over a three-day period. Too big a backlog of calls hinders the momentum of the overall programme. Also, some numbers may be wrong or may not be supplied at all. Therefore, if contact could not be established, the elusive prospect would receive a personalised letter which invited the recipient to make contact with Rover. This was not the preferred option as it lacked the persuasiveness of a live conversation.

Individual contact advices would then be sent to dealers informing them about the qualified hot leads who required test drives. Simultaneously, a letter was sent to the prospective customer confirming that a dealer would make contact with them. A Freefone number was provided in case there were any problems.

Dealer involvement

A fundamental difference between this, and previous programmes, is that the dealer had no contact (or even awareness) of the potential customer until Rover had obtained agreement from the prospect to take a test drive at that particular dealership. All that remained was for the dealer to confirm the appointment and lead the transaction through to a successful conclusion where the prospective customer purchased a Rover car.

It can be seen that unlike the earlier Catalyst programme which was designed to have dealer involvement, and sometimes suffered as a result, Conquest '91 put minimal demands on the dealer. It eliminated the administrative chores, and played to the dealer's strength – selling cars.

This approach also eliminated the need to have a development team to familiarise dealers with the relative complexities of the earlier Catalyst programme. In addition, by maintaining control over the process, Rover could ensure that all communications were conducted with a level of professionalism in line with the desired total-quality image the marketing department sought to establish.

Follow-up phone call

Approximately three months after agreeing to take a test drive, those who did not purchase would receive another telephone call. This would check if:

- The situation had changed, in which case the database would be updated.
- The prospect was aware that the incentive vouchers were due to expire (normally they were valid for three months).
- Something had gone wrong, in which case an attempt would be made to rectify the situation.
- They had purchased, but had yet to claim their gift (something which was found to happen quite often).

Evaluation

At many different levels this programme can be rated as a success. Targets set in December 1990 were achieved by December 1991, even though the car market was further hit by the recession. As part of the Roverisation programme, image and quality became the important tools of differentiation from competitors. The whole Rover marketing strategy was based on this differentiation.

Rover has achieved a purchase-to-test-drive ratio of something like 50–55 per cent, compared with the industry norm which is believed to be in the order of 40 per cent.

In addition, the Conquest '91 programme has yielded valuable information regarding:

- Best media in which to advertise for leads.
- Best methods of trawling for quality leads.
- Most attractive incentives (through analysing gift take-up).
- Levels of incentives according to model purchases (through analysing gift take-up).

- The effectiveness of telemarketing (voucher returns).

All this information can be quantified and is specific to Rover.

Not surprisingly, Conquest '92 built on the successful programme started in Conquest '91. Rover continues to experiment with the generation of leads. For example, Rover has set up retail information centres in Leeds, Reading and at Liverpool Street Station. These are good examples of more proactive lead generation.

The more Rover understands about proactive, short-term trawling techniques, the less it will rely on rented data, and consequent accusations of inappropriate targeting – rented data cannot usually fulfil all the requirements of 'right message, to the right person, at the right time'. Rented lists will, however, probably always have a part to play and, as Rover becomes increasingly confident in describing the profiles of its best prospects, lists can be sought which reflect this more accurately. New types of promotions will also be considered. The priority is still to find the most effective methods of generating high-quality, qualified leads in a single 'shot'.

Meanwhile, for existing and newly converted customers, Catalyst provides a continuing means of keeping in contact and is a vehicle to build and maintain an ongoing relationship between the Company and its customers. The database is updated continuously, ensuring timely identification of individuals moving into the purchasing window.

Research shows that recipients of *Catalyst* magazine do not see it as direct mail, but as part of a relationship with Rover. Rover is only mailing customers what they want to receive when they want it. Information on products and the Company is seen by recipients as a legitimate and positive use of direct mail.

Case 2.5　Direct Line Insurance Plc

This case was written by Professor Derek F. Channon as a basis for classroom discussion rather than to illustrate effective or ineffective handling of an administrative situation.
© Copyright Derek F. Channon, 1993.

'Too-to-to-toot toot toot toot toot tooo', the little red telephone on wheels scooted across the car park to stop in front of a lady motorist who had just been involved in a minor auto accident. She picked up the phone to Direct Line to find to her delight that, unlike with most insurers, immediate authorization could be given for her car to be repaired. The busy little red phone, which had become the household symbol of Direct Line Insurance and featured in all the company's literature and advertising by 1992, was achieving widespread recognition as a symbol of low cost but personalized service to customers interested in purchasing motor and household insurance.

From its beginnings in 1985, Direct Line Insurance, in the year to September 1992, posted its sixth successive gain in gross premiums written to £213.2 million, an increase of 71.3 per cent on the previous year. Moreover, in a year where there was severe pressure on the motor insurance industry as traditional competitors made moves to diversify their delivery systems and regain market share, Direct Line still managed to achieve underwriting profits. Details of recent financial performance are shown in Figures 2.5.1 and 2.5.2.

Company history

In April 1985, Nikki de Jaeger, a telesales operator, sold Direct Line's first motor insurance policy and so started a revolution which was to transform the UK motor insurance market in a short time. Peter Wood, Direct Line's Chief Executive, believed that it was possible to develop a high quality but cheaper service to motorists than that offered by traditional insurance providers operating via brokers.

Mr Wood, a former employee of insurance brokers Alexander Howden with a background in operations, approached the Royal Bank of Scotland, one of the leading British clearing banks and the premier institution in Scotland, to provide financial support and the benefit of the bank's assured reputation. With initial funding of £20 million to back the development of the necessary infrastructure and investment in comprehensive IT systems, the new company obtained Department of Trade and Industry approval to enter the insurance industry in January 1985, eight months after the venture had begun.

Direct Line became a subsidiary of the Royal Bank Group and Mr Wood

	1989	1990	1991	1992
Gross premiums	66.0	84.1	124.4	213.2
Reinsurance premiums	(1.6)	(1.8)	(3.3)	(4.0)
Net premiums written	64.4	82.3	121.1	209.2
Increase in unearned premium	(9.9)	(10.1)	(25.9)	(53.6)
Premiums earned	54.5	72.2	95.2	155.6
Investment income	5.7	7.8	10.5	11.7
Realized gains less losses on disposal of investments	(0.1)	–	–	–
	60.1	80.0	105.7	167.3
Gross claims	36.2	54.1	73.0	117.5
Reinsurance recoveries	(0.5)	(4.4)	(4.6)	(2.3)
Net claims incurred	35.7	48.7	68.4	115.2
Expenses	19.9	26.2	32.6	44.7
Increase in deferred acquisition expenses	(0.5)	(0.5)	(2.5)	(5.4)
	55.1	75.4	98.5	154.5
Underwriting result – transfer to consolidated Profit and Loss Account	5.0	4.6	7.2	12.8
	60.1	80.0	105.7	167.3

Consolidated profit and loss account, 1989–92 (£m)

	1989	1990	1991	1992
Investment income	3.3	4.7	4.7	6.1
Realized gains less losses on disposal of investments	(0.1)	–	–	–
Underwriting result	5.0	4.6	7.2	12.8
Other income	1.5	1.3	2.4	4.0
	9.7	10.6	14.3	22.9
Other expenses	(1.8)	(1.3)	(0.6)	(1.8)
Profit before CEO bonus	7.9	9.3	13.7	21.1
Provision against investments charged (1989 released)	0.2	(0.2)	–	–
Bonus payable to CEO	–	–	(1.6)	(6.0)
Profit before taxation	8.1	9.1	12.1	15.1
Taxation	(3.2)	(3.3)	(3.1)	(4.4)
Profit after taxation	4.9	5.8	9.0	10.7
Dividend paid/proposed	(1.8)	(2.7)	(2.4)	(4.5)
Retained for the year	3.1	3.1	6.6	6.2

Figure 2.5.1 Parent company revenue account, 1989–92
Source: Annual Reports

	Consolidated				Parent company			
	1989	1990	1991	1992	1989	1990	1991	1992
Fixed assets								
Tangible assets	10.9	20.7	26.5	30.2	5.8	7.6	17.4	18.4
Investments								
Listed investments	11.5	11.3	32.4	33.2	11.5	11.3	32.4	33.2
	22.4	32.0	58.9	63.4	17.3	18.9	49.8	51.6
Amounts paid in advance on new building	–	–	–	–	10.0	10.0	–	–
Deferred acquisition expenses	2.8	3.3	12.9	18.3	2.8	3.3	12.9	18.3
Current assets								
Debtors	11.0	18.0	24.1	28.4	9.7	16.9	22.8	27.1
Short-term deposits and cash at bank	73.7	84.5	109.5	194.2	63.9	80.0	109.3	194.2
	84.7	102.5	133.6	236.6	73.6	96.9	132.1	221.3
Creditors: Amounts falling due within one year	(8.7)	(12.3)	(15.3)	(27.2)	(5.2)	(8.1)	(8.2)	(15.3)
Net current assets	76.0	90.2	118.3	195.4	68.4	88.8	209.0	206.0
Creditors: Amounts falling due after more than one year	(1.9)	(1.8)	(2.5)	(0.4)	–	–	–	–
Provisions for liabilities and charges								
Deferred taxation	–	(0.5)	(0.4)	(0.4)	–	(0.2)	(0.3)	(0.3)
Insurance funds	(39.3)	(49.4)	(75.3)	(128.9)	(39.3)	(49.4)	(75.3)	(128.9)
Outstanding claims	(22.1)	(29.8)	(37.5)	(61.1)	(22.1)	(29.8)	(37.5)	(61.1)
Net assets	37.9	44.0	74.4	86.3	37.1	41.6	73.5	75.6
Capital & reserves								
Called-up share capital	35.0	38.0	62.0	67.0	35.0	38.0	62.0	67.0
Profit & loss account	1.3	4.6	15.6	21.8	0.5	2.2	14.7	21.1
Investment reserve	(0.2)	(0.4)	0.8	1.5	(0.2)	(0.4)	0.8	1.5
Property revaluation reserve	1.8	1.8	(4.0)	(4.0)	1.8	1.8	(4.0)	(4.0)
Shareholders' funds	37.9	44.0	74.4	86.3	37.1	41.6	73.5	85.6

Figure 2.5.2 Parent company and consolidated balance sheets year end 30 September 1989–92 (£m)

Source: Annual Reports.

sold his 25 per cent shareholding to the bank in return for an earnings formula based on the performance of Direct Line. In return, Mr Wood was given a free hand to develop the business without central interference. In 1992, Mr Wood was appointed to the board of the Royal Bank of Scotland Group.

This early period was spent in frantic development of the sophisticated communication and information systems behind Mr Wood's idea. It was intended from the beginning to develop a direct telephone and mail distribution channel, bypassing the brokerage route and thus saving the commission charged.

After the sale of the first policy there followed a period of test marketing prior to a national launch in September 1985. The business grew rapidly, driven by carefully monitored tactical press advertising and marketing to the customers of the Royal Bank. By 1992 Direct Line had grown dramatically and was the market leader in Direct Insurance, with some 670 000 policy holders of its motor policies. This represented an increase of 89.9 per cent in premium income over 1991, while at the same time the company's expense ratio had reduced from 19.6 per cent to an industry low of 14.5 per cent. In a traditional broker-based insurance company around 38 per cent of overall expenses were commission and a further 17 per cent related to claims handling. In Direct Line some 12 per cent of expenses were variable. The remainder of the expense base was more or less fixed. Substantial scale economies would thus occur if volumes were increased up to the point where capacity constraints occurred. Nevertheless, since much of Direct Line's fixed costs were based on computing and communication while labour costs were limited, substantial experience effects were possible.

In addition to its expense ratio advantage, Direct Line had been highly successful in driving down its claims ratio to less than 70 per cent compared with an industry average of over 80 per cent. This reflected the superior risk profile of Direct Line's motor portfolio. In motor insurance Direct Line had become one of the leading participants, with a national market share of 4.5 per cent, and many analysts believed that the company would shortly become the leading motor insurer in the UK. Further, Direct Line enjoyed a customer retention rate of 85 per cent compared with the industry average of around 50 per cent.

In October 1988, adopting similar tactics to those used to develop the motor market, Direct Line launched its Home Insurance product and by 1992 had over 206 000 policies in place. Like the motor industry, the split between structure and contents was some 60/40. Moreover, around 30 per cent of those with a household policy also had a motor policy with the group. Direct Line were also extremely careful in insuring properties with contents value at over £40 000. Such properties were physically inspected and strict security requirements laid down before policies were put in force.

Growth also brought regional expansion with the opening of a sales and claims office in Glasgow, followed by the opening in June 1990 of a further regional centre in Manchester to service Northern England and Wales. The original Croydon office continued to service the Southern parts of England and Wales but the company moved to a new, custom-built head office in Croydon in 1992. In November of the same year Direct Line also opened a new regional office in Birmingham to service customers in the Midlands.

As planned, the company came into profit in its third full year of trading and further profit gains were made in subsequent years. Details of the growth of motor and home policy holders and in premium income are shown in Figure 2.5.3. Direct Line had not, however, achieved the same superior performance in its household business as it had in motor. The company's claims ratio had tended to be consistently higher than the industry average although these figures fluctuated substantially from year to year due to weather conditions, theft, subsidence and the like. In 1992, while the industry average claims ratio for all property was 61 per cent Direct Line could only achieve 73.4 per cent.

While most insurers lost money in underwriting motor and household, they hoped to recover their position from profits made from investments. As a result, such investment portfolios consisted usually of a mix of property, equities and fixed securities. By contrast, Direct Line was extremely conservative in its investment policy. Investments were therefore relatively liquid and risk free, being held mainly in cash and deposits (£194 million at September 1992), government securities (£22 million) and owner occupied freehold properties (£18 million). Declining yields resulting from falling interest rates had therefore reduced investment income. However, the company argued that it did not wish to take risk on both sides of the balance sheet, in both insurance and investments, and hence its policy was to take only risk free investments.

The dramatic success of Direct Line had spawned imitators. The nearest competitor was Churchill Insurance. Started by one of the co-founders of Direct Line and operating in a very similar manner, this company had been acquired by Winterhur, one of the leading Swiss insurers. Other recent 'direct writers' included Topdanmate from Denmark and Gan-Minster from France.

The main casualties of Direct Line's success, however, were the British composite insurers who historically had sold their policies through independent brokers. Direct Line, by bypassing the brokers, saved their commission, which averaged 25–40 per cent of premiums written. The success of direct writing had spurred the composites to reply with a variety of strategies. Some had responded by tightening their links with those brokers who obtained the best quality business for them. Further, they had stopped accepting policies from brokers whose business produced higher than average claims. One problem in this strategy of broker

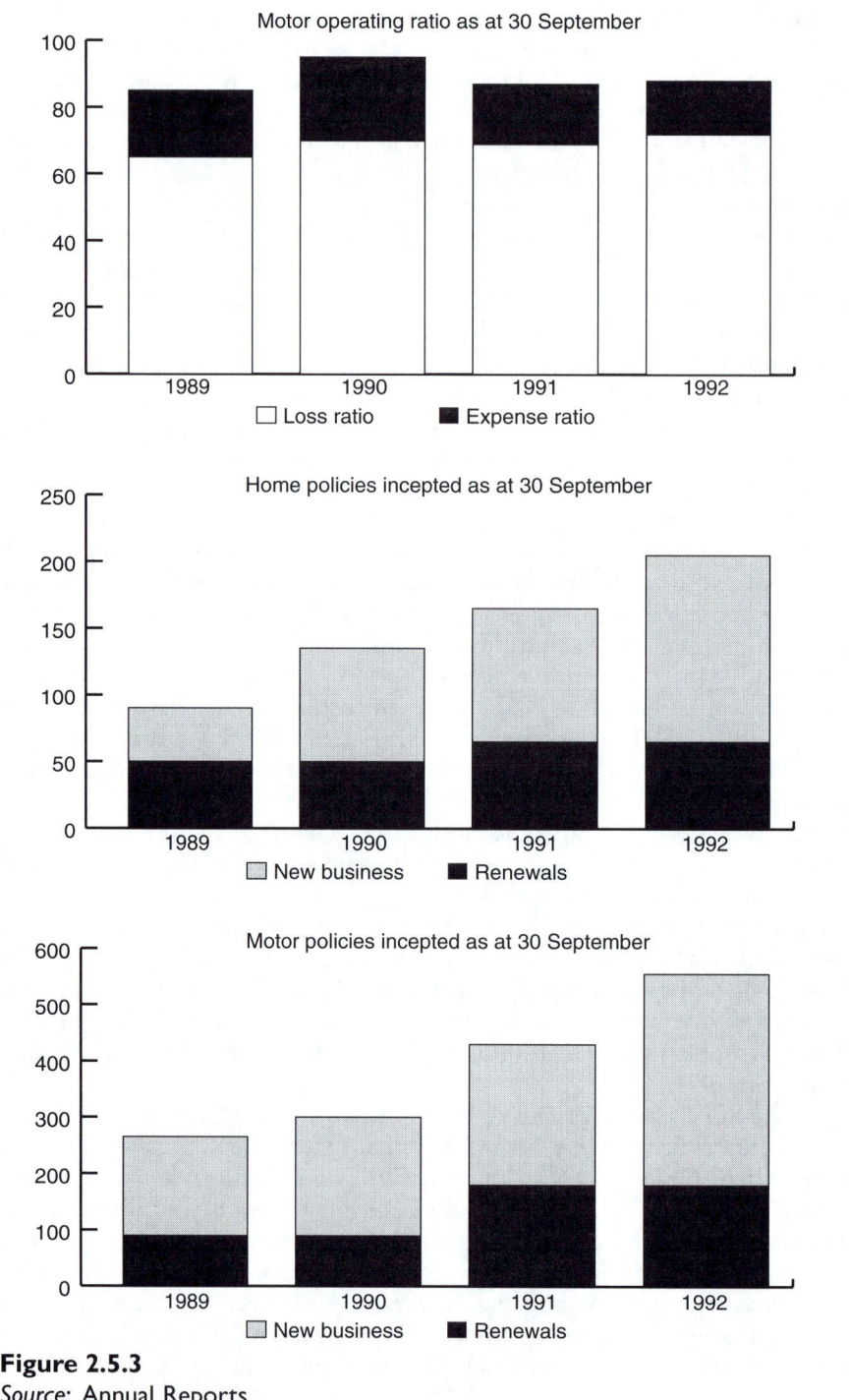

Figure 2.5.3
Source: Annual Reports.

selectivity had been the lack of sophisticated management information systems in many insurance companies, which made it difficult for individual companies to clearly identify the source of specific segment/broker profits or losses.

Other companies had attempted to secure their channels of distribution by integrating forward into broking by acquiring interests in major broking groups such as Swinton, AA Insurance Services and Hill House Hammond. Three companies – General Accident, Royal Insurance and Eagle Star – established their own direct writing operations between 1988 and 1990. However, these companies were anxious not to disturb their traditional broker-based channels and thus did not capitalize on the parent company's image by identifying closely with their subsidiaries. During 1991, for example, General Accident was faced with a brief boycott from brokers angered by its promotion of direct sales. The continued success of Direct Line and the other direct writers was forcing a change of attitude amongst the composites and greater recognition of the new channel seemed an essential strategy for the 1990s. 'They have to fish or cut bait. They have to choose one way or another,' stated one leading strategic management consultant, Mr Clive Bannister, insurance specialist at Booz Allen Hamilton. 'If they continue to walk down the middle of the road they will get run over.'[1]

The UK motor insurance market

Throughout the 1980s the UK motor market had been expanding slowly as a result of rising consumer affluence. The number of private cars and light goods vehicles expanded from 16.3 million in 1980 to 22.1 million in 1991. There had been a slight swing towards private cars in the late 1980s. This resulted from a decline in the attraction of 'company cars' due to the changes in tax treatment and a rise in the number of two-car families. This trend was expected to continue and even accelerate in the 1990s.

Premium income for motor insurers rose at a rate greater than GDP but claims tended to rise even faster. As a result, most motor insurers made a loss on underwriting, endeavouring to make these up with investment profits. In 1991, of approximately 600 motor insurers in the UK, only Direct Line amongst the majors made a net profit on underwriting alone. In addition, the industry had done little to reduce its expense ratios. This was partially due to the fact that nearly 40 per cent of expenses were related to commissions to intermediaries but also resulted from insurers failing to grapple with the issue. The largest motor insurer was the Norwich Union

Table 2.5.1 Leading insurance companies in the UK private car market (1991)

Company	Gross earned premium (£m)	Market share (%)	No. of vehicles	Average operating ratio		
				Claims (%)	Expenses (%)	Total (%)
1 Norwich Union	330.1	8.3	1078	96.3	22.4	118.7
2 General Accident	306.8	7.7	1302	84.3	29.1	113.5
3 Sun Alliance	275.4	6.9	1692	84.4	30.1	114.5
4 Eagle Star	258.2	6.5	925	80.2	28.3	108.5
5 GRE	247.2	6.2	960	97.6	25.2	122.8
6 Royal Insurance	219.6	5.5	1222	85.3	28.9	114.2
7 Cooperative	156.6	3.9	660	92.2	25.0	117.2
8 AGF Insurance	142.9	3.6	717	79.4	37.7	117.0
9 Commercial Union	128.0	3.2	486	82.6	28.4	111.0
10 Cornhill	121.4	3.0	542	82.7	25.8	108.5
17 Direct Line	65.4	1.6	336	68.4	23.0	91.5

Source: Gerrard Vivian Gray.

in terms of gross premium income written. Details of the leading insurers are shown in Table 2.5.1.

By the end of the 1980s there was a clear sign of a structural shift in the distribution system for motor insurance. The biggest intermediaries were beginning to more actively market their product. The Automobile Association, which had adopted a traditionally aggressive approach, was being followed by the major banks, and most recently by leading brokers which had been acquired by insurance companies. These latter included Hill House Hammond, acquired by Norwich Union, and Swinton, now owned by Sun Alliance. In addition, led by Direct Line, direct writing had expanded dramatically. The main insurers, although initially reluctant, had been forced to respond by offering their own direct writing alternatives despite their fear of alienating their traditional intermediaries. Finally, there had been a strong move to build direct business with affinity groups.

Direct writing was estimated by the Association of British Insurers to account for around 13 per cent of private car motor premiums in 1991. However, this figure included business written directly by branches, direct mail operators and affinity group business. Actual figures for the leading direct writers are shown in Table 2.5.2 for 1991 and 1992. The leader by far was Direct Line. This was followed by TIS, a subsidiary of Royal Insurance; Churchill, a company created by executives who originally were part of Direct Line and now owned by Swiss giant Winterhur and the direct sub-

Table 2.5.2 Market share of the direct response writers in the UK private motor industry

	1991			1992		
	Premium (£m)	Market share (%)	No. of policies in force (000)	Premium (£m)	Market share (%)	No. of policies in force (000)
Direct Line	84	2.1	411	159	3.5	670
Churchill	35	0.9	179	70	1.5	300
TIS (Royal)	44	1.1	292	60	1.3	314
GA 121	37e	0.9	175	48e	1.0	180
Eagle Star Direct	16e	0.4	55	38e	0.8	115
Smaller operators	15e	0.4	75	30e	0.7	125
Total	231e	5.8	1187	405e	8.8	1704

Source: Gerrard Vivian Gray. *Note:* e = estimate.

sidiaries of General Accident and Eagle Star. Direct writing was the fastest growing delivery system for motor insurance.

The UK household insurance market

There were few good available statistics on the household insurance market. However, the basis for such insurance was house ownership. It was estimated there were some 23 million homes in the UK, two-thirds of which were privately owned. On 75 per cent of houses either owner occupied or privately rented there was a need for structural insurance. A lower percentage of households had contents insurance coverage.

Premium income growth proceeded steadily during the 1980s but claims grew rapidly in the early 1990s due to weather losses and by dry weather leading to a sharp increase in subsidence claims. There were also substantial increases in claims for theft.

The sale of household insurance was concentrated in the hands of the building societies. The leading insurers were all linked to particular societies and the top seven insurers held some 59 per cent of the market in 1992. In addition, company agents held a 27 per cent market share, and direct marketing operations had achieved a 9 per cent penetration by 1991, again led by Direct Line. Other direct writers were still mainly at the entry phase of the market.

The Direct Line approach to insurance

The aim of Direct Line was to provide customers with high quality cover, backed by high standards of personal service at low prices. The company's research showed that this proposition was the opposite to the perception most people had of traditional insurance companies. Direct Line aimed to achieve this lower price, higher quality service by operating at lower cost not only eliminating the role of the broker, but by the innovative use of technology.

The use of advertising

Direct Line reached its customers by the use of high profile advertising coupled with extensive direct marketing. In its early years this was largely via the use of national press advertising. In 1990, however, Direct Line moved much of its advertising to television and introduced the little red telephone on wheels responding to a number of motor and household insurance situations. Conceived of by Mr Wood and his team, the TV campaign was awarded the BDMA Royal Mail Direct Marketing Award for the most innovative campaign of the year. Commenting on the success of the campaign, the originating agency, David Wilkins Advertising, stated:

> The lively use of the very simple device of the Red Phone has achieved a much greater level of enquiries than press alone. It also had to persuade a broader audience, which had perceived direct insurance operations as being 'fly by night' or financially unsound. The Red Phone has achieved the added benefit of making the method of response – the phone – virtually its own. It has become an immediately recognisable and memorable symbol for Direct Line.

As a result of the campaign, quotations from areas with TV support, compared against those without, were dramatically higher. Moreover, tracking data showed that awareness of Direct Line increased from 27 to 46 per cent, while perceptions of Direct Line as a forward-looking company rose from 20 to 40 per cent, as a reputable company from 11 to 26 per cent and for high quality service from 7 to 15 per cent. The judges of the BDMA award described the results of the campaign as 'fantastic'. The response rates to Direct Line advertisements were carefully monitored to identify for management their economic effectiveness. All potential customers seeking quotations were asked where they had heard of Direct Line. The number of quotations per ad, whether TV or press, was thus monitored, as were conversion rates from quotations to policies, although this latter figure was also influenced by the telesales operators. Nevertheless, management was confident that it had a reasonable view of economic effectiveness of Direct

Line advertising, including advertising decay. This data was immediately available on Direct Line's sophisticated MIS system.

Direct Line backed its advertising by the aggressive use of public relations with a view to seeking editorial coverage. Commenting on Direct Line's approach to home insurance for example, Mr Wood noted that, having introduced home insurance in 1987:

> We decided that 1990 was to see a major thrust from the company to acquire more household business. We launched a forceful press and TV campaign in early June aimed at challenging the dominant position held by building societies in the home insurance market. At the same time we made a formal submission to the Office of Fair Trading asking them to consider whether the customer really has a free choice in the selection of his household insurer. Our initiative gained widespread, quality coverage on TV and in the press and has prompted a growing number of consumers to direct their own complaints to the Office of Fair Trading.

In its advertising Direct Line guaranteed to cut household insurance costs to new policy holders provided houses were purchased during the previous five years via mortgages from 17 of the 20 leading building societies and had been continuously insured through them. The campaign and the related press coverage prompted a defensive response from many of the leading building societies which had previously actively promoted household insurance through themselves to mortgagors and on which they received commissions of some 30 per cent from the insurance companies. As a result, the number of quotations and acceptances of household insurance via Direct Line increased dramatically. Commenting, Mr Wood added, 'While the objective of all this activity was to increase our sales we, nevertheless, believe in a fair deal for the consumer and in the consumer's right to choose the product which best meets his or her requirements at the right price.'

The provision of personal service

Mr Wood believed that much of Direct Line's success could be attributed to the company's focus on high quality customer service. Unlike traditional insurers, who had little or no direct contact with policy holders, Direct Line encouraged customers to talk to them by telephone. During the financial year to 30 September 1990, Direct Line received 1.7 million telephone calls. By 1992 this number had grown to 4.4 million. Not only had the company attracted many new policy holders but the customer retention rate was substantially higher than most other major insurers.

Customers seeking a quotation were greeted by a telephone system which could prioritize incoming calls, direct them to a free operator and identify to the telesales operator the type of call he or she was receiving. The operators were based in each of Direct Line's four operating centres and received extensive and continuous training in telesales techniques to ensure that customers and potential customers received friendly, courteous and efficient service.

On receipt of a call, operators would effectively complete a proposal form via a computer screen which would prompt the operator through the data required. Within the software, automatic back-up systems would identify for the operator specific models of automobiles and the like, which had been preprogrammed to cover underwriting risk. This database was being continuously updated as new underwriting experience developed. After completing data entry, the computer would generate automatic quotations, for example dependent upon driver background, area and model of car, with alternatives covering different degrees of risk. When a quotation was made, if accepted by the customer, immediate cover could be granted if required, or deferred to a time convenient to the customer. Quotations remained in effect for a given period. The whole process took usually no more than three minutes for motor insurance. In the event of acceptance, the computer file was transmitted to a laser printer which operated day and night to ensure that all policies issued during the previous day were sent out to customers within 24 hours. Details of all quotes made but not accepted were also stored on the central database. This data was used to compare success and failure so as to structure marketing efforts. Failed quote prospects were also sent an unsolicited direct mail quotation on the first anniversary of their original telephone enquiry. Only 10 per cent of the Direct Line Portfolio was for non-comprehensive insurance compared with an industry average of 29 per cent. In addition, Direct Line had a strict policy of not accepting fast or exotic cars or young drivers.

Claims were also dealt with as rapidly as possible, as Mr Wood believed strongly that customer service did not end at issuing policies but rather began there and that the real quality of an insurance company lay in its speed in settling claims. In the case of smaller motor claims, Direct Line telephone claims department could issue immediate approval to motorists. In other, more significant cases, Direct Line's own engineers aimed to settle as quickly as possible and the company rather than the repairers guaranteed the quality of repairs if customers chose one of a carefully selected list of automobile repairers around the country.

For both motor and household policy holders Direct Line provided 24-hour helpline numbers. In motor these were staffed by friendly experts in accident and claims procedures, plus a free Accident Recovery Helpline to ensure that help was quickly provided following an accident. Vehicle recovery costs and emergency breakdown care were also provided as options on Direct Line policies.

Home policy holders were also provided with a free Household

Emergency Helpline providing 24-hour access to qualified tradespeople identified by Direct Line in the event of an emergency such as burst pipes, storm damage, broken windows or damaged locks.

In the consumer *Which?* magazine independent survey of some 100 motor insurance companies in 1990, Direct Line was one of only three companies achieving a top rating for its speed and efficiency in claims handling. The company was proud that this position had been maintained in subsequent years. After the Lockerbie air disaster in Scotland, Direct Line paid out emergency aid to policy holders in less than 24 hours.

The telesales quotations and claims service were also transparent to management via the Direct Line integrated MIS system. This reported in real time the number of quotations made by area, by operator and the conversion rate by operator, thus providing a clear view of performance in real time. Moreover, operators with lower conversion success rates could be individually counselled to improve performance. Successful operators or units could also be rewarded via bonuses or special rewards.

The pursuit of efficiency

Direct Line was an innovative user of technology to help keep down the cost of premiums. For example, the company only accepted payment by credit card or direct debit. In this way payments could also be received electronically, with the minimum of paper, which kept staff levels and, therefore, costs to a minimum. A combined effect of the level of personal service and low cost delivery systems meant that Direct Line, despite product/services copiers from the traditional service industries, had continued to show substantial growth throughout the 1980s and early 1990s. Moreover, via its integrated MIS system, the company constantly monitored its customer base to identify claim abuse, extra policy marketing opportunities and the like and adjusted premiums charged based on the level of overall actuarial risk.

The major products of Direct Line

By 1992 Direct Line insured some 4.5 per cent of all cars in the UK and was amongst the top five of motor insurers (out of some 600). Moreover, the company's swift and efficient claims service enhanced the Direct Line image. Direct Line, apart from offering standard motor insurance, had extended its product offer by allowing customers to add Motor Legal Protection cover for a fixed rate fee. This meant that if a policy holder should ever need assistance in recovering uninsured losses following an accident that was not their fault, the matter could be pursued through the Courts on their behalf.

Further, Direct Line made breakdown cover available to motor policy holders at prices much lower than traditional organizations like the RAC and AA. The company had also expanded its policy holder base by purchasing the motor book of Refuge Assurance, made up of some 7000 policy holders. Mr Wood considered this purchase as merely a dry run to test the risk and return of other insurers' motor books, because he felt that other leading composites would be getting rid of their usually unprofitable motor books.

> It was only four days' new business, in current terms, so I couldn't get excited about it ... It occurred to me that we would never take this route again because it's easier and better to grow your own business. There are probably 15 million motorists in the UK who pay for private motor insurance. Of them, ten million will have over three years' worth of no-claims bonuses. I want 30 per cent of that total.

In household insurance Direct Line was better able to quantify the effect of low prices. As a result, Direct Line had been able to guarantee purchase at lower prices for home insurance compared to those purchasing through building societies. Direct Line was also able to guarantee that the company's premium rate would not increase in the following year. By 1990 Direct Line had achieved cover on £7 billion of property and contents. By 1992 premium income had grown to £35.7 million and the number of policies in force had grown from 166 000 in 1991 to 206 000.

Direct Line also provided insurance for the Royal Bank of Scotland loan clients and credit card consumers. The service was in this case provided via the Royal Bank and covered over 350 000 accounts in 1990, though it was not as successful as motor or household businesses.

Early in 1993, Direct Line was developing further financial service products to offer on a direct sales basis. Mr Wood commented,

> We're doing a prototype now on home loans and deposits to our customers. It should be operational in four or five months' time. Longer term, we are looking at mortgages. However, I'll only give people home loans if they come through our core products. That really streams out some bad costs. It'll be simple – if you want one of our products, you have to have our home insurance.

Company organization

By 1992, the number employed by Direct Line had grown to 1086 people in its four regional centres, up from 787 in 1991. These centres were based in Croydon, Glasgow, Manchester and most recently Birmingham. Very roughly, there were around 520 in telesales and 330 involved in teleclaims. The company employed 47 in-house engineers and these evaluated 90 per cent of claims on motor business rather than using loss adjusters, as was normal in the industry. Household claims were evaluated by loss adjusters but Direct Line used its own engineers to conduct security surveys on high value properties. There were some 130 employees engaged in information processing, with around 40 of these being involved in systems development. In addition, some systems work was subcontracted out. Of the remainder, 40 employees were engaged in accounts, 7 analysing underwriting and 3 were involved in marketing. While premium income had soared in recent years the number of employees had grown at approximately half this rate. The company was still led by Peter Wood, who had originally conceived the idea of Direct Line. He was ably assisted by a small executive team many of whom had been with the company since its conception. In 1992, Mr Wood became the highest paid executive in the UK when he received a bonus of some £6 million for Direct Line's performance in the past financial year. This sum was based on a formula developed when the Royal Bank of Scotland bought out Mr Wood's original 25 per cent of Direct Line's equity.

Future prospects

Mr Wood, commenting upon the future prospects of Direct Line, stated,

> Despite its short history Direct Line has already claimed a position among the UK's leading personal lines insurers. During the 1990s many of our competitors suffered losses, in some cases substantial; some of them have responded with increases in premium rates on both their private motor and household buildings' accounts.
>
> We are aware of the problems facing the general insurer ... However, our underwriting success of the past year and the continued development of our infrastructure have put us in a strong position for the future and we consider that such an environment presents Direct Line with the opportunity to grow both its motor and its household accounts.
>
> This we plan to do, at the same time continuing to give our policy holders maximum value through the tried and tested combination of low costs and

high quality service. If all things stay the same, Direct Line will be making £100m in a couple of years. The only things you can't control are the competition and the weather. It could be worth more than the Royal Bank before too long.

Reference

1 Lapper, R. (1992). Direct Line Rings the Changes, *Financial Times*, 4 December, 23.

Case 2.6 Relationship marketing: The RSPB – a bird in the hand

Written by Helen Mitchell, based on an original project written by Karen Farquharson, Stephen Barlow, Aiden Cotter, Matthew Meredith, Malcolm Stainforth and David White.

A new chief executive

As her car turned into the large walled entrance of the Royal Society for the Protection of Birds (RSPB) and weaved its way along the gravel road leading towards 'The Lodge', Barbara Young couldn't help but notice the large number of visitors enjoying the trees and birds in the early April sunshine. Most of them were families or pensioners who had come to spend a day in the beautiful wooded estate that belonged to the charity.

Seeing so many people reminded her of the meeting scheduled for that afternoon with the marketing director. They were due to discuss the progress of the membership subscription drive and the level of membership donations. The RSPB was a thriving organization but they had a long way to go to meet planned targets for growth. Since she had joined the Society, she was only too aware of the importance of the members for continuing the success of the last century.

In January 1991 she had been appointed as the chief executive of the RSPB. Having previously spent 20 years in the NHS in a variety of roles, the newly appointed CEO was looking for 'a job that was worth doing' and one that she would care about passionately. She had a reputation for directness and a capability that had earned her a high profile and respect from the various public bodies for which she had served.

The announcement followed the retirement of Ian Prestt, an eminent ornithologist who, in his 15 years as Director General of the charity, had used his and the RSPB's influence to wage a ceaseless international campaign for the conservation of the environment.

Prestt had shaped what was generally acknowledged to be a professionally run organization, which was well positioned to move ahead into the 1990s. Under his leadership there had been rapid growth; in 1975 there were 275 000 members, this number had then trebled to 827 000 by 1990, bringing an annual income of £22.4 million. As well as this healthy membership base, Young also inherited from her predecessor a successful retailing operation, growing links with the financial services industry and also many substantial investments.

However, many different issues now confronted her. At the time of the appointment Young was quoted as saying, 'many challenges face the

Society and it is too big to be managed as a spare-time activity by people who really want to work on conservation'.

Her arrival marked the beginning of a new era for the RSPB and one which would bring a fresh focus on modern management techniques to the Society's business.

History of the Society

The RSPB was formed in 1889 and had an auspicious start, with 25 000 paying subscriptions being received by the end of its first decade. This helped to finance the first official 'watcher', who was appointed to protect breeding pintails at Loch Leven. The seal of royal approval arrived in 1904, when the Royal Society for the Protection of Birds was officially formed by a degree of charter. Personal recommendation from the Royal Family came in a letter from Queen Alexandra, who publicly supported the Society, and later that year King Edward VII went as far as to refuse a gift of plovers' eggs from a visiting dignitary.

Over the next 90 years the Society began to have great effect on legislation and environmental issues concerning wild birds. Shortly after the charter was granted the Importation of Plumage (Prohibition) Bill was introduced to parliament and after much hard lobbying the Bill was passed in 1921.

The majority of funds raised were reinvested in the purchase of land to create reserves, as efforts focused not only on the protection of birds but also the environments which sustained them. This work was rewarded in 1947 by the return of avocets to breed at Havergate and much excitement was generated as ospreys were sighted in Scotland after an absence of 50 years. The popularity of the cause and membership of younger members also grew; the Junior Bird Recorders Club, the forerunner of the Young Ornithologists Club (YOC), was formed at this time.

As the Society entered the 1960s public support for their cause grew rapidly and following the Torrey Canyon oil spill, that killed 100 000 birds off the Cornish coast, the RSPB raised enough funds in three years to buy four properties for conservation. In 1961 the move from London to the present headquarters, the picturesque 'Lodge' at Sandy, Bedfordshire, was not only a response to rising costs but also to enable major expansion.

Fund-raising appeals were now on a major scale. In 1977 the 'Save a Place for Birds' appeal raised over £1 million and membership of the charity now stood at 300 000 members. This success was to be repeated in the centenary year, 1989. During this year the RSPB successfully lobbied the government to abide by European Community directives to protect special sites associated with the movements of two million migratory birds. The launch of the 'Action for Birds' campaign also addressed a wide range of wildlife issues. These activities did much to bring environmental and conservation issues to the attention of the public through extensive press, television and radio coverage.

The RSPB entered the 1990s as the largest wildlife conservation charity in Europe, with over 827 000 members. It manages one of the largest conservation areas in the UK with over 130 nature reserves, covering more than 93 000 hectares, including important habitats such as lowland heath, wet grassland, estuaries and reed beds. Most importantly, the reserves help to protect 63 of the 77 most rare or threatened breeding birds in the UK.

Throughout the Society's history its objectives have always been to gain support from both members and the general public to achieve a healthy environment, rich in birds and other wildlife. However, although it deals with wider issues of conservation, its stated aim has always been to concentrate on the importance of birds and especially birds needing special protection.

The 'Future Directions' plan

Young recognized that the charity was a dynamic and successful organization but it needed careful planning and guidance to manage growth. It was important to recognize the need to balance the different aspects that make up the organization, to allow it to meet the challenges of the next century. To this end, the development of a new corporate strategy was one of the first tasks that she addressed.

The Society's vision of its future direction concentrated on positioning itself as the foremost bird conservation body as well as continuing to develop expertise in broader nature and environmental policies. However, wild birds would always be at the core of the Society's mission.

Young and her team began by identifying the core corporate values of the organization. These were that the RSPB should always strive to be:

Visionary	Innovative and inspire others.
Achieving	Provide real improvements to the environment.
Determined	Committed and do not go away.
Reasoned	Actions and views are based on sound knowledge.
Open	Respect and listen to others and actively seek partnerships.
Prudent	Resources must be used wisely.

Once these had been identified a new strategy plan was developed, called 'Future Directions – the RSPB'S Corporate Strategy for 1992/97'. The plan unveiled a new strategy which covered all aspects of the organization. It focused on both the 'ends objectives' and the strategy for income generation, without which the 'ends' were unlikely to be achieved.

'Ends objectives'

These centred upon the 'traditional' concerns of the Society: birds and wildlife and establishing and conserving habitats in the UK. However, larger emphasis than before was placed upon the importance of international work in Europe and Africa and the main concern was to 'take up a serious position in the global conservation community'. The newly prepared mission statement embodied these concerns:

> The RSPB believes that the beauty of birds and nature enriches the lives of many people but also that nature conservation is fundamental to a healthy environment on which the survival of the human race depends. The RSPB therefore strives for the conservation of wild birds and the environment on which they depend, primarily in the UK but increasingly in Europe and elsewhere in the world.

International development

Realization of this goal came in 1992 under Young's guidance as the RSPB's focus turned to global concerns and, together with other bird and habitat conservation organizations across the world, they formed a partnership called 'Birdlife International' to address conservation issues on a global basis. Under this banner the Society now funds projects to document the most important areas for birds in Europe, the Middle East, Asia and Africa. The proposed plan for expenditure is for a trebling of funding for international conservation over 1992–97.

A major contribution to this work comes from careful investment in research; advanced techniques using satellite imagery and electronic tracking provide vital information on bird movements and migrations. An indication of how important this has become is evident from the RSPB's expenditure of £2 million on research and monitoring in 1994/95.

Staff development

The plan stated that in order to be seen as the foremost bird conservation organization, the recognized gap between what they actually did and what the public perceived they did must be addressed (see Figure 2.6.1). Every mode of communication to the public would be used to clarify its position.

Achieving this goal would be impossible without the commitment of the staff and volunteers, and the 'Future Directions' strategy recognized the

What is the corporate strategy?

What do we need to achieve for birds, conservation and the environment over the next few years? What sort of RSPB do we want for the future?

These are the questions that the Management Board and Council have been considering with the senior managers. The result of our planning is Future Directions: the RSPB's Corporate Strategy for 1992/97.

The corporate strategy is a vision of the future direction for the RSPB. It defines what we need to do and achieve in a planned way by working together over the next five years. By having a clear vision of the RSPB's future direction, every function and department can see how it contributes to achieving the vision.

The corporate strategy sets direction but is flexible enough to adapt to changing conditions. It will allow us to grasp opportunities which arise unpredictably knowing that they are in tune with our overall direction. The strategy is evolutionary rather than revolutionary in that it builds on the excellent progress already made. But it does mean real change.

The conservation movement as a whole faces unparalleled opportunities and risks in the environmentally aware 1990s. At the RSPB we are well resourced with skills and finance. We have a wide and positive public profile and are well placed to make the most of the next five years.

The RSPB's mission

In preparing to review our direction for the next five years, we sharpened the definition of the RSPB's mission – really our reason for being. The following is a statement of our mission:

The RSPB believes that the beauty of birds and nature enriches the lives of many people but also that nature conservation is fundamental to a healthy environment on which the survival of the human race depends. The RSPB therefore strives for the conservation of wild birds and the environment on which they depend, primarily in the UK but increasingly in Europe and elsewhere in the world.

The vision

What is our vision of the future direction of the RSPB?

We aim to be the foremost bird conservation organisation as well as having considerable expertise and standing in broader nature conservation and environmental policy.

Birds will continue to be our start point. Given their visibility, popularity and value as indicator species for the health of the whole environment, birds provide an excellent focus for a wide range of conservation work. We believe the RSPB will achieve most for conservation by keeping wild birds at the core of its mission while continuing to develop work on broader conservation and environmental issues. We restate our commitment to all wild birds and the wider countryside, though we will continue to concentrate on priority species and habitats in order to target our resources most effectively.

The strategic objectives

What objectives do we plan to have achieved at the end of five years? And how will we know how successful we have been?

Conservation in the United Kingdom

We've set ourselves some very specific objectives, called 'ends' objectives, for safeguarding and enhancing priority species, habitats and sites and for the re-creation of the wider countryside. We plan to increase resources by over £4 million to achieve them. We will complete work already begun on species and habitat action plans and, in particular, we will increase the scope of our

Figure 2.6.1 Future Directions: the RSPB's corporate strategy for 1992/97.

policy work, especially that with a European Community dimension. We will continue to see purchase of reserves as one of the important ways of protecting sites and habitats and will acquire land to a level of £2 million per year, and more in exceptional circumstances. It will be important to collaborate with other conservation organisations where possible to get maximum impact.

International conservation

We will treble our expenditure on international work over the next five years. In particular we will develop our already substantial work in European conservation and will take on an increasing role in Africa.

Again we are setting, in collaboration with the BirdLife International network of bird conservation organisations, some specific 'ends' objectives for safeguarding priority species, habitats and sites. We will adopt the same framework as in the UK for choosing priorities in the rest of Europe and Africa and will help support selected BirdLife International priorities elsewhere in the world as funds permit. We will work as part of BirdLife International, the global network of bird conservation organisations, and play a key part in its development.

As we take up a serious position in the global conservation community, we will have to develop a view on global policy questions such as trade and aid.

Other important areas of work

There are similar objectives for all the other important areas covered by the strategy; in human resources, in infrastructure and accommodation, in information and financial systems.

Growth

To achieve all these objectives and meet the growing challenges to birds and conservation, the RSPB must grow. Our plan is to increase the RSPB's growth rate so that in six years' time we will have increased our spending by over £15 million a year and will be growing at a rate that would double our income in the five years after that. This is a major increase in our growth rate and will depend on how quickly the country recovers from recession. However, the current recession does not invalidate our growth plans. It may simply take us a year or two longer to achieve them. A detailed marketing strategy is being developed, to increase the membership, to get more help from each member, and to develop new sources of support.

Positioning

The strategy requires the RSPB to be seen as the foremost bird conservation organisation with expertise and standing in broader nature conservation and environmental policy within the UK, but increasingly in the rest of the world, especially in the rest of Europe and Africa.

But we know there is a gap between what we do and what the public thinks we do. As we grow, we must continue to measure the gap and manage it. We must test how much the gap between perception and reality impacts on our effectiveness in delivering our conservation objectives and in raising the funds on which they depend. We must manage positively the gap in presenting our work to the public, in order to decrease it while retaining the support of our loyal membership. Our position should be 'mission led but market modulated'.

There is no simple solution to achieving the public positioning that the strategy requires. But every mail-shot, every conservation policy, every appeal, every edition of Birds magazine, must contribute to clarifying our position to the membership, the public and other key audiences.

Youth and education

One of the most important groups we seek to influence and inform is young people. We will continue to implement the already agreed strategies for our educational work, and for our youth work through the Young Ornithologists' Club.

Figure 2.6.1   (*continued*)

Membership

The most important group is the membership. Members give the RSPB their time through volunteering, financial support and popular support of our work. For all these reasons we will aim to harness the support of as many people as possible. In addition, we will increase the opportunities for volunteering in line with the recently published volunteers' strategy.

Our members are vital to our work and we will commit more resources to finding out their opinions of the RSPB and our work, to developing improved 'member care' and to communicating well with the membership.

Values

As we grow we mustn't lose sight of the values that have been our strength in the past. These values include the high standards we adopt in our dealings with our supporters, and with the communities with which we work, conducting our activities in a business-like and environmentally conscious way and our fundamental insistence on well researched rational argument.

Implementation

How will the corporate strategy be translated into real change and positive action? Implementation of the strategic objectives will take place through the RSPB's annual planning and budgeting process. Work programmes will outline the annual steps we are taking towards achieving the corporate strategy.

The team

There is much to be done. This summary highlights some important aspects of the corporate strategy. A fuller version is available on request.

There's a role for all of us to take to make it happen. The strategy can only be achieved if we all work corporately towards meeting its objectives.

We look forward with excitement to the commitment we know you have to the RSPB's future. But the RSPB must also make a commitment to you. The success of our work depends ultimately on the quality and enthusiasm of you, the staff. So we plan to invest substantially in training and developing staff to help deliver the objectives we all believe in.

Together and with a membership that is stronger, more aware and active, we will pursue our future direction with boldness and make a real difference to the challenges that lie ahead for birds, the environment and all of us.

Chief Executive
February 1993

RSPB, The Lodge, Sandy, Bedfordshire SG19 2DL.
Tel: 01767 680551 Fax: 01767 692365

RSPB Northern Ireland Office, Belvoir Park Forest, Belfast BT8 4QT.
Tel: 01232 491547 Fax: 01232 491669

RSPB Scottish Headquarters, 17 Regent Terrace, Edinburgh EH7 5BN.
Tel: 0131-557 3136 Fax: 0131-557 6275

RSPB Wales Office, Bryn Aderyn, The Bank, Newtown, Powys SY16 2AB.
Tel: 01686 626678 Fax: 01686 626794

Figure 2.6.1   *(continued)*

importance of their contribution. A training and development programme was announced that would be an on-going process and aimed to empower staff to take customer-focused initiatives. Entitled 'Full Power', it aimed to underline the importance of every individual's role in communicating the message of the RSPB to the outside world. All participants would be encouraged to:

- examine the way in which the Society communicates its messages to the outside world
- identify 'audiences', both internal and external
- use clear communication skills in dealing with these audiences
- identify different opportunities to get messages across effectively
- apply these principles to improve the way the RSPB communicates.

The RSPB and other charities are different from other types of industry in terms of staff and staff recruitment. They embrace the whole spectrum of the labour market, from unpaid part-timers to full-time, highly paid executives. As some now operate highly sophisticated sales operations, it has been more recent practice to employ talented and experienced executives and to pay the market price for that skill.

Most employees are passionate about birds and highly committed to the RSPB cause. As a result, recruitment is usually a case of trying to identify the best candidate from large numbers of applications for each job. This is in addition to the increasing number of volunteers that give their services free to help in the shops and reserves each year.

Policy development
Policy issues also became broader under the outline of the plan. It recognized a need to develop policy on a more global basis and now covered such issues as trade and aid. Environmental and conservation issues would also increase their scope and concentrate particularly on the EEC dimension. The RSPB has always had a wide range of policy interests and has great expertise in actively campaigning and getting wide publicity for policy issues. For example, MPs, MEPs, EC officials, civil servants, peers and political advisers are constantly briefed and lobbied to promote changes to UK and EC legislation.

In order to campaign effectively, membership levels are vital to the Society; it has more members than the Labour, Liberal and Conservative parties combined and MPs are known to listen much more carefully to an organization that has influence on a large number of his or her constituents. The Society relies on members' endorsement of the work that is done; renewed memberships reflect this approval and give added strength to their lobbying power.

In order to gain maximum influence the Society must also be seen to be involved in decision making. Young has been a Committee member of the Secretary of State for the Environment's 'Going for Green' initiative since 1994 and a member of the Department's 'Round Table on Sustainability'. Political lobbying is an important part of the CEO's role and one which she handles with relish. In an interview with the *Glasgow Herald*, Young was quoted as saying that she was 'not convinced that Mr Major is very pro-bird'.

However, this does not hinder the Society from aiming to achieve the ends objectives within the 'Future Directions' strategy plan! Most obstacles faced come from within the sometimes restrictive environment of the charities industry itself.

The RSPB and the charity 'industry'

Charities in the UK
The charity 'industry' in the UK accounts for some 3 per cent of gross domestic product. Each year the British public gives nearly £5 billion to over 170 000 charities. These range from large, well-known concerns like the National Trust and Oxfam to more specific or local charities such as the Friends of Bristol Horses and the Confectioners Benevolent Fund.

The charities which benefit most from voluntary donations can be divided into five main groups:

1 *Medicine and health* – Imperial Cancer Research Fund, British Heart Foundation
2 *General welfare* – Salvation Army, RNLI, Barnardo's
3 *International aid* – Oxfam, Save the Children Fund, Red Cross
4 *Heritage and the environment* – National Trust, Worldwide Fund for Nature
5 *Animal protection* – RSPCA

The RSPB see themselves firmly in Heritage and the Environment but are often perceived as an Animal Protection charity. (See Tables 2.6.1–2.6.3 and Figure 2.6.2 for detailed information on competitors.)

The profile of the average charity supporter in the UK is predominantly AB (a National Readership Survey classification in which As are 'upper-middle-class' and from higher managerial, administrative or professional occupations and Bs are middle-class and from intermediate managerial, administrative or professional occupations), middle-aged, read the *Daily Telegraph* and are more likely to live in the South of England. The gender split is 50 : 50. However, within this very broad generalization there are strong variations. People giving to environmental and wildlife charities tend to be much younger than the average and not so well off. People who give to 'fluffy' animal charities tend to be female, have pets of their own

Table 2.6.1 Total charity voluntary income, by type, 1993–95

Source of income	1993		1994		1995		% change 1993/1995
	(£m)	(%)	(£m)	(%)	(£m)	(%)	
Household donations	1450	41.0	1476	39.9	1310	31.6	−10.0
Legacies	628	17.7	635	17.2	640	15.4	+1.9
Government grants	1213	34.3	1325	35.8	1417	34.1	+16.8
Corporate contributions to the community	248	7.0	264	7.1	272	6.6	+9.7
National Lottery – Charities Board	–	–	–	–	260	6.3	n/a
National Lottery – other good causes	–	–	–	–	251	6.0	n/a
Total	3539	100.0	3700	100.0	4150	100.0	+17.3

Source: Henderson Top 2000 Charities/Directory of Social Change/Home Office/Mintel.

and be from slightly lower socio-economic groups than the average donator.

Giving money to charity is essentially a subjective act and the well-known British characteristic of a passion for animals helps to explain why animal charities thrive particularly well in the UK. For example, the Donkey Sanctuary (£5.3 million in 1991/92) raised more funds than Shelter (£4.3 million in 1991/92) and Mencap (£3.6 million in 1991/92).

Another inherent advantage for animal charities, in terms of corporate sponsorship, is that some companies feel uneasy being associated with causes such as dementia or mental health and are happier supporting

Table 2.6.2 Ranking of top five conservation/heritage charities by voluntary income, 1996

Charity	Voluntary income (£m)
National Trust	75.9
RSPB	23.9
WWF	15.5
National Trust (Scotland)	6.2
Woodland Trust	4.9

Source: Charities Aid Foundation Report 'Dimensions of the Voluntary Sector' (1996).

Table 2.6.3 Ranking of top 30 charities by voluntary income

Name of charity	CAF ranking 1995	1993	Total voluntary income (000s)	Other income Trading (000s)	Sales of goods & services (000s)
National Trust	1	1	75 929	9100	9038
RNLI	2	3	58 231	560	–
Cancer Research Campaign	3	6	55 970	954	–
Oxfam	4	2	53 735	2070	38
Imperial Cancer Research Rund	5	5	45 644	1201	–
Save the Children Fund	6	4	41 355	–	–
Salvation Army	7	9	38 841	875	23 946
British Red Cross	8	15	38 763	194	20 715
NSPCC	9	10	35 315	794	1343
Help the Aged	10	8	34 829	194	1907
Barnardo's	11	7	33 915	1168	–
Cancer Relief Macmillan Fund	12	13	32 329	316	–
RSPCA	13	11	32 053	125	1658
British Heart Foundation	14	14	28 483	100	–
Marie Curie Cancer Care	15	20	27 434	57	–
RNIB	16	16	26 060	–	5158
Guide Dogs for the Blind	17	17	24 884	593	–
ActionAid	18	22	24 646	(218)	85
RSPB	19	21	23 983	1958	2223
Christian Aid	20	12	22 622	388	–
SCOPE	21	19	20 816	114	1347
Tear Fund	22	23	18 727	96	–
People's Dispensary for Sick Animals	23	25	18 674	1420	–
Institute of Cancer Research	24	18	18 283	–	–
Charity Projects	25	32	16 198	4081	–
CAFOD	26	30	15 659	–	–
Worldwide Fund for Nature	27	26	15 560	461	–
Royal British Legion	28	27	15 511	152	7842
Arthritis & Rheumatism Council	29	24	14 875	83	–
BBC Children in Need Appeal	30	–	14 056	1607	–

Source: Charities Aid Foundation Report 'Dimensions of the Voluntary Sector' (1996).

animal and children societies. Some members of the public also believe that health and medical research should be a Government responsibility and often refuse to donate to these types of charity.

Competition
Charities do compete with each other for funds and the charities that the RSPB regards as the closest competitors are the National Trust, the World-wide Fund for Nature and other wildlife charities. However, the industry

National Trust

Facts
- Membership stands at 2.29 million.
- Own 241 hectares, 550 miles of coastline and 245 properties.
- Spend £9.9 million on routine property repairs and maintenance.
- Million paying visitors per year to houses, gardens and land.
- Have high tech, interactive computer centres in over 30 properties.
- 32 000 volunteers give over 1.86 million hours of their free time per year.
- All property is inalienable – it can never be sold or mortgaged in the future.

Aims and objectives
- The Trust's role is to: 'preserve places of historic interest or natural beauty permanently for the benefit of the nation'.
- They believe the reason for their success comes from: 'the dedicated people who serve our mission, whether paid or volunteer, whether donors of property or funds; and our great army of members', and also from 'our independence of the serving government'.

They have recently been taking a more holistic view of their role to get away from the 'stately home image'. Using language similar to the RSPB they stress both their habitat management role and focus on conservation.

Perception of the National Trust
- The majority of members see it as a guardian of habitats, animal and plant life.
- Only a third of members believe the Trust aims to encourage participation in its work by the local community, few see it as pioneering or innovative.
- Members see their subscription as being both a contribution to heritage preservation as well as an encouragement to visit.
- The public see the Trust as wealthy as they own wonderful and impressive places filled with treasures. The Trust goes to great lengths to stress that they are not rich and inevitably a lot of houses run at a deficit.

Future strategic directions
- The Trust will emphasize more strongly their conservation/environment work. More involvement with global and European issues – 11 of the Trust's regions now get financial aid from the EU.

Figure 2.6.2 Competitor profile

Worldwide Fund for Nature (WWF-UK)

Facts
- Membership in 1993 was 230 214, however, calculation of this figure has changed and now members and donors are counted together and stand at 242 939 on 3 September 1995.
- Founded in 1961.
- Total income in 1995 was over £21 million.
- Direct expenditure on conservation projects and fund-raising activities was £13.5 million.
- Legacy and appeals income steadily increasing but regional fund-raising and catalogue trading have shown losses recently and the WWF-UK admit that cost of attracting new members is proving to be inhibitive.
- WWF-UK are rumoured to have a falling membership base and a poor staff morale.

Aims
- 'WWF's ultimate goal is to stop, and eventually to reverse, the accelerating degradation of our planet's natural environment. Our vision is a future in which humans live in harmony with nature.' This will be achieved by:
  - preserving the abundance and diversity of genes, species and habitats
  - ensuring that any use of biological resources is ecologically sustainable both now and in the long term
  - promoting actions to minimize pollution and the wasteful exploitation and consumption of resources and energy.

Perception of the WWF
- The greenhouse effect and pollution of the UK rivers and seas are seen as the two main issues by their membership.
- A quarter of the membership believe that global influence is their greatest strength.
- Only 15 per cent believe that WWF have a strong public profile.
- Members believe that the charity should devote more resources to active promotion and publicity of their cause.

Future strategic directions
- Active involvement in influencing governments across the world, projects cover timber certification schemes, wildlife trade legislation, encouraging and providing environmental management systems.
- Emphasis and concentration on conservation issues.
- Development of education programmes in schools and communities.

Figure 2.6.2 (*continued*)

Greenpeace

Facts
- Founded in 1971, the total number of 'supporters' is now 279 000. 5000 of these are 'Frontline supporters' who pay £10 per month in return for video updates on campaign activity, a magazine and invitations to regular campaign meetings.
- Total income in 1995 was £6 627 000. 80 per cent of this was from subscriptions.
- Direct expenditure on campaigns and local group support was £3.5 million.
- Membership base is rising steadily, the charity has benefited from the recent high profile battle with Shell Petroleum.

Aims
- 'Greenpeace has always been ready to campaign against things which we believe are morally or ethically unacceptable.' Annual Report (1995).
- Greenpeace is in the business of elimination of problems, not their management.
- 'The optimism of the action is better than the pessimism of the thought' (Harold Zindler, an early Greenpeace activist).
- It aims for long-term change. Recent projects have included: campaigning for new manufacturing technologies by commissioning the design of a Greenpeace fridge and car, championing energy saving schemes, protecting fishing stock, attempting to stop whaling vessels and campaigning against genetic engineering in the food industry.

Perception of Greenpeace
- In the past it has been seen as a radical activist group, having links back to the 'hippy movement'; this was reinforced by media coverage of their physical attacks against whaling boats and illegal fishing fleets with their mother ship, 'Rainbow Warrior'.
- In 1996 Greenpeace is regarded as more mainstream and relevant for a wider number of people. A nationwide poll commissioned by the charity found that 61 per cent of respondents 'supported' or 'agreed' with Greenpeace.
- Greenpeace now works closely with many of the world's leading companies. Recent projects have included: group working with Body Shop, IKEA, Tesco and Lloyds Chemist Group to investigate PVC's effects on human health, the introduction of a green refrigerator with Calor Gas and developing solar power for domestic use with the Co-op Bank.
- Journalist Richard North (usually a strong detractor of Greenpeace) wrote recently, 'Courage and charisma have always been its hallmarks. It is now adding a degree of corporate savvy in a Branson or Roddick sort of way ...'

Future strategic directions
- To concentrate on the twin missions of defence of nature and the reform of industrialism.

Figure 2.6.2   (*continued*)

RSPCA

Facts
The Royal Society for the Prevention of Cruelty to Animals
- Membership in 1995 – 34 000 adult members, 60 000 group members and 19 000 younger members of the Animal Action club, but overall 330 000 financial supporters.
- Founded in 1824 by a small group of animal activists, it is the world's oldest animal protection agency.
- Total income in 1995 was over £37 million.
- Direct expenditure on animal projects and fund-raising activity was £40 million.
- There are 305 inspectors investigating animal cruelty. It is the largest non-governmental law enforcement agency in the UK.
- In 1995 the Society's 10 regional communication centres received over one million calls from the public.

Aims
- 'Whether lobbying for change in the parliaments of Westminster or Strasbourg, providing food, shelter and treatment for abandoned dogs and cats or injured wildlife, tracking down dangerous dog-fighters or prosecuting animal abusers in court, the Society can be found just about anywhere the welfare of animals is at stake.' Annual Report (1995).

Perception of the RSPCA
- A much loved, traditional, British institution.
- Public awareness of the Society's work was heightened by the BBC television's *Animal Hospital* series, which was watched by over 11 million people. This highlighted the wildlife aspect of their work, which was not previously well known.
- Recent advertising campaigns have been highly controversial, almost in the style of 'Greenpeace' campaigns, and the Society is now being seen as more active in campaigning for change.

Future strategic directions
- Active involvement in influencing British and European governments.
- A focus on the problem of stray and unwanted animals, not only by building shelters but by ensuring animals are neutered to help avoid unwanted litters.
- Campaigning on a variety of issues: whaling, animal experiments, driftnet fishing, oil spillage disasters and the fur trade.
- To halt transportation of farm animals to the continent and to campaign for a supermarket labelling scheme called 'Freedom Food'.
- To initiate animal welfare training across the world.

Figure 2.6.2 (*continued*)

is unusual in that hostile competition is very rarely practised. The RSPB and most other charities are aware that their members are also members of other charities and that the very nature of giving to charity demands a level of integrity from them.

The level of cooperation between charities is surprising for businesses of their size. Charities involved in direct marketing meet on a bi-monthly basis to discuss when mailings are being planned and so coordinate them to avoid clashes of interest. This was especially useful when the RSPB and the National Trust celebrated their centenary years and mailings were at a peak. This meant that other charities could hold back on appeals and send them when they were likely to be more successful.

Another example of this cooperation was seen in the launch of a charity CAF (Charity Associations Fund) card account in 1996. Members of the public can place all the money that they give to charity in one account, receive a larger covenant on the total amount and then choose to direct the funds to the charities they wish.

Legal requirements on charities

With the large amounts of money collected by charities, combined with the diversity of different charitable organizations, regulation is becoming increasingly more stringent. Charitable status is sought by organizations because of the favourable tax status it confers. To achieve this status every charity is required to register with the Charity Commission of England and Wales. This body is then responsible for the supervision and regulation of that charity. The Commission will draft and propose any new laws and guidelines it feels fit and acts as a regulatory body to oversee charity activities.

In recent years the charities sector has been under increasing scrutiny. In 1993 864 charities were investigated by the Charity Commission for maladministration, malpractice and breaching various charities laws. The activities of charities are also becoming the concern of the public, who give their money based on trust. Unfavourable publicity on such matters can unnerve the public and change their perception of charities.

Just such a spotlight recently fell on Guide Dogs for the Blind, who were found to be holding over £100 million in reserves but spending only £28 million a year on training and care for guide dogs. The money in reserve represented 16 years of operating costs (£6.25 million is the annual operating cost) and was in stark contrast to Oxfam, for example, who currently hold only seven weeks' operating costs in reserve. This created bad publicity for the charities industry as the generous British public do not expect their donations to sit in deposit accounts.

Charitable status also means that the benefits cannot be for the members themselves. In the case of the RSPB, the nature reserves are for the benefit of the birds, with the enjoyment of the members being seen as entirely incidental.

However, some changes are in the pipeline. Heavy lobbying by the National Council for Voluntary Organisations (NCVO) has successfully persuaded the government to repeal the Trustee Investments Act of 1961, which put restrictions on investment of assets into gilts and equities in order to protect charities from 'foolish decisions' by trustees.

Charities welcomed the proposal. It will open up new investment options and allow them to make use of modern portfolio theory in managing their funds. Charities have always believed that they have been disadvantaged by this law; in the past a £2 million investment in the stock market at the time the Act was passed would have increased sevenfold by 1996, whereas the same investment in bonds would have yielded only a £750 000 return.[1]

Political activity

To achieve and retain their charitable status, charities are not permitted to partake in any direct political activity. Charities such as War on Want and Oxfam are often investigated by the Charity Commission on the political nature of their activities. Charities have to resort to 'lobbying', although the fine line between this and direct political activity has resulted in further guidelines and rules from the Charity Commission.

The public purse

There is little evidence that a recession affects individual donations to charities. Over the last 10 years, during one of the most prolonged recessions in recent history, arguments that the British public are suffering from 'charity fatigue' appear unfounded. Indeed, some of the most popular charity campaigns have been launched in this time, Live Aid, Children in Need and Red Nose Day being successful examples.

It appears to have had a greater effect on company donations. Many of the UK 'blue chip' companies reduced their yearly donations. By 1996 this had fallen to a yearly total of £160 million or 0.2 per cent of profits, the lowest level of corporate donations for five years. This was defended by companies such as BP and Marks and Spencer, who pointed out that they make contributions in other ways by seconding staff to charities or providing skills or accommodation instead. Company liquidations did not improve the situation. The collapse of Barings Bank meant £12 million less for charities in 1995/96.

However, the launch of the National Lottery, in 1994, appeared to bring a far more serious threat to charities' long-term revenue. A survey undertaken in May 1995 by the NCVO, seven months after its launch, showed a

£71 million reduction in charitable donations. If that trend continued, the voluntary sector was expected to lose £212 million by the end of 1995. The percentage of people donating was also seen to fall, from 81 per cent in November 1994 to 67 per cent in March 1995. However, the Lottery does redistribute funds back to charities, although the allocation is fixed at 5.6 per cent of each ticket sold. This means they are due to receive only £155 million, leaving a shortfall of £57 million in 1995.

There also appears to be confusion amongst the general public on how much of the money they spend on the Lottery is actually going to charity. Research by the Charities Aid Foundation found that people on average thought 20–22 pence went to charities. This view was encouraged by the first saturation advertisements, which promoted the Lottery as benefiting 'good causes' by 28 pence per ticket. This, in fact, refers not only to charities but to sports, the arts, National Heritage and the Millennium fund. The board which distributes the funds also can determine the type of charity to which the funds are attributed. The first payments in October 1995 were restricted to 'causes to help the poor'. This prompted a reply from Stuart Etherington, Chief Executive of the NCVO, who said in May 1995, 'while helping disadvantaged people is clearly worthwhile' it would have been helpful if the board had 'given some indication as to when other causes, like animal welfare and medical research, would be eligible to apply'.

Street collections and the purchase of raffle tickets are the biggest source of income that has been affected. These forms of fund-raising are used by large and small charities and did account for 10 per cent of the total value of charitable donations by the general public. Charities that run lotteries themselves have been severely affected. Many of them, such as Tenovus, the cancer charity, and Arthritis Care, were forced to abandon their lotteries in 1995 despite revenue running into millions of pounds in previous years. The RSPB has also been affected to a degree, as their raffle revenue has been reduced since the launch of the National Lottery. In financial terms this amounted to just over £100 000, which is less than 1 per cent of their total income. Thus the effect of the Lottery is less profound on charities that rely on memberships, bequests and subscriptions.

However, latest figures show that the gloomy prophecy of 1995 appears to be premature as the total revenue taken by charities in 1995 was up by 2 per cent on 1994. The study by the Charities Aid Foundation in May 1997 reported that the public have in fact increased donations over that time, despite a drop in corporate donations and revenue from charity shops. One theory proposed for this increase is that the huge publicity generated by the Lottery has had a beneficial effect and has raised charity profiles in the UK.

The study also showed that a large percentage of the growth has come from the newly formed Area Health Trusts, who have made concentrated efforts to increase fund-raising and have seen a 12 per cent rise over the year. What remains to be seen now is the breakdown of revenue between the various charities and whether certain categories have suffered, perhaps due to the focus on 'poverty' charities by the Lottery Board and the new competition from the health sector.

It is clear that, since 1991, the charity and fund-raising market has changed rapidly, bringing with it far-reaching implications for future income generation strategies.

Income generation and marketing in the RSPB

Income generation
The emphasis on the need to grow conservation in the UK and internationally, as outlined in the strategy plan, required a matching growth in income. The RSPB has always faced a need to balance the budget between income and expenditure: increased expenditure has to be matched by a combination of an increase in membership levels, higher donations and legacies or other methods of fundraising.

Table 2.6.4 Statement of income for years ending March 1991–96 (£'000s)

	1991	1992	1993	1994	1995	1996
Membership subscriptions	7219	9301	9814	10 529	11 139	12 037
Legacies	5446	7150	6788	8397	9300	9390
Fund-raising:						
Appeals	1913	2643	2606	2289	3121	2250
Mail order and shops	1588	2033	2057	2184	1678	1599
Raffles	319	634	591	687	1212	1122
Members' groups	284	349	386	289	335	273
Collecting boxes, admissions and general donations	543	551	590	706	529	571
Total	4647	6210	6230	6155	6875	5815
Business support and trusts	1775	1855	1989	1794	2364	2164
Films, consultancy and incidental income	639	783	699	860	875	779
Grants	560	705	936	1337	1628	3219
Financial – interest, dividends and investment profits	1096	1015	893	1018	665	693
Income from land	719	781	928	745	634	678
Total	22 101	27 800	28 277	30 835	33 480	34 775

Source: RSPB Annual Reports, 1992–96.

Table 2.6.5a Statement of expenditure for years ending March 1991–96 (£'000s)

	1991	1992	1993	1994	1995	1996
Direct conservation:						
Acquisition of new reserves	3439	2124	2545	2409	1975	2418
Management of reserves	3821	4103	4665	4948	5448	6386
Species and habitat protection (including research):						
Regional	2445	3242	3703	4043	5130	5191
National	1308	1685	2295	2452	3090	3402
International	610	875	1268	1416	1656	1843
Publications, films and education	3440	3809	4053	3777	2869	3240
Total	**15 063**	**15 838**	**18 529**	**19 045**	**20 168**	**22 480**
Costs of income generation:						
Recruitment of new members	1864	1712	2274	2445	2125	2356
Fund-raising costs	1211	1211	1674	1991	2158	2334
Mail order and retail shop costs	674	788	1113	1318	1220	1327
Support services:						
Operating costs	987	1476	1599	1599	1526	–
Capital system purchases	723	347	451	451	360	–
Transport and office services:						
Property and maintenance costs	1043	860	1024	1024	1099	–
Print, post and telephone	729	834	810	853	1023	–
Capital equipment	652	267	353	292	265	–
Total	**4134**	**3784**	**4237**	**3644**	**4273**	**4325***
Management, finance and personnel	1360	1535	1636	1636	2030	2162
Irrecoverable VAT	507	279	324	324	183	249
Total	**24 813**	**25 128**	**29 787**	**30 832**	**32 736**	**35 233**

Table 2.6.5b Operating deficit/surplus for years ending March 1991–96 (£'000s)

	1991	1992	1993	1994	1995	1996
Total income	22 101	27 800	28 277	30 835	33 480	34 775
Total expenditure	24 813	25 128	29 787	30 832	32 362	35 233
Surplus/deficit	(2712)	2672	(1510)	3	1118	(458)

Source: RSPB Annual Reports, 1992–96.

*1996 accounts were presented in a different format and not broken down by auditors, Price Waterhouse. This figure is a Cranfield estimate.

The statements of income and expenditure account in Tables 2.6.4 and 2.6.5a and 2.6.5b illustrate the performance achieved from 1992 to 1996. The surplus appears to fluctuate wildly from year to year but this reflects the nature of some of their sources of income and the projects on which it is spent. As a non-profit making organization the RSPB has to divide its focus not only on income attraction but also on allocation of those funds. Purchase of conservation areas may be more successful in some years than others. In 1992 the Society was unable to complete on a number of trans-actions and had a large surplus of funds whilst in 1993 they committed themselves in excess of income. Legacies can also vary considerably and it is impossible to have any notice over when and how much individuals will leave to the Society.

However, an increase in income has recently come from a new division, RSPB Sales Ltd, which was formed in 1993 in order to widen commercial activities. Under this banner it is allowed to profit from sales that are not normally allowed under its charitable status. By covenanting RSPB Sales Ltd's operating profit to the Society, it is able to recover the related income tax. Income from this division in 1994 stood at £963 000 and is expected to increase steadily over time.

This is a welcome addition, but it was apparent to Young that the exist-ing marketing strategy would have to be altered if sufficient income was to be generated to fund the impetus for growth. She set a target expenditure of £38.2 million and a target of 1.2 million members by the year 2000 in order to meet the planned 'ends objectives'.

Marketing in the RSPB – 'Towards 2000'

In Young's first meetings with Frances Hurst, the Marketing Director, and Karen Rothwell, Strategic Coordinator, much effort was focused on devel-oping a marketing strategy that would concentrate upon increasing both membership levels and donations. It was also recognized that a high per-centage of RSPB income came from the membership product: the Society needed to avoid over-reliance on this source and develop new areas of fund-raising. A confidential document 'Towards 2000' was produced. It laid down a strategy in which the role of the marketing department was seen as paramount, some key aspects of which follow. Its overall mission was to generate income which would be consistent with the policy aims of the Society:

- to uphold the values within the 'Future Directions' document
- to operate within best charity fund-raising practice

- to reflect the RSPB's positioning and personality
- to achieve an acceptable return on investment.

It concentrated on two objectives:

1 To build on the RSPB's recognized strength with its existing membership and to develop new and better methods of recruiting and retaining members to ensure that total membership grows.
2 To develop new sources of funds by testing new fund-raising mechanisms within the following areas:
    a Regional fund-raising networks. Evaluating potential face to face fund-raising opportunities using regional fund-raising offices based around the UK.
    b Developing a legacy marketing programme.
    c Optimizing lottery opportunities. The removal of restrictions on lotteries arising from the Lottery Act provides another means of developing other innovative ways of approaching the public for funds.
    d Establishing a high level network, to identify people of influence who support the RSPB's work and build relationships with them.
    e Coordinating new grant opportunities. To coordinate grant applications more effectively and ensure that potential new sources of income are sought through the most appropriate channels.

As a response, the marketing department was reorganized in order to be able to respond most effectively to the new demands (see Figure 2.6.3):

Marketing Development. Responsible for identifying and developing new income streams to create a 'better balance of funding'. Marketing operations are divided into a number of functions, as seen in Figure 2.6.3.

Corporate Marketing. This section is responsible for developing partnerships with commercial organizations with the aim of raising money. The fund-raising can take the form of sponsorship, joint promotions, grants, trusts, product endorsements.

Regional Marketing. Concerned with 'face to face' fund-raising and membership recruitment via the Regional Fund-raising Officers working with the reserves and members' groups, together with generating income from reserves.

*Brands Management.** Responsible for developing and maximizing opportunities for the RSPB by developing brand plans for the RSPB, YOC and Birdlife International.

*Market Research.** Provides a central management tool for the whole Society by providing and interpreting market information and key statistics.

---

*Brands Management and Market Research work cross-divisionally.

Trading. Retail, Mail Order (catalogues), Members' Group sales and catering are brought together to maximize income from trading. The aim is to extract the greatest potential from members and non-members alike. Suppliers are recruited by competitive tendering, although some moves are being made to appoint approved suppliers.

Database Marketing/National Recruitment. The RSPB's computer systems are in urgent need of upgrading and are considered obsolete. At the beginning of 1996, the database capability is restricted to relatively unsophisticated analysis. The software cannot analyse defections or complaints raised by members, although every complaint is personally dealt with.

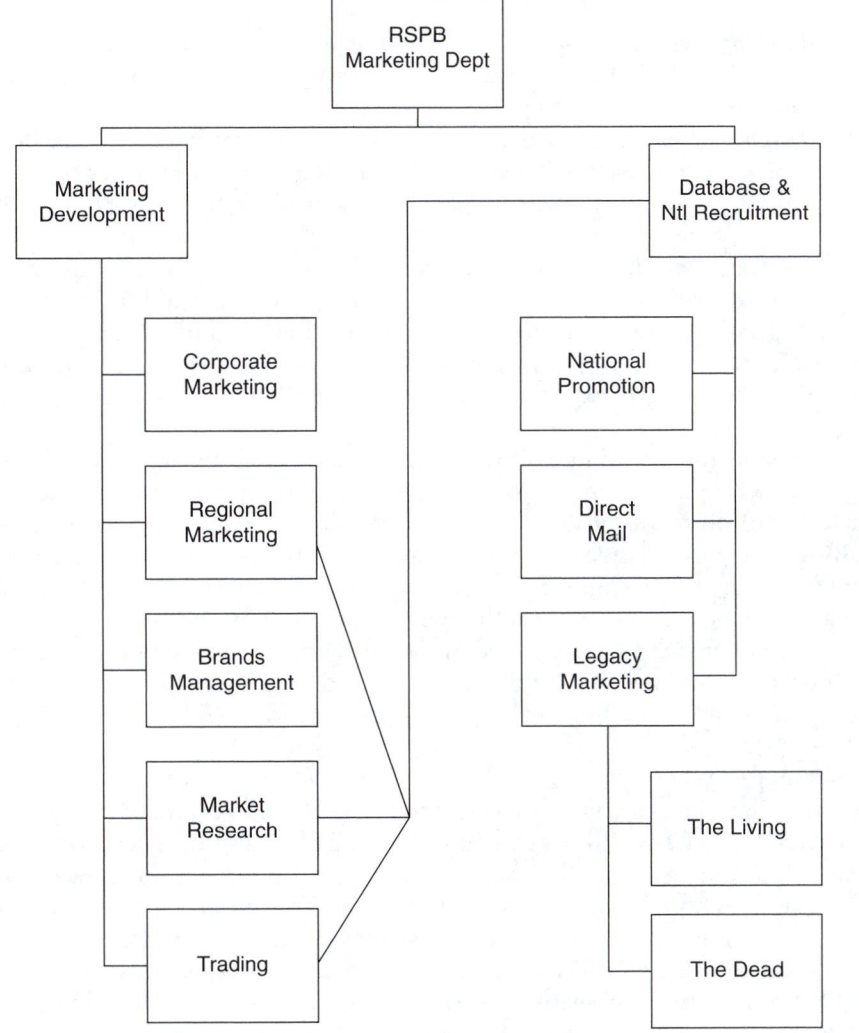

Figure 2.6.3 RSPB marketing department structure.

National Promotion. Their aim is to recruit members with a tight cost per member budget. To achieve this target recruitment information is mailed to rented lists of non-members and advertising is placed in national and regional media. The names of non-members who come into contact with the RSPB are also collected onto the database for mailing with recruitment information.

Direct Mail. Responsibilities are to retain members and to build relationships with them, to raise funds from appeal mailings, lotteries, covenant and financial products, including the RSPB Visa Card.

Legacy Marketing. The objective is to maximize the RSPB's legacy income from members whilst also working within the boundaries of corporate positioning. The work is divided into two areas, called appropriately 'the Living' and 'the Dead'.

> *The Living.* This involves generating interest in legacy donations through advertisements in *Birds* magazine and through mailing of the legacy prospectus to supporters.

> *The Dead.* Income from existing bequests is monitored and outstanding income (approximately £5 million per annum) is regularly chased; occasionally the Society are involved in administering or contesting wills.

The second objective of 'Towards 2000' was immediately addressed by the appointment of 10 regional fund raisers, whose agenda is to increase involvement with local companies, individuals and organizations. Their role is to raise awareness of the RSPB's work and increase donations by organizing charity events and functions in their respective regions.

As well as those outlined above, other areas of commercial fund-raising potential were investigated. Joint ventures have been set up with manufacturers of bird care products in which the RSPB receives funds for carefully vetted endorsement of products such as bird seed or garden feeding tables. Commercial sponsorship is also actively being sought, although great care has to be taken with the types of industry involved. Oil companies have expressed a desire to be involved with the charity but this could seriously jeopardize the RSPB's integrity and is not consistent with the policy aims of 'Towards 2000'.

The donor pyramid

When undertaking the issue of membership growth, as outlined in the first objective, Hurst and Rothwell began by analysing the past membership marketing strategy. This had been accomplished by using the concept of 'the donor pyramid', which provided the basis of the marketing strategy prior to the publication of the 'Towards 2000' document (Figure 2.6.4).

The aim of the pyramid was to develop a strong supporter base in which RSPB membership was seen as a series of steps that members take towards becoming committed long-term, high-level, donor members.

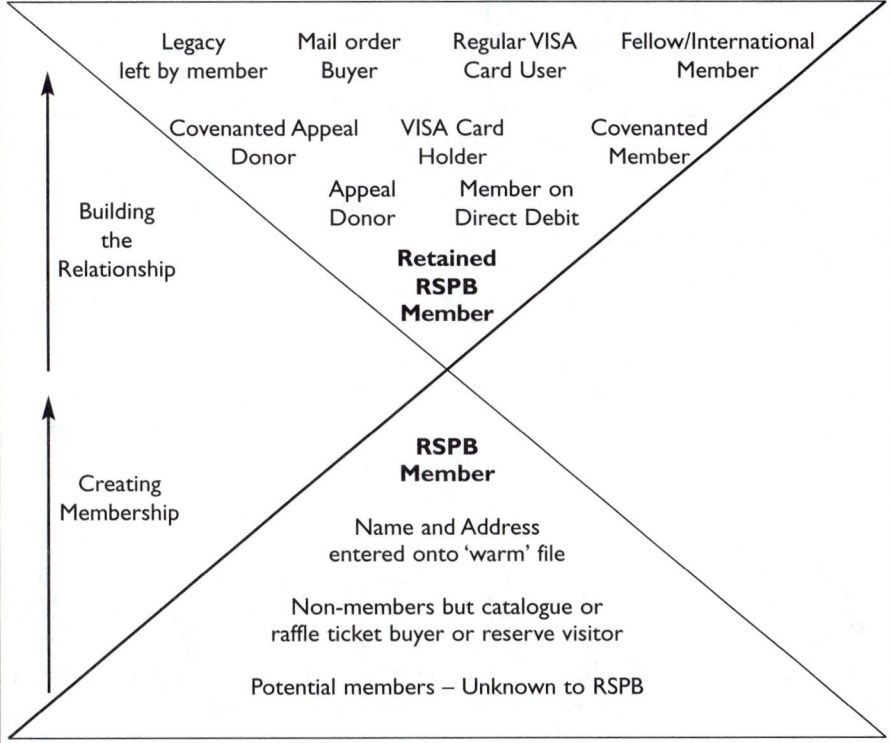

Figure 2.6.4 The donor pyramid.

The first step was to encourage non-members to become members, as shown in the lower half of the donor pyramid in Figure 2.6.4. This had been the main focus of their marketing efforts. The pyramid was based on the premise that individuals could be upgraded to greater levels of involvement. Although it was recognized that not every person will reach the highest level of the pyramid, once someone had joined programmes were developed to increase their level of commitment and support to the Society in order to facilitate their progress up to the next stage and build a deeper relationship once a prospect becomes a member of the Society.

However, this encouraged the neglect of those who didn't rise up the stages and fall-out of members was common. For example, an old-age pensioner with little disposable income may pay a concessionary rate and could never afford to donate. Under the pyramid approach he or she would receive appeal communications that were not relevant to them and they would not respond. They would eventually come to be regarded as not valuable to the Society and be neglected, yet they could suddenly leave their whole estate to the RSPB upon their death.

Hurst and Rothwell realized that the pyramid was useful for describing the membership hierarchy in terms of contribution and involvement but did very little towards developing a comprehensive marketing plan. In fact, the upgrade programme was based on a hierarchy of financial contribution as perceived by the RSPB rather than by the members themselves. In reality, individuals were more likely to support the Society in a variety of ways which would not necessarily fit in with the pyramid hierarchy in which a member has only one identity.

As a result of a review of the donor pyramid approach, it was soon recognized that the Society ran the risk of becoming too focused on trying to upgrade individuals from one level to another and were concentrating too heavily on recruiting new customers to fill the gaps left by lapsed memberships. More importantly, the RSPB were in danger of alienating their members, about whom they knew very little.

Market Research

In order to address these concerns, a market research survey was undertaken in which over 1000 members were surveyed by telephone. The results gave some insight into members' motivations and interests and helped to identify five distinct membership segments. The following list summarizes the characteristics of each segment identified.

1 Active Birdwatchers (21 per cent)
   - They're out and about the countryside and keep notes of what they have seen, in the UK and abroad.
   - They have very light television viewing habits.
   - Keen on conservation and protecting rare species, they visit reserves regularly and like to get their hands dirty, helping out with projects.
   - Tend to be working, middle-aged men.
   - Most likely to have children at home, a strong group for family memberships.
2 Basic Birds (31 per cent)
   - A passive group who watch birds for their beauty, either in the garden or on television, they have an inward, home-based focus in life.
   - They tend to be of retirement age and have little interest in conservation and green issues.
   - Equal male/female split.
   - High concessionary membership.
3 Basic Birds Plus (20 per cent)
   - Another passive group with a garden and television focus. They tend to be 'greener' and younger than Basic Birds and have stronger conservation interests.
   - They joined for the social aspect and to learn more about birds and tend to be up-market women with younger children at home.

- They have been members for the shortest time and support many other charities.
4 RSPB Enthusiasts (26 per cent)
- Those interested in birds of all types.
- Twitchers or keen garden birdwatchers.
- Interested in green issues and outgoing – they tend to be middle-aged (or older) women who enjoy socializing, visiting reserves and meeting others.
5   Distant Donors (2 per cent)
- Not really interested in birds.
- Low interest in environmental and green issues.
- Wealthy and active charity givers.

The membership research emphasized the severe limitations of the present database technology, as no detailed analysis could be undertaken on these segments until it was updated. Until this was possible, efforts were directed towards developing a database strategy for the future.

Database development
The stated aim of 'Towards 2000' was to ensure overall growth of the membership base and to achieve increased revenue. To this end, a database management model (adapted from database management seminars attended by staff) was adopted and is outlined in Figure 2.6.5. Whilst in the RSPB's case 'customer' refers to member, the objectives closely mirror those of the Society.

Hurst and Rothwell identified two priority strategies. Their overriding concern was not recruitment but to grow the membership base by retaining new and present members (shown as Strategies 2 and 3 in Figure 2.6.5). The value of prioritizing these was underlined by further analysis of the present membership database.

Retention
Previous research had shown that members who join and pay by direct debit have a much higher retention rate: 99 per cent of members renew their membership at the end of the first year compared to only 42 per cent for those who initially subscribe by cash or cheque. The large difference in retention rate meant that the marketing department must concentrate on trying to switch new members to direct debit after they have joined. If a member is subscribing by direct debit at the end of the first year, the Society is much more likely to build a profitable long-term relationship with them.

At present, the incentive to move to direct debit is three months' free membership. Other promotional items in the past have included bird tables, books and videos but the gifts are always positioned to reinforce the

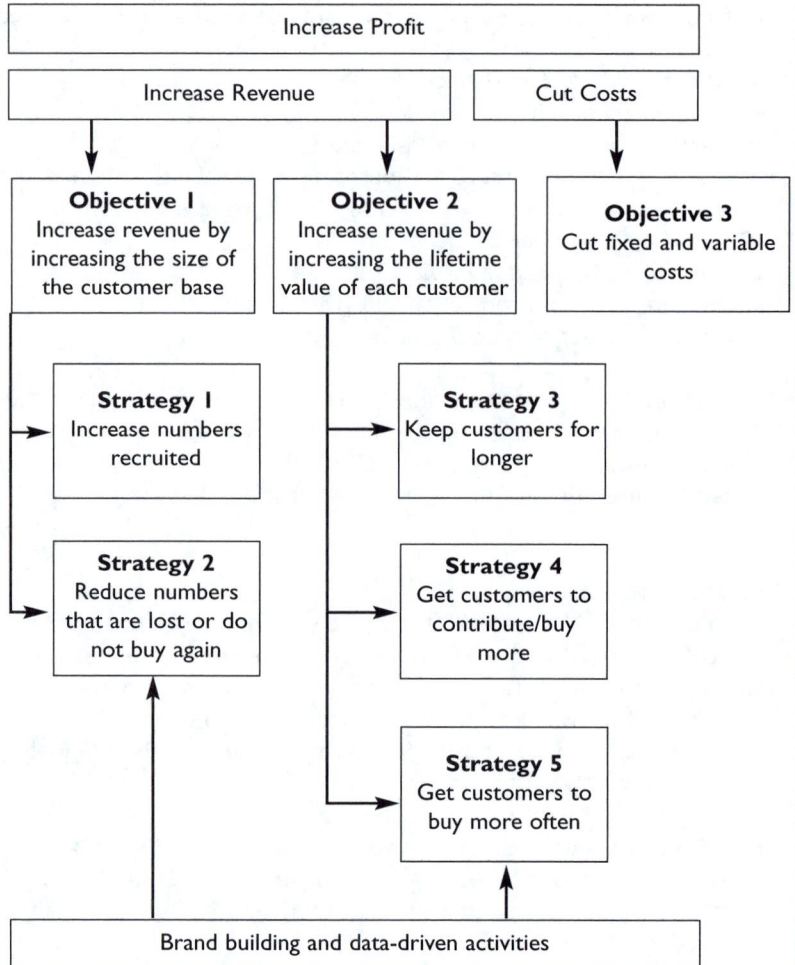

Figure 2.6.5 RSPB strategic option model.

main focus – the birds. This reflects the fact that the majority of the membership are primarily interested in birds and not the broader field of wildlife conservation.

These incentives have proved to be valuable to gain new members. The Society loses on average 10 per cent of its membership each year, including 4 per cent through death, and needs to recruit over 60 000 new members each year, just to cover this loss.

Cost of recruitment is a vital consideration as it can vary greatly, depending on the method of recruitment. A direct mail campaign costs approximately £13 per member to recruit; however, if a member responds

Table 2.6.6 Marketing costs analysis 1996, an analysis of recruitment and marketing costs for each membership category over a ten-year period

(------------------------Based on a continual ten-year membership------------------------)

Membership type	Recruitment cost	Welcome pack	Total R. cost	Cost of 'Birds'	Cost of 'Birdlife'*	Cost of renewal	Total costs of ten-year membership	Approx. income over ten years	ROI
				$(0.96p \times 40)$	$(0.40p \times 60)$	$(0.54p \times 10)$		(Membership fee at 1996 rates $\times$ 10)	
Single	£35.41	£4.06	£39.47	£38.40		£5.40	£83.27	£220	2.6
Joint	£35.41	£4.06	£39.47	£38.40		£5.40	£83.27	£270	3.2
Family	£35.41	£4.06 + 1.61	£41.08	£38.40	£24.00	£5.40	£108.88	£320	2.8
Fellow	£35.41	£5.40	£40.81	£38.40		£5.40	£84.61	£450	5.0
Concessionary**	none	£4.06	£4.06	£38.40		£5.40	£47.86	£120	2.5

*'Birdlife' is exclusive to family memberships; **The RSPB do not actively seek to advertise concessionary/subsidized memberships and no money is spent on trying to recruit them. The marketing costs analysis assumes: an average membership of ten years; that recruitment costs are constant over this time; reminder costs are incorporated into renewal costs; legacies and donations are not taken into account; subscriptions and all other costs are at current rates (1996); advertising and administrative costs are included in recruitment costs as an average.

Table 2.6.7   Total of member households (000s)

	1988	1989	1990	1991	1992	1993	1994	1995	1996
Start of year	361	377	395	460	508	500	506	508	514
Lapsed memberships	33	29	33	50	69	65	59	61	55
Renewed memberships	328	348	362	410	439	435	447	447	459
New memberships	49	47	99	98	61	71	61	67	74
Total at end of year	**377**	**395**	**461**	**508**	**500**	**506**	**508**	**514**	**533**

via a television advertisement the average cost is almost £40 per member. If the membership fee for a single person is £22 a head, clearly the recruitment costs and retention patterns are crucial to the Society's finances as break-even points are not met for one or two years (see Table 2.6.6 for detailed marketing costs). If these expensively recruited members do not renew their membership, losses can escalate. (See Tables 2.6.7–2.6.13 for detailed membership data.) It became increasingly evident to Hurst and Rothwell that retention of both existing and new customers was their chief priority.

Table 2.6.8   Projected membership growth as targeted in 'Towards 2000', 1993–2000

	1993	1994	1995	1996	1997	1998	1999	2000
Memberships:								
RSPB memberships*	506	508	514	540	573	615	659	703
YOC memberships	38	38	39	41	44	47	48	49
Total	544	546	553	581	617	662	707	752
Members:								
RSPB members*	726	734	749	811	859	922	989	1055
YOC members	125	130	144	145	152	157	158	161
Total	851	864	893	956	1011	1079	1147	1216
New recruits:								
RSPB	71	61	67	90	92	103	105	105
YOC	10	12	12	15	16	16	15	18
Retention rate:								
RSPB	87	87	89	90	89	89	90	91
YOC	67	68	73	70	70	70	70	70

*Memberships refer to the number of member households, members are the individual members of the RSPB.

Table 2.6.9 Recruitment of new members by category (%)

	1990	1991	1992	1993	1994	1995	1996
Single	18.2	8.4	36.1	40.9	38.5	38.9	39.8
Joint	68.4	78.0	41.7	39.3	38.0	36.7	34.1
Family	4.4	3.9	7.5	10.5	12.4	15.7	13.3
Fellow	0.1	0.01	0.2	0.1	0.1	0.05	0.04
Life	0.3	0.01	0.2	0.1	0.1	0.05	0.04
Concessionary	8.5	9.3	11.5	5.4	4.2	3.0	2.1
Others*	0.1	0.38	2.8	3.7	6.7	5.6	10.62

*Others are mainly members who have lapsed in the past and are rejoining after an absence. They are not strictly 'new' members but are classed under this category as they are not renewing an existing membership. These figures are approximate.

Table 2.6.10 Recruitment of new members by category (000s)

	1990	1991	1992	1993	1994	1995	1996
Single	17 886	8061	22 049	29 040	23 496	26 100	29 500
Joint	67 241	75 019	25 452	27 946	23 202	24 600	25 300
Family	4321	3788	4595	7485	7591	10 500	9900
Fellow	139	39	96	61	43	40	30
Life	279	152	110	55	50	40	30
Concessionary	8397	9024	6991	3830	2606	2000	1500
Others	780	2000	1800	2700	4200	4000	7800

The membership, research also initiated many questions about the methods previously used to market to the membership database. Little attention had been paid to targeting information or communications by activity or interest but had been directed towards the pyramid hierarchy which was based on response to appeals and products.

Table 2.6.11 Retention of new members by membership type

Type of membership	Average % retention
Single	58
Joint	67
Family	61
Fellow	75
Concessionary	74

Table 2.6.12 Retention of new members by
method of payment (1994)

Method	Average % retention
Payment by cash	33.5
Payment by direct debit	99.2
Average retention	62.9

Table 2.6.13 Retention of all members recruited by
direct debit (1994–96)

Year	%
1994/95	48
1995/96	67

The membership experience

At present, membership subscription fees vary according to the membership category type that each person belongs to. Table 2.6.14 gives details of the 1996 membership charges and the benefits that each member will receive over the year. Prior to Young's arrival and 'Towards 2000', communications were geared to an individual's location on the inverted pyramid. For example, once a member had agreed to a direct debit subscription, communications would then be geared to encouraging the member to take out an RSPB Visa card or become a covenanted donor and rise to the next level of the pyramid, this was regardless of the individual's specific interest in the charity.

Table 2.6.14 RSPB membership fees and benefits (1996)

Membership type	Membership fee (£)	Joining incentive	Welcome pack	Free visits to reserves	Birds magazine	YOC** membership
Life fellow	500*	Yes	Yes	Yes	Yes	No
Fellow	45	Yes	Yes	Yes	Yes	No
Joint	27	Yes	Yes	Yes	Yes	No
Family	32	Yes	Yes	Yes	Yes	Yes
Single	22	Yes	Yes	Yes	Yes	No
Concessionary	12	No	Yes	Yes	Yes	No

*The Life Fellows fee is a one-off payment. They and Fellows gain from other benefits, such as copies of the RSPB Annual Review and an extra 'behind the scenes' newsletter; **Young Ornithologists Club.

Both Hurst and Rothwell were becoming increasingly aware of what they called 'charity fatigue'. Indeed, within the charity industry many fund-raisers were becoming concerned by the public perception of appeals and donation requests. Research was showing that individuals felt that as soon as they donated to a certain charity the floodgates opened. They were bombarded by requests for funds and eventually this resulted in a cut-back on overall charity-giving as they felt guilty that they just were not able to give anymore.

There was a high risk of donor fall-out from following this strategy, so the Society developed a new approach of communicating to their current members. Rothwell saw the RSPB's target as achieving the status of 'being the most trusted in the hierarchy of giving' and rewarding members for loyalty to the charity.

Branded relationship marketing

In a review of their marketing strategy in 1995, it was agreed that the RSPB would develop the concept of 'branded relationship marketing'. This approach recognized the fact that individuals support the RSPB in a multitude of ways. This was in marked contrast to the donor pyramid approach that encouraged relegation and minimum contact with members who failed to upgrade to higher levels of involvement or donation.

The 'branded relationship' approach takes the basic principles of the pyramid but improves its effectiveness by analysing membership types and devises different communication plans for different groups. These will change over time, depending on what each individual member does and does not respond to. As database and information capabilities improve, more distinct and sophisticated membership segments are expected to emerge and relationships with those members can be improved.

The profiles of the key member groups will allow the marketing department to develop brand positioning statements for each of them. These statements, once agreed, will help to define clear strategies on how to pitch most effectively for each group, as well as develop guidelines on adopting a tone and style in communications that is as relevant to them as possible.

The next step is then to look at which groups are more likely to represent the greatest potential for growth in terms of recruitment and revenue. Once new database technology is in place, work will begin on analysing the retention rates, transactions and donation levels of each of the new segments. This exercise will give the Society a much clearer view of:

- the size of the market within which they could find new supporters
- which other charities were likely to be competing for the same supporters
- the level of membership service required from each group.

Of the five segments, Basic Birds and Basic Birds Plus have initially been identified as priority segments for new memberships but obviously all segments are important when considering overall membership retention. The Society also wants to focus attention on referrals in the future through a 'member get member' promotion.

However, work on these priorities will be slow as the RSPB's computer systems are in urgent need of upgrading and are considered virtually obsolete. At the beginning of 1996, the database capability was restricted. Only relatively unsophisticated analysis was possible as the database holds only basic information such as name, address, type of membership and renewal date. The software cannot analyse defections or complaints raised by members, although every complaint is personally dealt with.

Heavy investment is planned for the next financial year (1996/97) to address this problem and once this has been completed RSPB plan to shift their direct mailing emphasis to 'warm names' – their own enquiries and visitors, rather than relying too heavily on recruiting new members from expensive rented lists, as in the past.

The future

Over 1996 the Society plan is to move towards a full 'branded relationship marketing' programme. It is hoped that this approach will make membership of the RSPB more relevant for individual members and give them more value. Likewise, by concentrating on growing segments that provide greater input to the Society, they hope to increase financial and voluntary contributions to their cause.

Any increase in revenue will mean that the RSPB can concentrate on serving their other customers – the birds! Projects can be costly but hard work and patience is rewarded. In May 1996, under the supervision of the Society, golden eagle chicks hatched in the Lake District to the only breeding pair of eagles in England and although this is a small triumph Young sees each one as symbolic. To her the two customer bases are symbiotic, 'If we get the environment right then we get things right for people.'[2]

References

1 *Financial Times* (1996). Charities expected to move assets into equities, 14 May, 11.
2 Little, A. (1996). A bird lady soars high. *Glasgow Herald*, 2 March, 28.

Chapter 3

The supplier and alliance market domain

The supplier and alliance market domain

The 'Supplier' and 'Alliance' markets or market domains are the focal point of this chapter. Whilst it may sometimes seem semantic to treat them separately – in the relationship marketing model they both need to be viewed as partnerships – there is a subtle distinction between the contributions they can make to the establishment of a successful relationship marketing strategy. Here we define them as follows:

- *Supplier markets*. Suppliers or vendors are the providers of physical resources to the business. Sometimes these resources will be augmented by services but typically they will be characterized as the upstream source of raw materials, components, products or other tangible items that flow on a continuing basis into and through the customer business.
- *Alliance markets*. In a sense, alliance partners are suppliers too. The difference is that typically they will be supplying competencies and capabilities which, more often than not, will be knowledge based rather than product based. They may well provide services and often these alliances will have been created in response to the perceived need to outsource an activity within the company's value chain.

In the mid-1980s, the Austin Rover car manufacturing company had well over 1000 suppliers with whom it had arms'-length, often adversarial, relationships. Ten years later a transformed company, now called the Rover Group, had fewer than 500 preferred suppliers with whom it had the closest possible relationships.

The UK-based high street retailer BhS at the beginning of the 1990s purchased clothing products from 1000 suppliers. By the end of the decade it was working closely with only 50 strategic suppliers.

These two examples reflect the significant change that has taken place in the way companies view their supplier base. Similar dramatic changes have occurred in the use of alliances with other organizations in order to import resources, capabilities and expertise into the business instead of trying to keep everything 'in-house' as previously was the case. Hence British Airways seeks a strategic alliance with American Airlines to enhance its access to North American markets. Andersen Consulting, IBM and Ryder team up to provide a global, information-based logistics solutions capability. Unipart and Yotaku-Giken combine their manufacturing and technological expertise to bring innovation in catalytic converters to market.

These new style relationships are markedly different from the previous focus on vertical integration, whereby as much of the value-added in the final product as possible was brought into the same legal ownership. Ford in its early days in North America owned the steel mills that made the steel for its cars, as well as most of the factories that made the components. Courtaulds established a vast, vertically integrated business in textiles with the

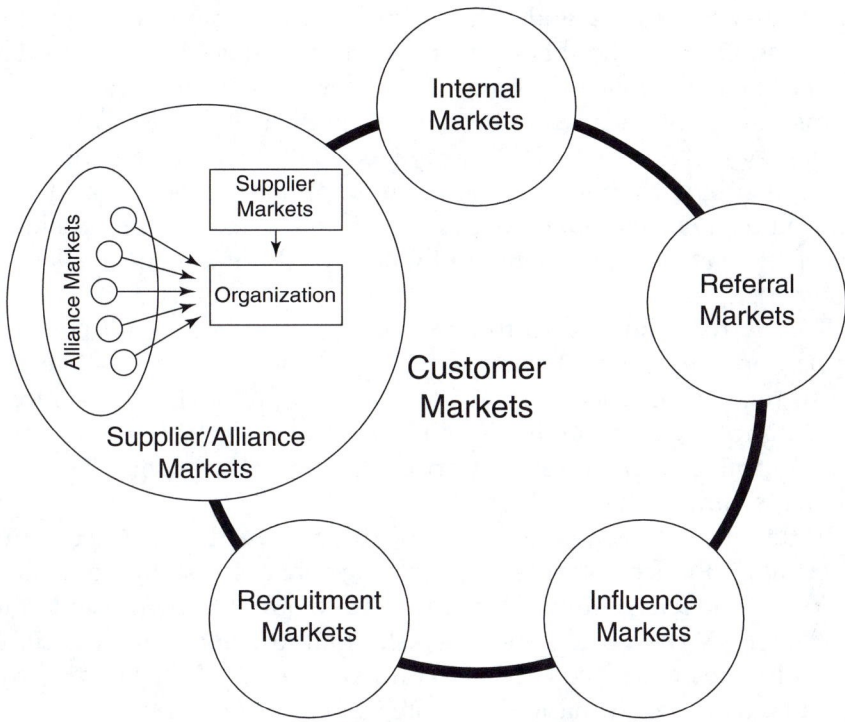

Figure 3.1 The supplier and alliance market domain.

capability to control the value chain from synthetic fibres to finished garments.

Now, the emphasis is upon virtual integration, by which is meant a confederation of organizations combining their capabilities and competencies in a closely integrated network with shared goals and objectives.

Figure 3.1 brings together the concept of vertical supply relationships and horizontal alliance partnerships as a closely coupled network within the Six Markets model.

It could be questioned whether there is a need to make a distinction between alliances and suppliers. Increasingly leading firms seek to create alliance-type relationships with suppliers and to manage in a spirit of partnership. Alliances have often been seen as a means of importing a capability or a resource into the business, for example the strategic alliance that existed for some years between Rover and Honda which provided Rover with much-needed new products and Honda with a platform for developing its European business. On the other hand, suppliers have usually been thought of merely as the providers of goods and services and where often an 'arms'-length' stance was adopted by the firm.

However, it is now widely recognized that the management of the relationship with both alliance and supplier partners has to be viewed as a critical business process. The strength of any business will increasingly be determined by how well it manages these relationships.

It is perhaps helpful to think of alliances as horizontal partnerships and suppliers as vertical partnerships – 'horizontal' in the sense that an alliance partner can be viewed as playing a value-creating role within the firm's value chain and 'vertical' in the sense that suppliers can be seen as an extension of the firm. The basic model is summarized in Figure 3.2.

In this model, sometimes termed 'the extended enterprise', suppliers and alliance partners link with the core organization to enable more cost-effective, timely and innovative offers to be presented to customers. Virtual integration seeks to gain the benefits that accrue to companies who focus on core competencies whilst providing the advantages of coordination and integration that can flow from vertical integration.

As a result, it is coming to be recognized that the management of these interlocking networks of organizations – and in particular the relationships between them – is vital to competitive success. Thus we would argue that one of the central elements of a relationship marketing strategy has to be the way that supplier and alliance 'markets' are proactively managed.

This chapter will explore some of the emerging trends in supplier

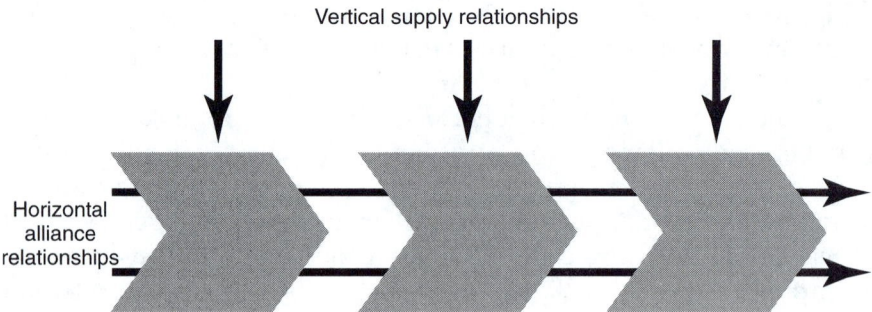

Figure 3.2 The firm's value chain.

and alliance relationships, in particular the trend towards strategic outsourcing and supply chain management. It also examines the critical ingredients for successful alliances and partnerships.

Strategic outsourcing

For many years companies have bought goods and services from other companies and thus it might seem that there is little new in the idea of outsourcing. However, in the past the tendency was to subcontract rather than to outsource. The difference is more than semantic. In the conventional model, organizations would seek to subcontract those activities that could be more cheaply or efficiently performed by others. There was little strategic thinking underpinning such decisions. Economists have sought to explain this behaviour by reference to 'transaction costs'.[1, 2] Under this model, companies will retain activities in-house if the external price plus the transaction costs are greater than the internal cost of performing this activity. Transaction costs reflect any inherent risk, the need for specific assets and the costs of negotiating and managing the subcontract arrangement.

It has been argued[3] that if an external firm can find a way to cover the transaction costs such that the external price is now less than the internal cost plus the transaction costs, then the basis for a relationship exists. However, we would suggest that strategic outsourcing decisions should be made on a broader basis than simple economics alone.

Porter[4] has suggested that the basis for competitive advantage comes through focusing upon that part of the value chain where the firm has either a distinctive cost or value advantage. A cost advantage implies that for one reason or another the firm is able to perform that activity at less cost than others can. A value advantage means that the firm can perform the activity in a distinctly different way and thus create superior value for customers. Hence the argument has been made that organizations that do not have either a cost or a value advantage in specific parts of their value chain should outsource those activities to firms that do have such an advantage.

Thus, the basis for strategic outsourcing now becomes firmly rooted in the search for competitive advantage in the marketplace.

If widely adopted, this idea will fundamentally change the nature of the competitive framework. It will take us from a world where individual firms compete against other individual firms to a world where networks or supply chains compete against other networks or supply chains.

Supplier development

The conventional arms'-length approach to suppliers often meant that customers were putting themselves at a significant competitive disadvantage. Firstly, it has to be remembered that the customer pays for the supplier's costs. In other words, the price we pay for goods and services inevitably is impacted by the upstream costs. Furthermore, many of these upstream costs are incurred because of downstream actions! For example, a manufacturer who does not share information on usage of parts with the supplier of those parts and who makes frequent changes in requirements at short notice and still expects just-in-time deliveries is clearly going to create costs for the supplier. These higher costs are generated through the additional inventories that the supplier needs to carry to buffer against unpredictable customer demands and also the cost of 'schedule instability' in the supplier's operations.

Conversely, customers who work more closely with suppliers, who bring those suppliers into their planning and strategy formulation processes, will benefit not only from lower costs but will also typically be able to create a more responsive supply chain that can meet final demand in a more flexible and timely manner.

What is also apparent is that those companies who work more closely with a limited number of strategic suppliers also benefit through supplier-led innovation both in processes and products. In many industries a growing amount of innovation is supplier originated. For example, in the car industry, much of the improved technology in the finished product, such as ABS or engine management systems, was devised by upstream suppliers rather than by the car assembly companies themselves. By being a part of a close relationship with suppliers such manufacturers can gain leadership in their chosen technologies and markets. The same is true for quality improvement, in that suppliers who believe that they are in a long-term relationship with committed customers tend to be more prepared to focus on quality improvement, which leads to further benefits of reliability of supply and lowered costs – the so-called 'cost of quality'.

Recognizing that significant benefits can accrue from closer relationships with suppliers, many organizations have now embarked upon programmes of 'supplier development'.[5, 6] The simple idea that underpins the concept of supplier development is that by working more closely with suppliers significant improvements in their performance can often be achieved. The improvements can come in the form of lower costs through greater operational efficiency, higher levels of product quality, lower inventories, the adoption of 'paperless' transactions through electronic data interchange and joint product development.

One of the first companies in the UK to see supplier development as a critical business process was Nissan. When they opened their first European vehicle assembly plant in the North East of England in the late 1980s, they sought firstly to minimize the number of first-tier suppliers they did business with – under 200 – and secondly to establish cross-functional Supplier Development Teams to work closely with those suppliers on product and process improvement. These teams comprised a mix of Nissan engineers and specialists whose goal was to seek out and exploit opportunities for creating a 'seamless' supply chain that would enable Nissan to gain significant advantage in cost and quality terms. It was claimed at the time that Nissan had, as a result of this approach, a £600 per car cost advantage over other European car manufacturers.[7] Not surprisingly perhaps, this idea was rapidly adopted by others in the industry!

Another manifestation of the idea of supplier development is the creation of supplier associations. Hines[6] has defined a supplier association as:

> a mutually benefiting group of a company's most important sub-contractors, brought together on a regular basis for the purpose of co-ordination and co-operation as well as to assist all the members to benefit from the type of development associated with large Japanese assemblers such as kaizen, just-in-time, kanban, U-cell production and the achievement of zero defects.

The idea is to bring together suppliers to form a 'club', the purpose of which is to benchmark and learn from each other as well as from the customer. Whilst the evidence of the efficacy of these supplier associations is mixed, it does seem that if the traditional adversarial approach can be replaced by a genuine concern to achieve 'win–win' outcomes then the basis for partnership can be established.

It is perhaps this fundamental prerequisite of mutual trust that is the hardest step in the journey to what has sometimes been called 'partnership sourcing'. Lamming[5] has drawn some lessons from the Japanese that suggest that 'trust' is not some woolly, amorphous concept but instead is based upon a very clear understanding of mutual commitment:

> Partnership is based upon commitment, trust and continuous improvement. The fact that Japanese assemblers do not have written long-term contracts, relying instead on short-term stipulations for deliveries but very long-term (next vehicle model) involvement of the supplier, based upon mutual trust, is an indication of the very different business culture that partnership requires.

Supply chain management

As the critical role of suppliers and alliance partners comes to be increasingly recognized, the need for formal processes to manage the supply chain emerges. These processes are in effect an extension of the internal linkages that create the smooth flow of information and materials within a single organization into the other parties in the supply chain. Supply chain management has been defined[8] as: 'the management of upstream and downstream relationships with suppliers, distributors and customers to achieve greater customer value at less cost'.

The key to the achievement of this more responsive and cost-effective marketing process is buyer–supplier integration. What this means in effect is that rather than the totally separate decisions on critical issues such as production schedules, inventory levels and distribution plans that typify the uncoordinated chain, there is instead a single 'end-to-end' plan to manage the pipeline as a whole.

Whilst many companies have begun the journey to supply chain integration, few have made it to the destination. It must be recognized that there are some significant hurdles to be overcome in the transition from a 'stand alone' organization to supply chain partner. Stevens[9] has identified four stages in this transformation process:

1 *The baseline organization.* This organization operates the classical system of management, with the motivation of profit maximization and a high level of functional specialization. The company cannot adapt quickly to changes in the consumer market and has a low ability to exploit materials flow or market information.

2 *The functionally integrated company.* This organization has begun to erode the hierarchical structure and short-term financial focus by concentrating on customer service criteria and sales order processing. The major competitive advantage of this organization is in the distribution efficiency of the system and the collaboration between the sales function and the distribution function.

3 *The internally integrated company.* This organization has continued to restructure and align the activities of manufacturing and purchasing to create a systems approach to customer service. The company has reduced the number of administrative functions required and operates effective interfaces between departments to optimize information exchange and hence the overall performance of the company. The planning horizon has also extended from the short term to the medium term and involves a limited interaction with suppliers. At this point the organizational structure may become product focused and involve a high level of cross-functional management.

4 *The externally integrated company.* This organizational state involves the externalization of the alignment process and the integration of the supply base with the demands of the consumer in a transparent system of materials and information exchange. The company seeks deliberately to manage the interfaces between companies to generate a flexible and responsive system of long-term collaboration. At this point the company has completed the restructuring of its internal supply chain and has recognized the importance of external supply-chain management strategies and the need to synchronize the supply process. The company operates internal cross-functional management structures, which may be product related, and typically develop supplier networking groups.

Figure 3.3 illustrates these four stages to supply chain integration.

If supply chain integration can be made a reality, then the potential impact on the final market can be significant. The point of supply chain management, it must be remembered, is not just to seek out efficiencies and cost reduction opportunities, but rather to

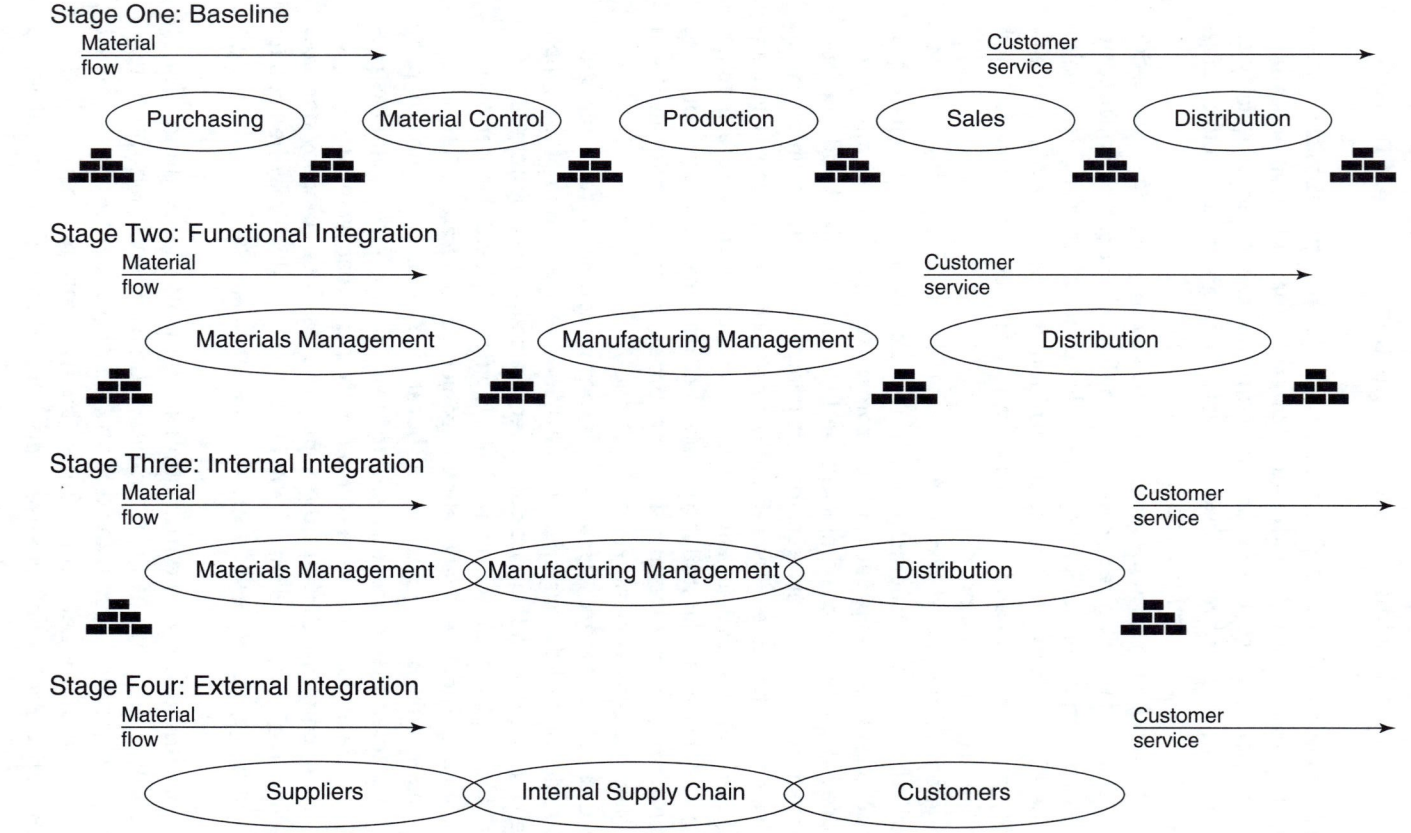

Figure 3.3 The progression to supply chain integration.
Source: Stevens, 1989.

create a more responsive and flexible stance towards customers and consumers. In today's marketplace, where time has become a critical success factor, the ability to move quickly in response to volatile demand is vital.

The achievement of a closely integrated supply chain ultimately, however, will depend upon the recognition by all the parties involved of the need to work on the basis of collaboration and to seek out 'win–win' strategies.

Creating win–win relationships in the supply chain

The benefits of cooperation rather than conflict in buyer–supplier relationships have been set out in some detail by Axelrod[10] amongst others. He demonstrates that cooperation can provide greater payoffs to both parties than can be achieved through the more traditional 'win–lose' scenario.

In the same work Axelrod describes the well-known example of 'The Prisoners' Dilemma', in which two outlaws are captured and can only escape a severe punishment by trusting each other not to confess. Because they are both offered lighter sentences if they convict the other (and knowing also that the other is inherently untrustworthy) the likelihood is that they will both confess and both be punished! Whilst this may be an over-simplification of relationships in a supply chain, it graphically highlights the issues.

Lewis[11] has studied in depth a number of companies who have strong partnerships with suppliers – companies like Marks & Spencer, Chrysler and Motorola – and has identified that typical benefits include:

- *On-going cost reductions* that can double those possible through market transactions.
- *Quality improvements* that exceed what individual firms can possibly do alone.
- *Design cycle times* 20 to 75 per cent shorter than those in traditional relationships.
- *Increased operating flexibility*, which in some firms has yielded an economic lot size of one – the ultimate in flexible manufacturing.

- *More value for the customer's customers*, including faster and better responses to new needs and opportunities.
- *Enhanced leverage with technology*, including earlier access to new concepts and more control over technological change.
- *More powerful competitive strategies*, gained when a customer adds its supplier expertise to its own.

One example of win–win thinking that is beginning to emerge in supply chains is the idea of Vendor Managed Inventory (VMI) or – a subtle variation – Co-Managed Inventory (CMI). The traditional approach to replenishment at each step in the chain has been for the customer to place an order on a supplier. Typically there would be no early warning of requirements from the customer and thus the supplier would have to carry inventory in the form of safety stock as a 'buffer' against this uncertainty. Similarly, the customer would also carry safety stock of the same items to guard against the possibility of non-supply. The result of this conventional 'arms'-length' approach was higher levels of inventory in the chain and paradoxically lower levels of service and responsiveness.

The idea behind VMI is that the customer no longer places orders on the supplier, but instead shares information on actual demand or usage on a continuing basis. Because the supplier now has 'visibility' of the rate of off-take lower down the chain it can plan and schedule production and transportation more efficiently, duplicated inventories are greatly reduced, service levels improve and the customer's cashflow is enhanced because they only pay for the product as they use it.

Co-managed inventory (CMI) is a further extension of the idea whereby the customer jointly plans with the supplier appropriate inventory levels taking into account promotional activity, specific local conditions, competitive activities and so on. In either case, VMI or CMI, what is happening is what might be termed the 'value-added exchange of information'. Value is created through more responsive supply as a result of customers providing suppliers with information on off-take or product usage.

The key to supply chain integration is shared information. By working to the same data on demand, inventories and marketplace trends, a much more cost-effective logistics process can be developed. Under the conventional model – where no information is shared – both the supplier and the customer had to carry inventory on a 'just-in-case' basis. The supplier carried inventory because they had no forward notice of customer requirement. The customer carried inventory because they knew from experience that the supplier may not always be reliable. When information is shared, then uncertainty is reduced and hence inventories can be dramatically cut.

The benefits of shared information go beyond cost reduction, however. There has been a clear tendency for companies to become increasingly mutually dependent as they start to link information systems together. The use of Electronic Data Interchange (EDI) to create an environment where 'Electronic Commerce' can eliminate documentation such as purchase orders and invoices leads inevitably to the supplier taking on more and more of the activities that previously were performed by the customer. For example, in retailing there has been a trend, particularly in North America, for the suppliers to become very actively involved in 'Category Management'. This entails the supplier assisting the retailer in making decisions on shelf-space allocation, on layout and merchandising and by managing the flow of product from the factory to the shelf.

The same phenomenon can be encountered in industrial markets, where suppliers of equipment and materials may take on the actual operation of that part of their customers' processes where they have specialist skills. Thus BOC, the industrial gases company, will build, operate and run gas production facilities at their major customers' premises.

These patterns of collaboration in the supply chain are gradually starting to change the shape of the competitive environment. It was suggested earlier that companies no longer compete against other companies as single entities but rather as supply chains or networks. Under this model a key determinant of success or failure in the marketplace is the extent to which the supply chain can be managed as an integrated network with shared strategic goals and closely linked processes to support those goals. It follows from this argument that an increasingly important source of competitive advantage will be the strength and the quality of the relationships between members of the network – both 'vertical' supply chain partners and 'horizontal' alliance partners.

Creating successful alliances

As we suggested earlier, supply chain partnerships may be thought of as 'vertical' relationships, alliances could be described as 'horizontal' relationships. The idea of an alliance is to bring skills or competencies into the business that may not exist or may be in need of strengthening. The move to outsource all those activities that are not considered to be 'core' to the business has been gathering pace in recent years, and this has given further impetus to the search for appropriate alliance partners. Increasingly the value-creating process is no longer confined to a single firm but instead is rooted in a confederation of firms each of which contributes specialist skills and capabilities. The value chain, in effect, now spans several organizations that work as partners in creating and bringing products to market.

Normann and Ramirez[12] have described this new business model succinctly:

> Increasingly the strategic focus of successful companies is not the company or even the industry but the *value-creating system* itself, within which different economic actors – suppliers, business partners, allies, customers – work together to co-produce value. Their key strategic task is the *reconfiguration* of roles and relationships among this constellation of actors in order to mobilize the creation of value in new forms.

Alliances can be formed between a few or many partners. The basis of an alliance is shared strategy determination – if this does not happen but instead the dominant party unilaterally determines the strategic direction, then it is unlikely to succeed as an alliance. More and more examples are beginning to emerge of alliances that are formed to exploit particular market opportunities or technologies. These alliances may be confined to a single product or market sector or may be more broadly based. An example of the latter is Airbus Industries, where a number of specialist manufacturers have combined their specific capabilities through a jointly owned company. An interesting example of the former is the joint venture between Daimler-Benz and the SMH watch company (the makers of Swatch). They have come together to create, produce and market the 'smart' car through a jointly owned organization – Micro Compact Car (MCC). What makes this alliance different is that other specialist organizations have come together to operate specific parts of the car manufacture and assembly process. Hence, Magna International, a Canadian company, welds the structural shell; Eisenmann – a German coatings specialist – paints it before it moves up the assembly line to where VDO – a German instrument manufacturer – builds and installs the cockpit. Other specialist companies supply and install complete modules for lighting and electrics, doors, engines and axles and so on. Each company employs its own managers and work force but they work as members of an integrated team in the purpose-built factory at Hambach in France.

Other alliances can be based upon shared knowledge. This type of alliance will often involve technology/R&D transfer or it could involve marketplace knowledge or access. The key to success in this type of alliance is to recognize that both partners are looking for different outcomes and that it may not be long-term. These types of alliance can be formed to exploit a technology or a market or to solve problems that may be beyond the financial means of one partner alone.

Quinn[13] has suggested some of the foundations for successful alliances:

- *Jointly developed goals and plans.* Many companies have found that openly sharing their overall goals for the project with partners, and then jointly developing explicit goals for the project team's management, can eliminate much confusion and ill feeling.
- *Avoiding niche collisions.* Each party needs to analyse carefully where each partner might intrude into the sacred territory of the other – geographical, demographic, customer, technology, or componentry niches of strategic importance – and block these off by corporate-wide agreements for a finite period.
- *Structuring the team.* Alliances need teams to manage the shared processes and these teams need to reflect the technical and business issues that the alliance was established to confront. They must be empowered to take decisions and hence require representation from senior management in their membership.
- *Clear communication links.* Having clear contact points at the technical-operating, project manager and top management levels – especially in the early stages – can avoid conflicts in information and impressions conveyed. Often, confusion comes as much from misunderstandings about whom to contact or ignorance of the partner's internal processes and decision rules as it does from the complexity of the information to be transferred.
- *Understanding cultures.* Creative companies, operating in adhocracies, generally find their programmes slowed by the bureaucracies of the more conservative partners. Cultural conflicts are especially acute with

foreign partners. But every culture (not just those in foreign countries) has its different language, style, internal politics and hidden decision rules, which, until understood, exact a high price.

- *A structural learning process.* Companies enter alliances because the partners can do certain things better than they can. From the beginning, therefore, it is helpful to codify and to transfer knowledge gained from the partner to those who can best use it. Periodically, representatives of all units on the alliance team need to be brought together to share results from the process and to design further steps to gain more knowledge.

It is clear that such alliances require a totally different management approach from that which used to prevail in more traditional 'subcontract' type relationships. In these 'network' or 'virtual' organizations the need to create a 'boundaryless' business with joint decision making, complete transparency on costs and a sharing of risks and regards is essential.

The organizational change that is implicit within this new style of relationship is profound. The need to fuse together processes across partners' boundaries – whilst still recognizing the essential independence of the entities – is paramount.

Figure 3.4 represents the transition that is required from the classic arms'-length buyer/supplier interface where the only contact focuses around the transaction (typically a salesperson and a buyer negotiating a deal) to a multi-level partnership.

In the partnership model there is an emphasis on joint teams, which themselves are cross-functional, working to establish 'seam-

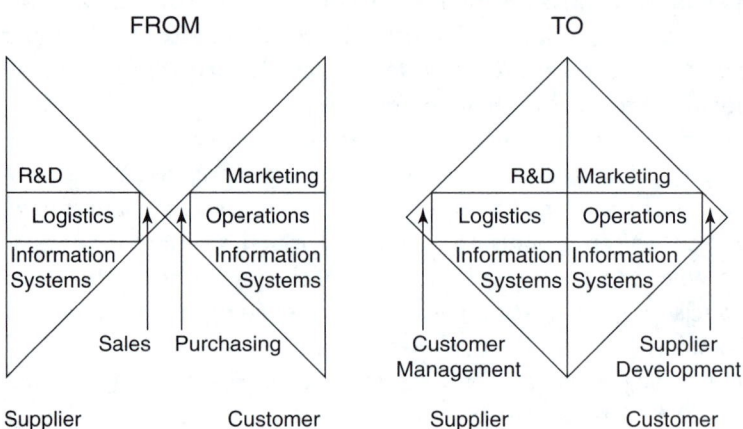

Figure 3.4 Creating closer relationships with supply chain partners.

less' processes to achieve mutually determined strategic goals. Because of the significant resources required to manage these team-based relationships, there is further pressure to reduce the supplier base and to seek out a limited number of key alliance partners. Inevitably these remaining partnerships will be of longer duration than would have been the case in the conventional, arms'-length way of doing business.

What seems to be emerging, therefore, is a radically different concept of competition. The fundamental paradigm shift that is taking place is that it is no longer the individual firm that competes with other individual firms, but rather it is networks of partners and alliances that compete with other networks. Companies that recognize this and that actively develop strategies for strengthening relationships amongst the members of their network will emerge as the winners in whatever markets they compete in.

References

1 Coase, R. (1937). The nature of the firm. *Economica*, **4**.
2 Williamson, O. (1975). *Markets and Hierarchies*, Free Press, New York.
3 Jarillo, J.C. (1993). *Strategic Networks: Creating the borderless organization*, Butterworth-Heinemann, Oxford.
4 Porter, M. (1985). *Competitive Advantage*, The Free Press, New York.
5 Lamming, R. (1993). *Beyond Partnership: Strategies for innovation and lean supply*, Prentice Hall Europe, Hemel Hempstead.
6 Hines, P. (1994). *Creating World Class Suppliers: Unlocking mutual competitive advantage*, Pitman, London.
7 Christopher, M. (1992). *Logistics and Supply Chain Management*, Pitman, London.
8 Christopher, M. (1997). *Marketing Logistics*, Butterworth-Heinemann, Oxford.
9 Stevens, G. (1989). Integrating the supply chain. *International Journal of Physical Distribution and Materials Management*, **19**, No. 8, 3–8.
10 Axelrod, R. (1984). *The Evolution of Cooperation*, Basic Books, New York.
11 Lewis, J. (1995). *The Connected Corporation*, The Free Press, New York.

12 Normann, R. and Ramirez, R. (1993). From value chain to value constellation: Designing interactive strategy. *Harvard Business Review*, July/August.
13 Quinn, J.B. (1992). *Intelligent Enterprise*, The Free Press, New York.

Chapter 3 case studies

The major theme of this chapter has been the need to establish a partnership style approach to suppliers and alliances as the foundation for a far more effective supply chain or network. The three cases that we have chosen for this chapter highlight specific aspects of the way in which strong relationships with network partners can contribute significantly to competitive advantage.

Case 3.1: Supplier relationships at Trico

This case was written by Professor Peter Hines of the Lean Enterprise Research Centre at Cardiff Business School, University of Wales.

Abstract
The case study focuses around the advantages and opportunities emerging from the establishment of supplier associations. Trico is a manufacturer of car windscreen wiper systems supplying the automotive industry. As a result of encouragement by one of their major customers, Trico sought to establish a more proactive approach to supplier development through the creation of a supplier association. The objectives and workings of the supplier association are described and some of the early results of its actions are presented.

Learning points
Some of the key learning points to emerge from this case are: the importance of supplier relationships in improving competitiveness for the downstream customer; the opportunity for benchmarking and process improvement through a willingness to learn from other members of the association; the way that gains have to be seen to be shared throughout the supply chain to ensure 'buy-in'.

Broad marketing and strategic issues:

- the challenge of global competition
- the trend towards supplier rationalization
- the role of product quality in gaining competitive advantage.

Relationship marketing issues:

- the growth of network competition
- the importance of supplier development as a key business process
- the ingredients for successful supplier associations.

Since the inauguration of the Trico Supplier Association, significant improvements in product quality, delivery reliability and cost have been achieved at each step in the supply chain. This has significantly enhanced the competitiveness of Trico and its network of suppliers.

Case 3.2: Cafédirect™: The building of a unique coffee brand

This case was written by Keith Thompson of Silsoe College, Cranfield University and Professor Simon Knox of Cranfield School of Management.

Abstract
This case study describes the formation and development of Cafédirect, which, as a virtual company consisting of four charity partners, successfully launched the Cafédirect ground coffee brand with the full support of the major UK supermarket chains. The primary purpose of this virtual enterprise was to bring together the skills and expertise necessary to enable Third World coffee producers to achieve a higher price for their product. To achieve this goal the alliance partners recognized the need to build strong supply chain relationships from the farmers through to the end consumer.

Learning points
Some of the key learning points emerging from this case are: the way in which a virtual company can be created through synergetic alliance partnerships; the importance of establishing appropriate relationships

both upstream and downstream in the supply chain; the need to balance supply chain efficiency with brand marketing effectiveness.

Broad marketing and strategic issues:

- the idea of the value delivery process
- the particular problems of marketing 'ethical' products
- the role of the retailer in consumer markets.

Relationship marketing issues:

- the nature of the 'virtual' organization
- using the network to provide differentiation in a commodity market
- the challenges of relationship management in an extended supply chain.

Cafédirect has now established itself as a recognized brand in the UK market, albeit still with a relatively low market share, given the growing consumer interest in ethical issues.

Case 3.3: Transvaal Nickel Mines

This case was written by Professor David Blenkhorn and David Holowack of Wilfred Laurier University, Waterloo, Ontario, Canada.

Abstract
This case describes some of the issues encountered by a company in the primary metals industry as it addresses the issues involved in initiating and implementing a relationship marketing strategy in an industry which historically has not spent much time analysing and working with the complete supply/value chain.

Learning points
Amongst the key learning points to emerge from this case are: issues involved in supplier collaboration and coordination in a complete supply/value chain; that the paradigm shift necessary to implement relationship marketing may be too great for traditional marketers in some firms; to illustrate how relationship marketing can be a differentiating factor in a commodity-based industry, where price reigns supreme.

Broad marketing and strategic issues:

- the need to understand the linkages between value chains and supply chains
- communicating the importance of supply chain cooperation, internally and externally
- the concept of supply chain competition.

Relationship marketing issues:

- the importance of internal marketing to establish a relationship marketing culture
- working with supply chain partners to present a value-added proposition
- how suppliers may need to take the initiative in establishing relationship marketing strategies.

At the time this case was written events were still unfolding and so the success or otherwise of the relationship marketing strategy is unknown.

Case 3.1 Supplier relationships at Trico

This case study was prepared by Dr Peter Hines, Lean Enterprise Research Centre, Cardiff Business School.

Pontypool lies at the threshold of economic prosperity. Its history of coal and steel is giving way to a bright future based on new innovative firms such as automotive component supplier Trico. This manufacturer of car wipers moved to Pontypool in 1992 from its previous London location. In order to succeed it needed a modern factory with world-class production facilities and methods, a new organizational culture and a better supplier base. This entailed building a new site and new relationships with its work force and suppliers.

A visitor to Trico's site will be struck by three things before reaching their respective meeting room. First, the number of quality awards from world-class companies displayed in the reception, taking up a complete wall. Second, the open-plan, uncluttered offices that exude an aura of calmness and efficiency. Third, at every turn of a corner, up-to-date notice boards displaying before and after Polaroid photographs of the surrounding work areas. It is hard not to be impressed.

It is usually not long before visitors are invited to look round the shop floor. The same sense of space, efficiency and order is the immediate impression. Notice boards abound with similar before and after photographs of the respective work areas. The same is true in the stores and tool-room areas. Again, the visitor can see that this company is doing the right things. But was this always the case?

This is their story.

How it all started

Three factors are responsible for what the visitor can see today. The first and most important of these was the move from London in 1992. This move allowed Trico the opportunity to rethink both its internal processes and the way it dealt with its suppliers. It enabled them to start the development of a new internal culture as well as a re-focusing of the product range and the make-or-buy decision process. This was fundamental to the change one sees today at Trico.

The second, some time later, was the launch of the Welsh Automotive Supplier Association by Rover and the Welsh Development Agency's Source Wales division. The third was their first meeting with Paul Morris, a consultant from a company called Supplier Association Partners, who was project managing the Supplier Association programme for Source Wales.

In late 1993 Trico, along with 22 of Rover's other strategic suppliers in

Wales, were invited to a Supplier Association meeting by the car maker. Trico's Purchasing Manager, Jim Taylor, and his boss, David Jones, did not, at this time, realize that this meeting was to have far-reaching implications for them and their company. Indeed, at this time they did not know what a Supplier Association was and were understandably wary of any new initiative that their main customer might be launching. They had in the past had far too many experiences with other customers of 'talking shops', 'flavours of the month' and 'initiative overloads' .

As it turned out, Rover were trying to get past the rhetoric of 'partnership words but adversarial actions'. They were attempting to move themselves and their supply chain into the next century as a world-class extended enterprise. As one of the Rover directors mentioned at this first meeting, 'we need to do more than just talk the talk, we need to walk the walk, we need to actually work together and implement real change'. Well, it all seemed to make sense to Jim and David, but were Rover and the Source Wales team actually going to deliver?

Some months passed and Trico were invited to more events. Discussions took place, plans were drawn up and both they and the other suppliers started to buy into the process and make significant improvements. Rover reported that members of the group had reduced quality failures by around 80 per cent whilst suppliers outside Wales had improved by only a few percentage points. Jim and David could see that the process was working. It was not long before Rover mentioned that their suppliers should start thinking about their own supplier development. The Rover Purchasing Director commented that 'we're not saying that you all need to rush away and set up your own Supplier Associations, that might not be the best vehicle, but we at Rover would be very interested in hearing about a better approach because we haven't found one yet!'

This set Jim and David thinking. Would the same process work for them and their suppliers? At about this time they attended a two-day Source Wales briefing run by Paul Morris. After a day and a half they were still not convinced that a Supplier Association was the right thing for them. Surely it would just end up being a 'talking shop'. It was then that Paul Morris started a discussion on how two types of hard and soft benchmarking could be used to focus suppliers and allow for mutual target setting. He also shared his experiences of problems in setting up such groups and how to avoid the dreaded 'talking shop' syndrome. Jim and David started to be convinced that this was for them.

Then Paul showed a ten-stage checklist (Figure 3.1.1) for Supplier Associations that ensured that there was not just a wish list for improvement, but a clearly defined process which will ensure that suppliers bought in were made aware of where they needed to improve, could become educated both in theory and practice and, what's more, could actually achieve change. This was what Jim and David were waiting to hear. They now

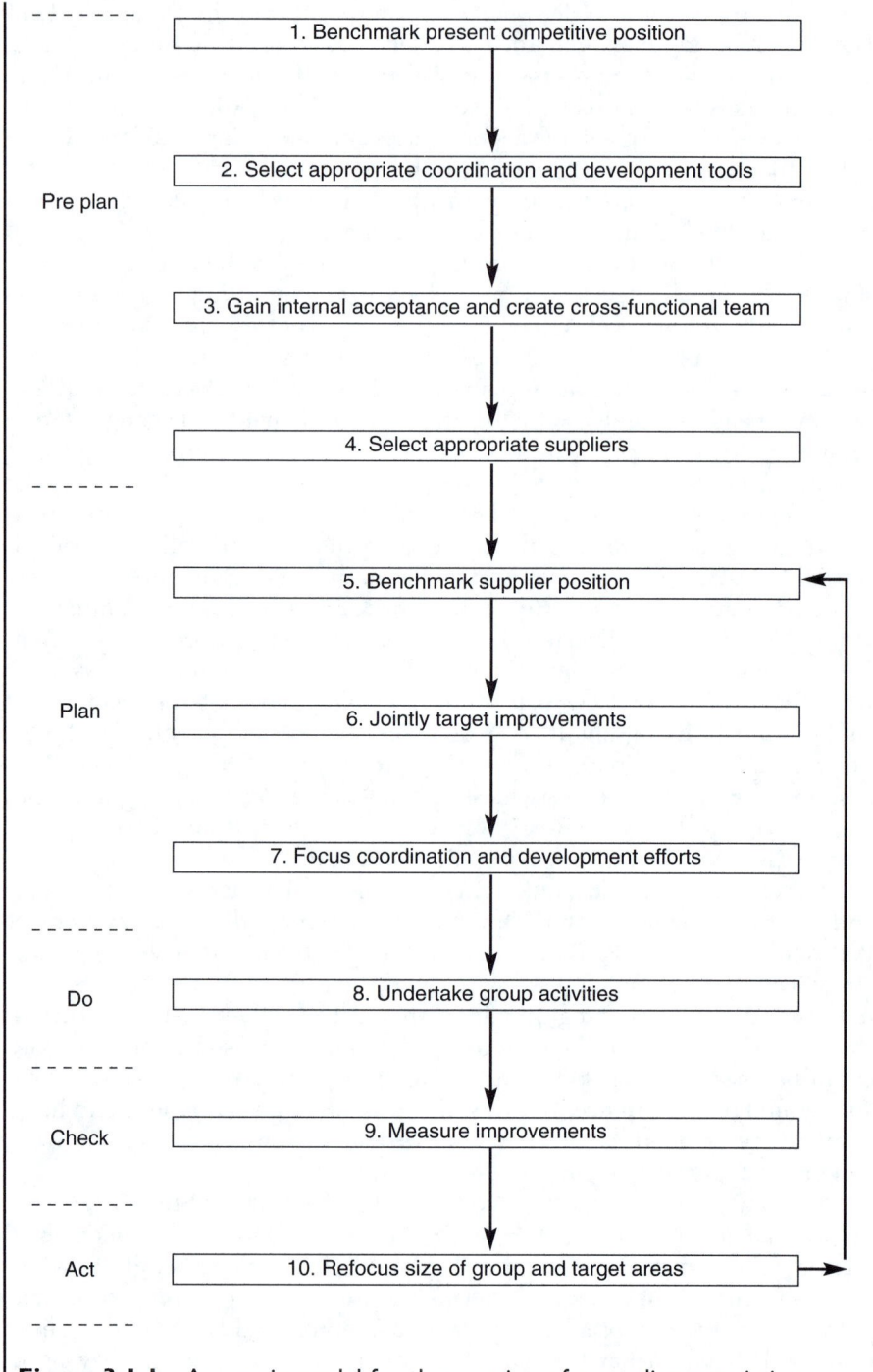

Figure 3.1.1 A generic model for the creation of a supplier association.

knew that the Supplier Association was for them. They took away the ten-point checklist and with Paul Morris started to plan their campaign.

Trico's supplier association

The concept agreed, a name had to be agreed for the forthcoming Supplier Association. Paul happened to mention in passing that there was already a Trico Supplier Association in existence. Jim and David were rather surprised, so Paul explained that the name Trico, as the firm who invented the car windscreen wiper, had been genericized in Japan so that in Japan Trico meant window wiper. As a result, due to the widespread use of Supplier Associations in Japan, another wiper maker had called their group the Trico Supplier Association. Undaunted, Jim suggested that the group needed a more unique name anyway. After some discussion the name 'Trico Quest Supplier Association' was chosen, with Quest being used as the group would be used to achieve **Q**uality and **E**xcellence through **S**upplier **T**eamwork. All that was required now was to develop an implementation strategy.

At this stage David gave Jim and Paul full authority to go ahead in doing this with his support for whatever they felt was the right approach. The approach, however, should be directed by the existing Trico mission statement:

> It is the policy of Trico to provide products and services that give total customer satisfaction. Customers are those who buy our products, our suppliers, our staff and all those with whom we have contact.
>
> We will treat each other with respect and strive for excellence in all we do to provide a high level of service to all customers – internal and external.
>
> We will maintain a culture based on continuous improvement in all we do and will train and fully involve every employee.

Like most mission statements, it contained an ideal that was difficult to match, one that was near impossible without a process to help. Jim and Paul set about developing such a plan, at least for the supplier relationships. This was critical, as around half of Trico's costs of sales were due to bought-in products from their suppliers. After several meetings in the autumn of 1994 a simple Gantt chart was produced to control this process (Figure 3.1.2).

The first part of the process was to develop a strategy and implementation plan. Issues here included areas such as which suppliers to invite, how many meetings to have, what to focus on, how to benchmark. Drawing on Paul's experience of helping to set up many other Supplier Associations, Jim put together a strategy document. This 21-page document explained

Tasks/Milestone	Nov	Dec	Jan	Feb	Mar	Apr	May	Jun	Jul	Aug	Sep	Oct	Nov	Dec	Jan	Feb	Mar	Apr
Develop strategy & implementation plan	←	→																
Supplier invitation conference			←→															
Performance benchmarking					←		→											
Inaugural seminar							←→											
Ongoing seminar programme								←→			←→			←→			←→	
Ongoing workshop programme									←→			←→		←→				←→
Networking									←									→
Other activities										←								→
Refocus for the future															←			→

Figure 3.1.2 Supplier Association implementation plan.

why the association was important to Trico and how it was crucial to their achievement of the mission statement. It went on to define the Trico Quest Supplier Association as:

> A mutually beneficial grouping of key suppliers linked together in a strategic alliance by their supply to Trico. A self help team brought together on a regular basis to share knowledge and experience in an open and co-operative manner with the purpose of members mutually approving and developing skills, systems and techniques, integrating processes and eliminating wastes.

The document explained why the association was being developed, the commitment Trico was giving and what they expected the suppliers to commit to (Figure 3.1.3).

The process for implementation was drawn up in the strategy document and was to include the use of benchmarking, four management seminars per year, with supplementary workshops run as required. All of this was designed to be within a networking framework. Lastly, the document outlined the types of mutual benefits to be sought. The benefits fell very much in line with the commitments being promised by Trico and the suppliers.

In January 1995 a group of nine suppliers were invited to the Supplier Invitation Conference. These suppliers, although only about 10 per cent of the total number, were responsible for 80 per cent of Trico's purchases. They ranged from plastic and rubber raw material and parts makers, to metal processors and packaging suppliers. During the whole day conference Trico's strategies were explained by the top management team, the concept of the Supplier Association was explained and a proposed working framework was presented by Jim based on the strategy document.

After lunch the suppliers were split into groups and asked to give their views. There was, of course, a good degree of scepticism from the suppliers and some concern about what the work would achieve. However, with the guidance of Paul, Jim was able to persuade all the suppliers to buy into the process and to agree to take part in the two types of benchmarking.

The first type, called Continuous Improvement Benchmarking, measured nine key attributes concerned with quality, productivity and delivery performance. This was designed to produce hard data which could be used for targeting of future performances. However, when this benchmarking was applied to the firms (including Trico themselves) a problem arose, namely, most of the firms were not recording some aspects of their performance and had trouble even understanding one or two of the questions. As a result, agreeing hard targets would be next to impossible. Consequently this was not done at this stage.

The second type of benchmarking, called Supplier Capability

Trico's commitment

Our operating practices are also receiving attention and we realise we have a role to play in supporting our suppliers; we are committed to:

- A better understanding of each other's needs
- Involve suppliers at the earliest stages of design
- Involve suppliers in internal project teams
- Encourage and promote timely and effective communication across all functions
- Listen and act upon criticism
- Strive continually to improve schedule stability
- Provide methods of measuring performance and improvement
- Provide benchmark positioning against best practice processes and practices
- Participate in and encourage joint continuous improvements
- Be open and honest
- Strive for quality and excellence in all activities
- Make the Supplier Association our vehicle to ensure teamwork

The suppliers' commitment

Trico will focus business with suppliers committed to continuous improvement and proactive relationship who share our goal of World Class status.

Our suppliers will be encouraged through our Supplier Association mechanism to:

- Work with us in an open cooperative way involving all functions
- Improve continuously supply logistics to deliver frequently to the point of use:
  - guaranteed quality
  - guaranteed quantity
  - on time, every time
- Provide active involvement in our design processes to:
  - optimise materials, processes and tooling
  - reduce costs, delays and lead times
- Actively participate in target costing, value analysis and cost reduction programmes
- Contribute to effective and timely communications across all functions
- Develop a culture of continual improvement and elimination of waste from all processes
- Introduce EDI and CAD links where appropriate
- Be capable of growth and diversification to support future business needs
- Strive for quality and excellence in all activities.

Figure 3.1.3 The commitment statement.

Benchmarking, was also employed. This involved staff from Trico facilitating round-table management discussions at suppliers to scrutinize each aspect of the firm's internal processes. This was very time-consuming for Trico and took longer than expected. The method was used to find out whether each of the firms (again including Trico) had the processes in place to achieve world-class performance. Sadly, this was not the case. As a result, a matrix was drawn up showing in which areas the individual firms were deficient (Table 3.1.1). The different firms were assigned specific letters for this table.

These findings were brought back to the suppliers at the Inaugural Seminar held in June 1995. The general reaction to these findings was, 'Well, what do we do next?' Jim and Paul already had a plan in place but did not want to strong-arm the suppliers into something they did not want to do. Trico had suffered from their customers doing this over the years and they did not want to repeat that particular mistake. The result of the discussion at this seminar was a little inconclusive as a result, and the suppliers went away unsure as to what would happen. This was a make-or-break time for the association.

Shortly afterwards Jim brought the suppliers together again after talking to most of them on a one-to-one basis and getting their viewpoints. At this next meeting the House of Lean Production (Figure 3.1.4) was put forward

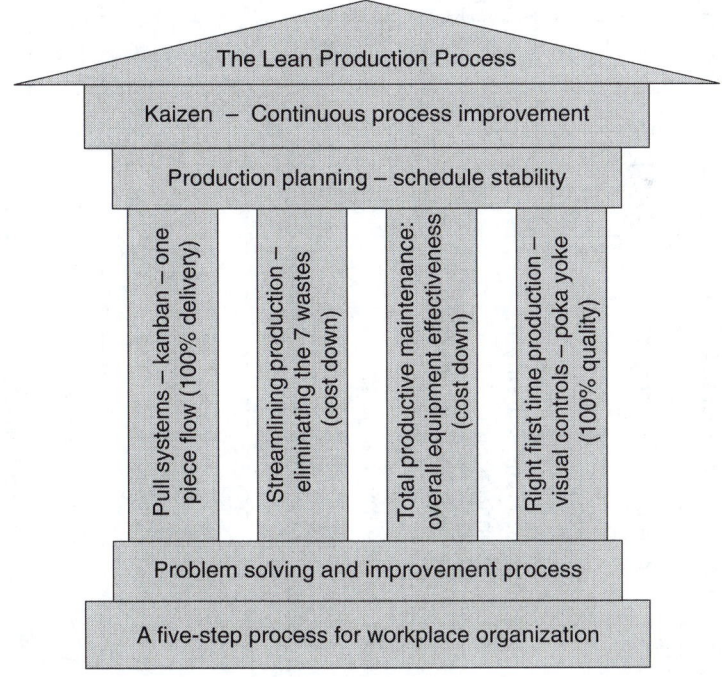

Figure 3.1.4 The House of Lean Production.

Table 3.1.1 Trico Quest: Common areas for development matrix – the enablers needed

Development subject	A	B	C	D	E	F	G	H	I	J	Total
Housekeeping	∗		∗	∗	∗	∗	∗	∗	∗	∗	9
Performance measurement, visual management	∗	∗	∗		∗	∗	∗	∗	∗	∗	9
VE/VA streamlining production, 7 wastes	∗		∗	∗		∗			∗		5
Employee involvement/ continuous improvement	∗	∗		∗				∗	∗		5
Equipment effectiveness/set-ups		∗		∗							2
Kanban			∗	∗	∗	∗		∗	∗		6
Problem solving	∗	∗	∗		∗				∗	∗	6
Performance appraisal		∗	∗		∗			∗			4
Supplier evaluation and measurement		∗	∗		∗	∗		∗		∗	6

as an operating framework to implement the various enablers identified in the development matrix. This system was suggested by Paul as a compilation of his views as to how world-class performance had been created in firms like Toyota.

After some more discussion, the suppliers bought into this as a viable operating framework. As a result, a plan was drawn up to implement the change. At this point Jim was feeling rather relieved and announced to the group that he was not going to ask the group to undertake any activities that Trico were not willing to undertake themselves. In order to start implementing a Lean Production system in buyer and suppliers alike a 5S* housekeeping programme was suggested. 'Once this firm bedrock is in place', Jim suggested, 'we can go on to develop a kanban[†] based delivery system.' At this point Jim nearly lost the suppliers and as Paul has commented since, 'you could see the shutters go up and the morale and self-belief of the firms disappear, they were clearly not ready to embark on a kanban programme without first developing more trust.'

The situation was turned around when one of the suppliers suggest they just concentrate on 5S housekeeping at this stage and worry about kanbans later. So this is what they did. At the end of the meeting a plan had been put together to implement a 5S programme (Figure 3.1.5).

The plan employed the Deming Plan–Do–Check–Act improvement cycle. When put into practice this cycle took around six months to complete, being led by Jim, drawing on Paul's technical knowledge. It consisted of an initial seminar for senior managers held at Trico's premises in August 1995 for the purpose of getting each firm to understand the basics of 5S and buy in actually to doing something. This was followed by a more detailed training workshop in October 1995 for operations staff. This was held at one of the suppliers' premises. Following this, each firm was asked to pilot a 5S project inside their company. All bar one of the firms did this. However, Trico took the lead and have made tremendous strides internally as a result.

It is this lead, not only in the 5S work, but generally, that so impresses the visitor as they tour Trico. For the 5S process, for instance, Jim Taylor not only leads the Supplier Association activities but also the internal activities

---

*5S is a system designed to create an organized workplace. As such it is the basis for any improvement activity. Organizations that seek to make improvements without using 5S or a similar programme are unlikely to succeed. The 5S's refer to the first letters of the Japanese words representing the five types of activities involved. The five activities are: clearing up, organizing, cleaning, standardizing and training and discipline.

[†]Kanban is the Japanese word for 'card' or 'signal'. It describes a system used to control the transfer of materials between the stages of an operation. In its simplest form, it is a card, plastic marker or even coloured ping-pong ball, used by a customer stage to instruct its preceding supplier stage to send more materials.

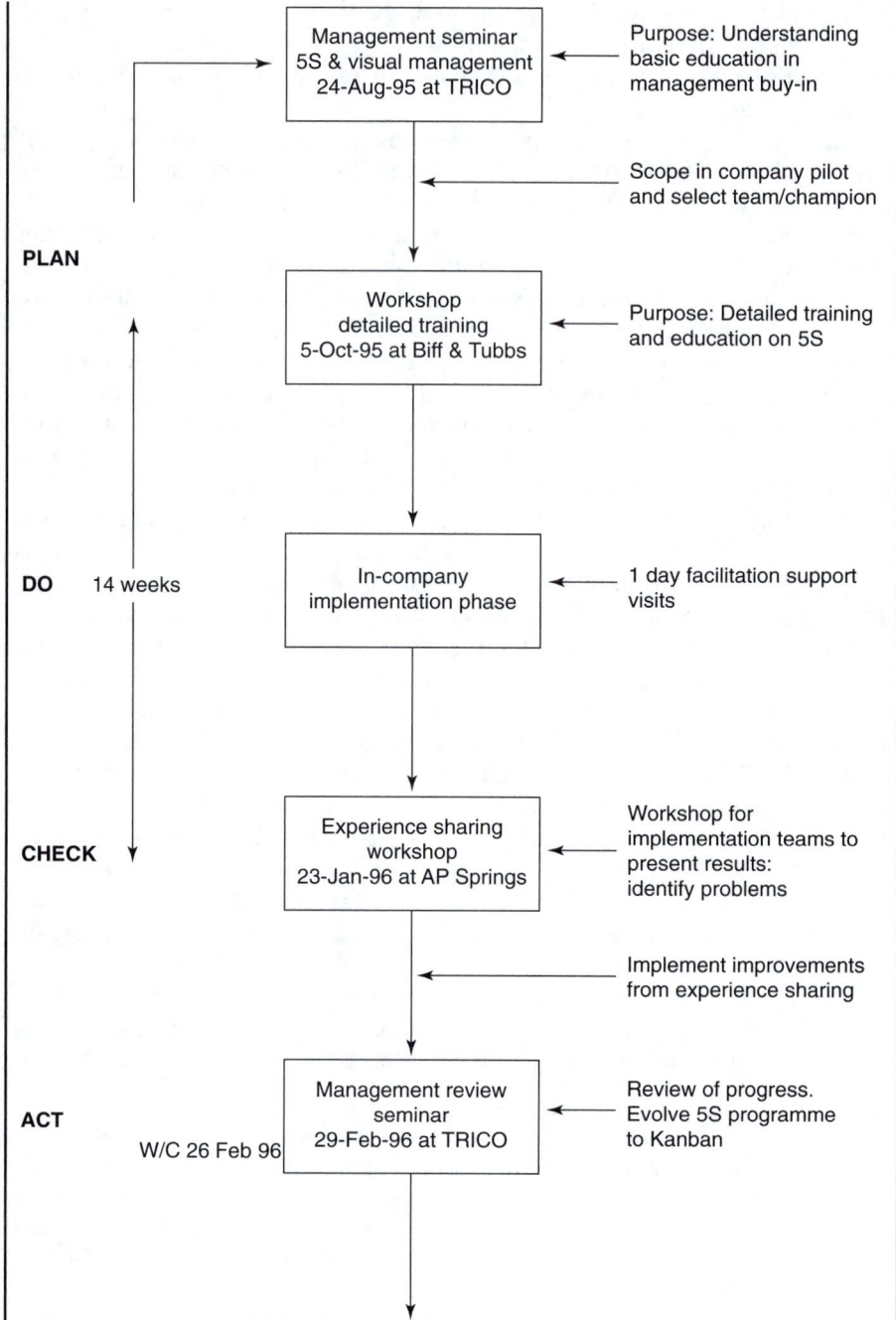

Figure 3.1.5 Implementation improvement process, theme for 1995/96: 100 per cent delivery by kanban replenishment.
Source: Suppliers Association Partners, 1996.

right across the company. To date 33 5S groups have been started, involving around a third of the 500 plus work force. The first of these to be started was in the purchasing office. Jim not only doesn't ask suppliers to do things before Trico, he doesn't ask the rest of the firm to do things before he and his team have tried them. The result is that the purchasing department, part of a large open-plan office, is extremely well organized. When asked if he had any information on the Quest group Jim pulls out two folders and within seconds can produce the exact piece of paper required. This was not the case before becoming involved in 5S. Jim mentions in passing that just after the 5S work started there was 'much peering over walls, and people came from right across the site to look at my desk – they thought I must have left!'

The same attitude and attention to detail is now apparent wherever you go inside the company. The 5S work, together with a number of other separate internal programmes, has transformed the organization and the culture of the work force. This is now a company that is capable of becoming a world-beater, and it has a group of suppliers to help it on the journey.

The experiences that Trico had in their 5S work, as well as those from the operational staff from the other firms, was shared at a workshop in January 1996. This progress was brought back to the senior managers at a further meeting in February 1996. At this seminar the relationship and atmosphere was quite different from earlier meetings. With the exception of the firm that had not really made any progress, everyone could see the benefits of working together. What's more, there were real tangible benefits. As a result, the suppliers were no longer afraid when Jim Taylor again suggested they implement a kanban delivery system. The reason is that the suppliers can see that Trico do not ask anything if they are not willing to do it themselves. They also do not ask the impossible and they stick to their word. As a result, a similar awareness raising, learning and implementation programme is being commenced for kanban implementation at the time of writing in spring 1996.

Impressions to date

Jim sits back in his chair and appears to echo the message given from Rover some months before as he suggests that the Supplier Association has been 'extremely beneficial [with] results now showing – the best vehicle we have seen for communicating to a group of suppliers'. Asked for an example, rather humble about his own achievements, he gives one from a plastic supplier. 'The quality of product has already improved, as they started their 5S activity in the warehouse. It soon resulted in a major time saving for the team leader who could then address various quality issues for product being shipped to us ... as a result quality has improved almost overnight.' Although not eager to give exact details, a simple tour round the factory reveals many empty stock shelves where stock used to be but has already been taken out of the system, with several other racks already having been removed completely. Also, a glance at one of numerous charts on the purchasing office walls shows the cumulative yearly price changes for suppliers being a below zero figure. Things seem to be under control.

In addition, the process is spreading. Once news of the early success spread, the American sister company decided to launch their own group. Already a 25-strong Supplier Association has been formed. In addition, one of Trico's Quest members is starting its own group, fired by the enthusiasm and success from working with Trico.

Summing up his impression of the work, David Jones comments, 'I didn't come on board that quickly at first because I was not totally convinced, [but I realized that] by and large we needed a vehicle to create improvement and that the days of rattling cages and threatening suppliers were ended. What's more', he adds, 'we enjoy it and we encourage the suppliers constructively to criticise us as much as we do them, and they do!' He continues by stressing the internal improvements and relationships that have been built as a result of the work, 'we're trying to change the philosophy here, with the enthusiasm generated from the 5S programme here [at Trico], the main problem is fighting people off who want to take part!'

So what is next?

As mentioned above, the implementation of a kanban system with suppliers is the next major activity for the Trico Quest Supplier Association. After this they may re-benchmark and be in a position to set realistic targets. They also have to consider what to do with the rest of their suppliers, although a less comprehensive supplier development programme is already in place.

In 1992 Trico lay at the threshold of economic prosperity. Whether they achieved it depended on having a modern factory with world-class production facilities and methods, a new organizational culture and a better supplier base. The new Pontypool facility has provided the physical structure. Their Supplier Association has laid the foundations for a new relationship with their suppliers. It has also, rather unexpectedly, helped to provide a framework for significant culture change and internal dynamism and innovation.

The future looks bright for Trico and their suppliers.

Questions

1 Discuss the strengths and weaknesses of Trico's approach to building supplier relationships.
2 What should Trico do now with its Supplier Association members and other suppliers?
3 Is the existing relationship as employed by Trico suitable for other industries? Discuss what modifications would be required if their approach was adopted by:
   a an FMCG food manufacturer
   b an electrical distributor
   c a retail supermarket.
4 Can the Supplier Association be applied within a more traditional Relationship Marketing approach, i.e. a company forming a network group with a number of customers? Under what circumstances would this be appropriate or possible?

Case 3.2 Cafédirect™: The building of a unique coffee brand

This case was prepared by Keith Thompson of Silsoe College, Cranfield University and Professor Simon Knox of the Marketing Group, Cranfield School of Management, as a basis for class discussion rather than to illustrate effective or ineffective handling of an administrative situation.

Some of the 4000 Peruvian coffee producers that Luzmila Loayza visits can only be reached by donkey, even though most agents in this part of the world seldom bother to go further than the coffee processing plant. But then Ms Loayza is unusual; she buys coffee from small producers for an organization which processes, packages and sells it under the brand name 'Cafédirect' in order to give a fair return to small growers. According to Oxfam, these growers normally receive only about 10p for a standard 8oz pack of ground coffee selling in a supermarket for around £1.30. Over 80 per cent of the world's coffee is still grown on small plots farmed by families who depend for their livelihood on one of the world's most unstable commodity markets. Although news of 'frost in Brazil' may bring a *frisson* of excitement to informed supermarket shoppers, since it offers them a chance to 'win' by stocking up with coffee before price increases reach the supermarket shelves, for the growers the unpredictable rise and fall in market prices can be catastrophic.

For almost 50 years coffee production was controlled by quotas imposed through the operation of the International Coffee Agreement, a mechanism which effectively stabilized prices (see Figure 3.2.1). When in 1989 the agreement collapsed, the price of green beans fell so low due to uncontrolled coffee production that marginal producers were unable to even cover their costs (Figure 3.2.2, years 1990–1993). There was no point in them harvesting an unsaleable crop, so farms were neglected and families could not afford sufficient food, school books or medical treatment. With no resources to tide them over, survival frequently meant giving up the farm and moving to the city to scratch a living among the urban poor, putting intolerable strains on the economic and welfare infrastructures of the countries concerned. Others drifted in desperation to the illegal coca fields in order to support their families.

A unique organization is born

Charities, such as Oxfam, had long been buying coffee from small producers, having it processed and selling it through their own outlets up and down the country. Since distribution through charity shops could only

---

Trade represents 80 per cent of the income of developing countries, and coffee is second only to oil in importance as an internationally traded commodity. In South America coffee is of such economic importance that in 1940 a number of countries agreed upon export quotas. Each country was given a quota for the amount of coffee it could put onto the market each year. If the market price of coffee beans fell below a specified level, all of the quotas were cut until prices rose again. World wide coffee export quotas were adopted in 1962 when an International Coffee Agreement was negotiated by the United Nations. The agreement was subsequently renegotiated three times, involving 41 exporting and 25 importing countries. By the late 1980s the agreement started to run into difficulties. Coffee-growing countries began bartering coffee with countries outside the agreement. Consumer countries wanted an increase in the quota for higher grade beans in order to meet changing consumer preference. Then in 1989 the participating countries failed to reach agreement, and the world price of coffee plummeted.

Figure 3.2.1 The international coffee agreement.

reach a very small market, the reality was that such 'campaign'* quality coffee appealed only to the relatively few stalwart supporters who were prepared to make sacrifices in product quality and shopping convenience for a good cause. However, the collapse of the International Coffee Agreement and the problems caused by the consequent price fluctuations was the inspiration for some radical thinking.

In mid-1990, representatives of four organizations met to plan a new enterprise which would produce and market a brand of coffee which could compete successfully on supermarket shelves alongside the biggest brands in the business. Cafédirect, the company which emerged, has four partners (Oxfam, Twin Trading, Equal Exchange and Traidcraft), each with representation on the Board. It has no full-time employees, no production units and no distribution facilities. Sales became the responsibility of one person working four days a week, and brand management activities were undertaken by Media Natura, an agency which specializes in working for environmental and charitable organizations. The mission of the Cafédirect enterprise was to pass more of the proceeds back to the coffee farmers; instead of 10p they would get 40p for every 8oz pack of Cafédirect coffee sold. A further objective agreed by the partners was to stabilize the fluctuations in the price of coffee beans. A crucial aspect of this was contracting to pay growers a guaranteed $1.26 per lb to cover basic production costs and wages, regardless of how low the market price

---

*'Campaign' coffees have a poor reputation for flavour, quality and availability. For some time associated with support for the Sandanistas in Nicaragua, it is imported, processed and sold through charity outlets to the more dedicated charity supporters.

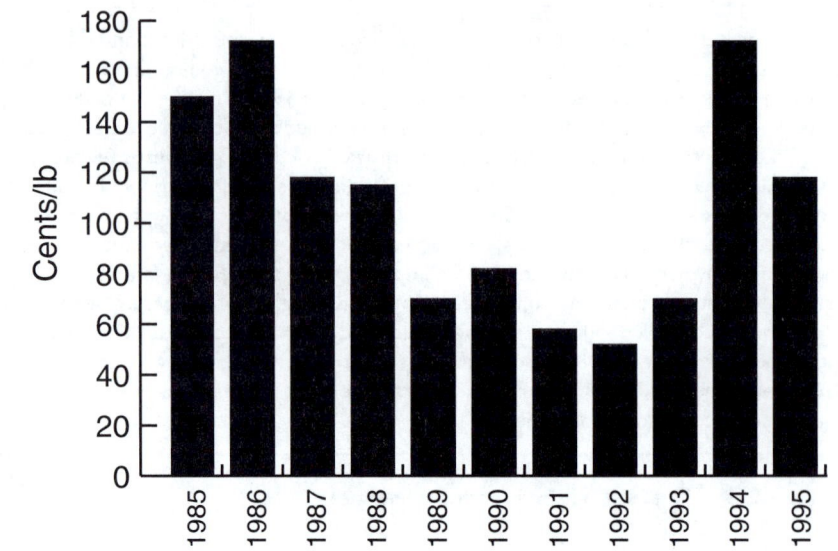

Figure 3.2.2   Market price of green coffee beans, Brazil.

drops.* With the security offered by a Cafédirect contract, farmers could make plans for the future, obtain loans, improve their production methods and provide more security and welfare for their families. Despite suggestions made by the big coffee processing companies that fair trade merely leads to overproduction and lower prices, experience has shown that few farmers risk planting new coffee trees since they take three years to bear fruit, and conditions in the international coffee market are too uncertain to make the investment viable. Instead, many farmers have sought to diversify into other products, like honey, black-eyed beans, lemon grass or cocoa.

In developing the Cafédirect brand, the group was planning to take on the giants of the industry, like Kraft General Foods, Paulig and Lavazza. What is more, they planned to do it with higher raw material costs, limited funds and none of the scale economies and marketing expertise enjoyed by their rivals.

They knew that they could not depend upon the fair trade message alone to convince buyers. In order to compete, they would have to deliver real value to the consumers and meet the needs of the supermarket

---

*Later, in 1994, the market price for green coffee beans briefly exceeded $1.26 per lb. Cafédirect responded to this by changing the contract so that coffee growers were paid the market price plus 10 per cent, with a *minimum* floor price of $1.26 per lb.

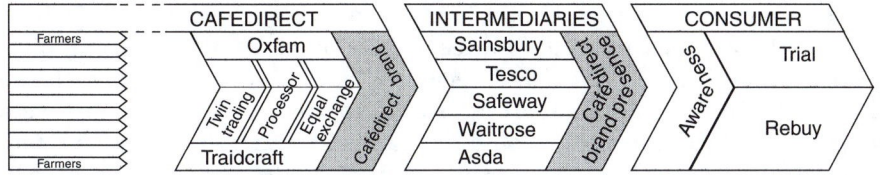

Figure 3.2.3 Cafédirect brand value chain.

buyers. They planned to do this by creating a vertically integrated network, linking small farmers, the coffee processor, supermarkets and consumers (Figure 3.2.3).

Each of the four partners has relevant expertise to contribute in building the value chain of the brand. The long experience of Twin Trading in advising small-scale producers how to meet customer specifications and export requirements would be invaluable in ensuring quality and consistency of supply. The *raison d'être* of Equal Exchange is to market foodstuffs from small-scale Third World producers. Finally, Oxfam and Traidcraft would bring their knowledge gained from their own retail experience, as well as their wealth of social, cultural and political intelligence from their contacts in countries all around the world, and their PR expertise in raising awareness of issues with the support of their many activists.

Identifying consumer value

It is a fundamental tenet of consumer marketing that corporate objectives are achieved by satisfying consumer needs more effectively than your competitors. Since Cafédirect had no marketing management, the task of planning how to provide this superior value fell to Bruce McKinnon of Media Natura, the advertising agency which was working for Cafédirect on a retainer basis.

Bruce knew that in order to identify consumer perceptions of what constitutes superior brand value, he would need a thorough understanding of changing needs, economics and the coffee buying process. Ideally, the value proposition presented by the brand should be unique and not easy to emulate. Most food purchases are still made by women, so the potential coffee buyers were likely to be female, aged 20–45 years (it was thought to be more difficult to change the behaviour of older buyers), in social groups A, B and C1 (having the disposable income needed for discretionary spending). In order to identify the target consumers more accurately, Bruce commissioned a market research study using standard questions from the Target Group Index omnibus research questionnaire.*

The study served to identify three groups of consumer: Selfish, Ethical and Semi-ethical. The Selfish category were considered unlikely to be interested in the plight of Third-World farmers at all, so they were excluded. Ethical consumers, on the other hand, were already committed, and so were quite likely to buy 'campaign' quality coffee from charity shops anyway. Semi-ethicals, the chosen target group, were found to be brand oriented, aware of fair trade issues, but largely uninformed.

The strategy for positioning the Cafédirect brand now became much clearer to Bruce. The brand values offered would have to reflect both functional and non-functional benefits: feeling good about supporting Third-World coffee farmers without the usual hassle and without sacrificing flavour or convenience, at a price that was acceptable to this target market.

Building consumer value

Market research showed that the main motives behind buying ground coffee were pleasure and discernment. This meant using high quality arabica beans, that were processed under strict quality control schedules, then roasted to perfection, ground and packaged to retain the flavour. The main responsibility for ensuring product quality fell to Twin Trading. Using their many years of experience, they were able to help the coffee growers' cooperatives to meet the demanding European customer requirements, and advise them on export practices. For instance, imports of processed coffee into the European Union attract a customs tariff, whereas green coffee beans do not. Twin Trading therefore advised that the green coffee should be imported into Britain. They hired a consultant 'cupper' to determine just the right quality and roast to suit British palates, and negotiated a contract with a processor in Kent to roast, grind and package the coffee under tight quality controls. Through Twin Trading (who liaise closely with the processor), Cafédirect retain control of quality from the farm to the supermarket, unlike their international rivals who buy green coffee beans on the open market.

---

*Target Group Index is a service provided by the British Market Research Bureau, a commercial agency which offers omnibus research. This is operated through a standard questionnaire covering a wide range of consumer products, which TGI send out to their database of 240 000 respondents at regular intervals. Clients can commission studies by either subscribing to standard questions, or by adding specific questions of their own, at extra cost. Cafédirect could not afford to add their own questions. Nevertheless, using standard questions they were able to gather valuable market data, from a large sample, at very low cost.

The market research also showed that consumers suffered some uncertainties about choosing and making ground coffee. This led them to seek the reassurance of a familiar brand which meant, in turn, that if the Cafédirect brand was to succeed, it would have to combine the fair trade appeal with a strong brand identity. Although the technical aspects of getting the flavour and quality right were within the Cafédirect team's capabilities, building a brand and getting shelf space in major supermarkets was really breaking new ground for them. Nevertheless, these were the minimum requirements; the 'hygiene factors' which have to be met merely in order to compete but which do not bestow any additional competitive advantage.

Establishing the superior value needed to acquire this competitive edge, which would then justify a premium price to cover Cafédirect's high raw material cost and diseconomies of scale, essentially had to be achieved through the fair trade message and a pack design which denoted quality and conveyed fair trade symbolism.

The promotional campaign to launch the brand

One reason for entering the ground coffee sector was because it was thought to be much less competitive than the instant sector and the brands less well established. Furthermore, since research had indicated that people are sometimes confused and usually uncertain when buying ground coffee, the introduction of Cafédirect may just give them the reason to choose one coffee brand over another. However, to achieve this, Cafédirect needed to inform their potential customers of the product's unique offering and to build up the brand. Recognizing the need to invest is one thing, finding the capital to do it is another; resources were very limited at Cafédirect!

The choice of promotional techniques available for this purpose was further limited by the nature of the value being offered to consumers. Although sales promotion may be relatively inexpensive, and perhaps even self-liquidating, special offers, coupons and contests were felt to conflict with Cafédirect's brand values. Money-off tactics left consumers wondering who was losing out.

Despite these constraints, by mid-1991 the Cafédirect team felt that their brand proposition was now sufficiently strong to make inroads in the ground coffee sector; the decision was taken to launch within a year and detailed launch plans were worked out. Prior to the launch date, stalls at the Good Food Show and the Global Partnership Exhibition were used to raise awareness of fair trade issues. Early samples of Cafédirect coffee were served free to railway passengers and leaflets distributed by Oxfam staff, who answered passengers' questions about fair-traded products.

At the product launch, the main thrust of the promotional drive was an advertising campaign with the strapline 'Fair trade, excellent coffee', which appeared on 1000 Ad Rail poster sites on InterCity and some local routes. One such advert featured a young child in its mother's arms with the message: 'You get excellent coffee. They get vaccines.' Another version featuring two older children proclaimed: 'You discover excellent coffee. They discover school.' Advertising agencies usually strive for a single definitive proposition, but Cafédirect had two important messages to communicate: the 'help the world' appeal had to be balanced with 'without sacrificing taste and quality'. Since the audience were defined as Semi-ethicals, they had to be told what the problem was *and* reassured about the product quality. A similar problem had been faced by Café Hag several years earlier, when they had to explain the health hazards associated with caffeine intake before they could put across the benefits of decaffeinated coffee.

Posters were chosen as the launch advertising medium because the volume and breadth they offer can create the impression of a big advertising spend. In reality, promotional funds were so limited that the 'campaign' consisted of only a few different executions. Post hoc advertising research indicated that the campaign had been successful, despite its lack of scale (Table 3.2.1).

Consumers responded well to the advertisements in terms of their propensity to purchase at a price premium, but they still had to be induced to take the product off the supermarket shelf. For that, a good pack design was critically important. Fortunately, a key part of Media Natura's service was cheap (and sometimes free) access to excellent creative people. Those given the task of designing the Cafédirect pack thought that most coffee packs had a 1970s look, which gave Cafédirect the opportunity to appear fresh and up-to-date by comparison. Colour was found to be an important discriminator; the final design was predominantly dark blue and gold because the combination suggested quality and was empathetic towards the values of the ground coffee consumer.

A picture may speak a thousand words, but research showed that con-

Table 3.2.1 Post hoc research results, Cafédirect poster advertising campaign*

	Cafédirect (%)	Norm (%)
Recognition	36	36
Liking	74	53
Clarity	48	n/a
Purchase incentive	43	31
Premium price acceptance	82	n/a

Source: RSL poster advertising research. Note: *Sample frame: travellers on the relevant railway routes.

sumers would not recognize a coffee tree or green coffee beans, so these were avoided, as was the depiction of a mug, which implied poor quality. The research also showed that as people are uncertain about ground coffee (what sort to buy and how to make it), they tend to become quite involved in the purchase. Consequently, many people do read the information on ground coffee packs with some care, unlike the labels on other products. So, if consumers could be persuaded to pick up the Cafédirect pack, they were likely to take in the fair trade message mentioned briefly on the front, and in more detail on the back (Figure 3.2.4).

---

cafédirect™
Four Fair Trade Organisations have worked with coffee producers to make this coffee available.

EQUAL EXCHANGE
EQUAL EXCHANGE is a workers' co-operative marketing foodstuffs from small-scale producers in the Third World.

OXFAM TRADING
OXFAM TRADING is the trading arm of OXFAM. Working with craft and agricultural producers in more than 40 countries.

TRAIDCRAFT
TRAIDCRAFT plc promotes fair and responsible trade with small farmers and manufacturers in developing countries, based on Christian ethical principles.

TWIN TRADING
TWIN TRADING works closely with producers, finding markets and guiding them through the process of meeting export requirements and specifications of customers.

cafédirect's promise is a high quality coffee for the consumer, and a higher income for the small farm producer. World coffee prices are unstable and have fallen dramatically in recent years. This coffee comes to you directly from the producers, and the price includes a premium for them.
cafédirect is part of a new movement towards fairer trading to create stability and justice for ordinary people around the world.

Figure 3.2.4 Text on the back of the Cafédirect ground coffee pack.

Charities like Oxfam and Traidcraft have been importing and selling fairly traded products through their own shops for many years. However, they reach a market which is severely limited by the nature of the outlets and the size of the concerned minority segment that they serve. The Fairtrade Foundation was established in 1992 with the aim of improving terms of trade with Third World producers by bringing fairly traded products into the mainstream and onto supermarket shelves. Retailers and manufacturers are encouraged to apply to use the Fairtrade mark on their products, which is awarded if the product meets criteria which 'set out standards of employment and/or terms of trade which ensure widespread benefits for the workers and other producers in the Third World'. The Foundation's standards require the payment of a 'social premium' on top of the normal price. But that is not sufficient in itself, the use of the social premium must be with the active involvement of the producers, and the social conditions must be good. Products licensed to use the Fairtrade mark pay a licence fee of 2 per cent on turnover up to £1 million per annum, and on a reducing scale thereafter.

Figure 3.2.5 The Fairtrade Foundation.

Supporting this fair trade message is the logo of the Fair Trade Foundation, which appears down the front and across the top of the pack. Cafédirect was one of the first products to be awarded this Fairtrade mark (Figure 3.2.5), which is considered essential supporting evidence in providing consumers with the necessary reassurance to make a preferential brand selection.

Building supermarket distribution

No matter how well presented or how good the product actually is, it simply will not sell if it's not readily available to consumers. Although both Traidcraft and Oxfam have their own shops, these distribution channels are not able to reach a mass market. It was essential that Cafédirect competed for shelf space in the major supermarkets. But shelf space is also crucial to the success of the big players and, since there is limited availability, a gain by Cafédirect is a loss to companies such as Paulig and Kraft General Foods. This was the job of the Cafédirect Sales Director, Lorna Young, who agreed to work four days a week on the account through Equal Exchange, her employer. The challenges facing Lorna were enormous, and her resources were minimal in comparison to those deployed by the account teams of her international rivals.

The first question Lorna Young had to address was, which supermarket chain would be most receptive to stocking yet another brand of coffee – particularly one with very little resources for advertising to 'pull through' sales?

The next question was, which supermarket chain would be the most accessible?

On the face of it, the Co-op supermarket seemed like a good choice since their current promotional platform emphasized the ethical nature of the whole organization, which they were able to reinforce through the historical roots of the Co-operative movement. The unique appeal of Cafédirect would complement this positioning very well. However, it emerged that the Co-op's response was coloured by plans to launch their own range of fairly traded products. Furthermore, the younger, ABC1 consumers crucial to success do not do much shopping at Co-op supermarkets, so the target group could not easily be accessed through Co-op outlets.

The supermarket group eventually selected for the 1992 product launch was Safeway. Not only does Safeway have the right consumer profile, built through an association with ethical consumerism over a number of years, it also has a reputation for supporting organic food and fair trade initiatives.

The choice of Safeway as the point of market entry had other advantages too. In Scotland, where the store group planned to test market Cafédirect, they have a single distribution depot. Furthermore, the Beverages Buyer was sympathetic to Cafédirect's cause and was willing to give help and guidance on matters such as delivery scheduling, tailgate heights and the division of responsibilities between supermarket and supplier. As Equal Exchange is also based in Scotland, Lorna Young was able to build relationships with Safeway managers on a day-to-day basis from her Edinburgh head office. These relationships were to prove invaluable to the success of the launch.

By the middle of 1996, the brand was stocked by all major British supermarkets and had achieved a national market share of 3.1 per cent (Table 3.2.2). This was a remarkable achievement in a market that is traditionally image-led and which usually requires multi-million pound

Kraft General Foods	15%
Sara Lee	14%
Paulig (inc. Lyons Tetley*)	13%
Lavazza	7%
Cafédirect	3%
Other brands	5%
Own label	44%
Total	**100%**

Table 3.2.2   Manufacturer's value share of the ground coffee market, 1996.
Source: IRI Infoscan, 12 months to May 1996.
*Paulig acquired Allied Lyons' ground coffee business in March 1994.

advertising expenditure as the cost of maintaining market share.

As it turned out, Lorna Young was not entirely alone in her quest for shop distribution. In the early days of the Cafédirect launch, Oxfam and Twin Trading activists called on supermarket managers to alert them to the issues and opportunities of fair trade products. The Fair Trade Foundation circulated hundreds of thousands of postcards to consumers so that they too could give them to supermarket managers with the message to stock fair trade products. After years of complaints about checkout queues filling the supermarkets' postbags, suddenly failure to support Cafédirect became the top complaint.

Other retail marketers were all too aware of this shift in consumer values and were working with their buyers to search out suitable products. Cafédirect was considered to be in the vanguard of the fair-trade movement, and ethically beyond reproach. Furthermore, it is no secret that many supermarkets like working with small suppliers because the imbalance of power means that they can get firms like Cafédirect to 'do it their way'. Publicly, supermarket buyers give three reasons for listing Cafédirect. Firstly, if it is not on the shelves it may just reflect badly on the store chain. Secondly, and more positively, it is seen to be the right thing to do and they hope that it will help to build consumer loyalty. Thirdly, with supermarket listings now multiplying, the brand is selling well. The latter is, of course, the real acid test.

Launch postscript

British supermarkets are giving the product a fair chance; the rest is really up to consumers and their willingness to repeat purchase the brand on a regular basis. Perhaps they will be prepared to do this. Research commissioned by the Co-operative movement indicates that, 'Consumers are willing to penalise retailers and brands which fail to meet their ethical standards and to reward those that do ... one in three consumers reported that they had boycotted a shop or brand in the past. Six out of ten are ready to do so now.'

However, now that Cafédirect is supplying most of the supermarkets, the buyer–seller relationship has changed and the call of the more strident activists to 'protest march down to your supermarket' has become counterproductive.

It is time for the Cafédirect team to reassess their marketing strategy and to build closer relationships across their entire network.

Questions

Cafédirect have formed a vertically integrated network to link Third World coffee producers into the markets of western Europe.

1 What are the features which this network needs in order to implement relationship marketing practices?
2 Identify the main relationships in the Cafédirect network and their relative strength and importance. Show these relative positions on the Relationship Map below.

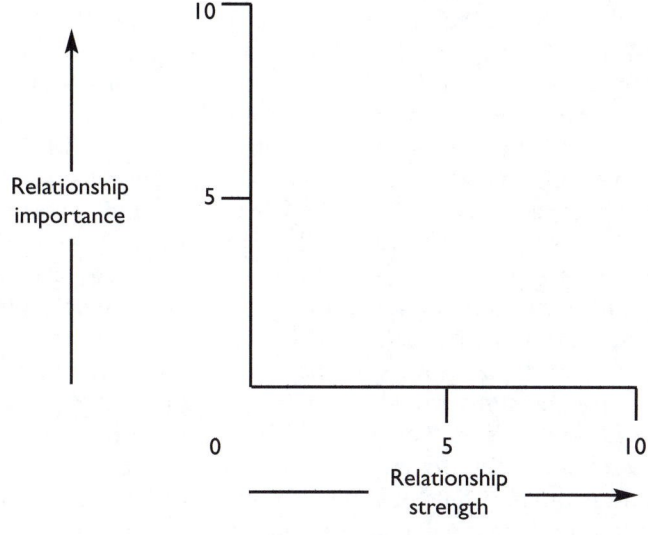

Cafédirect relationship map

3 Use your analysis to determine which are the priority relationships for Cafédirect. How can these relationships be further developed to build on the success of the launch?

Case 3.3   Transvaal Nickel Mines

This case was prepared by Professor David L. Blenkhorn, Wilfrid Laurier University, Waterloo, Ontario, Canada, and David W. Holowack, who, at the time the case was written, was an MBA student at the University of Western Ontario, London, Ontario, Canada. The case was written as a basis of class discussion rather than to illustrate effective or ineffective management practices.
© *Copyright 1996 David L. Blenkhorn and David W. Holowack.*

Jane Keegan, Marketing Director – Europe for Transvaal Nickel Mines (TNM) looked at the data on the pages sitting in front of her. Each of her nickel customers was rated in terms of annual volumes, gross realized revenues and Net Back Premium (NBP).* The year over year data indicated an erosion in the premium realized from almost all of Jane's customers; moreover, the results in Europe were also representative of all of TNM's other global markets. The trend was disturbing both from a corporate and from Keegan's personal perspective. For TNM, the US$0.02 decline in premium from 12 cents to 10 cents in the latest 12 months represented a direct reduction of US$4 million in the company's profit. For Keegan, the narrowing premium brought into question the need for, and the viability of, a global sales and marketing organization within TNM. The reason for this was simple. As the premium realized by TNM moved closer to zero, it became more cost efficient for the company to sell its commodity nickel products directly onto the London Metal Exchange (LME)** rather than incur the administrative overhead to sell directly to industrial users. Staff in TNM's sales and marketing function felt that their contribution to the company was greater than that indicated solely by the premium, especially in terms of market recognizance and understanding emerging trends in nickel con-

---

*The NBP equalled the net contribution to TNM after direct marketing, transportation and financing costs less the sales price available on the London Metal Exchange. For example, if TNM sold a pound of nickel for $4.00 while nickel on the LME was selling at $3.80 and if it cost 8 cents per pound to transport the product to that particular customer and if the financing cost related to the pound of nickel while in transit and to collecting accounts receivable totalled 3 cents, the NBP would equal:

Sales price	$4.00
Less:	
Cost to transport to customer	0.08
Cost to finance inventory in transit & A/R	0.03
Price available if sold on LME	3.80
Net Back Premium (also referred to as 'premium')	**0.09**

sumption. However, TNM's flinty eyed accountants remained unconvinced of these intangible benefits. The issue was particularly poignant to Keegan as she had recently moved to the Marketing Department after a successful five years as a production engineer in TNM's South African mine, smelter and refinery operation in Rustenburg. Prior to her shift into the marketing area the company had sponsored her in the one-year full-time MBA programme at Cape Town's International Graduate School of Business, following which she was transferred to TNM's European marketing headquarters in London, England. Keegan reflected to herself whether she had moved into the Marketing Department only to watch the department's dismantling, or was there some way to introduce innovation to TNM's marketing and sales efforts to bring more value to the function? She had taken a number of marketing courses during her MBA and felt that the emerging area of relationship marketing held the greatest potential for adding value to TNM's sales and marketing function.

The company

Transvaal Nickel Mines (TNM) was created in 1926 to mine and refine nickel deposits found in the Transvaal area of South Africa. With its head office located in Johannesburg, the company had branched out internationally and now had mines, smelters and refineries on three continents – Africa, Australia and South America.

The market

Primary nickel consumption in the Western world for 1995 was estimated to be 905 000 tonnes; 66 per cent of this nickel was used in the production of stainless steel.[†] The balance of the nickel produced was used in the production of super alloys and in electroplating applications. Super alloys are metal alloys formulated to withstand extreme conditions; common applications include jet engine turbine parts, automotive turbocharger blades and nuclear reactor tubes. It is in this area that high purity/quality producers like TNM can obtain higher than average prices for their nickel. Key areas of nickel consumption are Japan, the US, Korea, Germany and other parts of Western Europe.

[**]The London Metal Exchange is a terminal market where standard quality nickel can be bought and sold. The LME establishes the benchmark price for the majority of commercial transactions involving nickel.
[†]Nickel Development Institute, 1996.

The competition

TNM competes against Canadian, Australian and Russian nickel producers. Since the break-up of the Soviet Union in 1991, large quantities of Russian nickel have been entering the European market, trading at a 2–4 per cent discount to the price charged by Western world nickel producers. Many nickel customers have decreased purchases from suppliers like TNM, shifting to the lower cost Russian nickel. TNM has been forced to reduce prices to protect its remaining market share, thereby adversely affecting overall profitability. Most customers continue to purchase a certain baseline quantity of nickel from TNM or other Western producers in order to maintain an on-going business relationship. Generally metal industrial users recognize that Russian material does not represent a secure source of supply and therefore try to maintain ties with more reliable Western producers.

The relationship marketing initiative

Keegan's direct superior, Hans Volker, was Managing Director of TNM's European marketing operations. Volker had learned the metals business through working as a metals trader for the Austrian metals company Metalgaburger. Years of trading had honed his skill at assessing situations quickly and making split second decisions. When Keegan approached Volker with the idea of introducing relationship marketing Volker's retort was:

> There are no relationships in metals marketing – raise your price 1 cent a pound, you lose your relationship; drop your price 1 cent and you have more relationships than you know what to do with. This relationship marketing theory you describe may work in consumer goods but it has no place in the marketing and sale of industrial commodity products.

Volker then paused and, thinking for a brief moment about his own daughter, who was about Keegan's age, reflected:

> Listen Jane, we were all young and inexperienced once and we all have to learn the hard way – if you think relationship marketing will bring you higher premiums, try it, but in my opinion the only way you get higher premiums is to charge higher prices – it is not that complicated! And remember, no one lasts too long around here if their group doesn't make its targeted numbers – both sales volume and sales price!

Keegan thought to herself that real life certainly was not much like the strategic market planning systems that the marketing professors at the Graduate School espoused. In those models, senior management drove and supported innovation, while here in the real world the message appeared to be, 'innovate if you want but don't let it get in the way of your real work'. After further contemplation, she reasoned to herself that her plan didn't really have a downside – at least if she tried to implement relationship marketing she had the potential to win by gaining competitive advantage for TNM. If she didn't even try, she would not be in control of her own destiny. From her marketing classes she remembered that a key principle of relationship marketing was to add value to the players in the value chain.

The relationship marketing plan

Keegan decided that the most logical method to develop the relationship marketing concept was to target one key customer in a high profile end-use category. Additionally, she realized that she needed a sympathetic, receptive champion at that customer who would push the introduction of relationship marketing. After surveying her customer base she decided that Supermelt Vacuum Melters (SVM) represented a prime candidate for the relationship marketing initiative. Some of the reasons for this were:

1 SVM was a significant consumer of nickel – consuming over a million pounds per year.
2 SVM was a lead supplier of super alloys to the automotive/truck turbocharger industry – a growth market.
3 Nickel made up to 50 per cent of SVM's costs.
4 SVM's customers (i.e. automotive suppliers) had received training in developing partnership relationships and hence would more likely be familiar with some relationship marketing concepts.
5 A success in the automotive industry would increase the acceptance of relationship marketing in other nickel end-use segments like aerospace and the nuclear fabrication industry.
6 SVM's Managing Director, Dr Herman Bearing, was recognized as a leading innovator in the metal alloy industry.

A phone call to Dr Bearing quickly gained his interest and tentative support for Keegan's proposed relationship marketing initiative.

Supermelt Vacuum Melters (SVM)

The process used by Supermelt Vacuum Melters involved combining an exact mixture of high purity metals such as nickel, cobalt and chrome and melting these elements in a vacuum furnace, creating small ingots of specific alloys which are then cast by another manufacturer into turbocharger wheels, jet engine turbine fins and other items used under conditions of high temperature and stress. A small miscalculation in alloy composition could ultimately cause a jet aircraft to crash or a car or truck's turbocharger to explode. Since consistent high quality was so critical, only certain vacuum melters were certified by aerospace and automotive companies to produce the required alloys. SVM, being a leader and very well respected in the industry, was certified by all the major aerospace and automotive companies.

Keegan then reviewed what her company files contained about SVM and its customers. Historically TNM had spent little time and energy on understanding its customers' customers and information in this area was scarce. Keegan called Dr Bearing of SVM and explained that a key component of relationship marketing was to develop new ways to add customer value in the relationship. An understanding of SVM's value chain would assist in this effort. After a brief discussion with Dr Bearing, a supply/value chain was outlined (Figure 3.3.1).

The chain starts with the nickel miner, smelter and refiner (TNM), which sells nickel ingots to a melter such as SVM, which combines the nickel with other metals such as cobalt and chrome to produce very high quality alloys for specific uses. These alloy ingots are then sold to an investment casting turbocharger manufacturer, which then produces turbocharger blades to very exacting specifications, becoming one component in automobile or truck turbochargers, which are then sold to a transmission manufacturer and become part of an automobile or truck drive system. The transmission manufacturer is a first-tier supplier to an automobile or truck assembler, which puts the transmission in the vehicle. The vehicle, through the intermediary of a dealer, is then sold to the final user – the consumer.

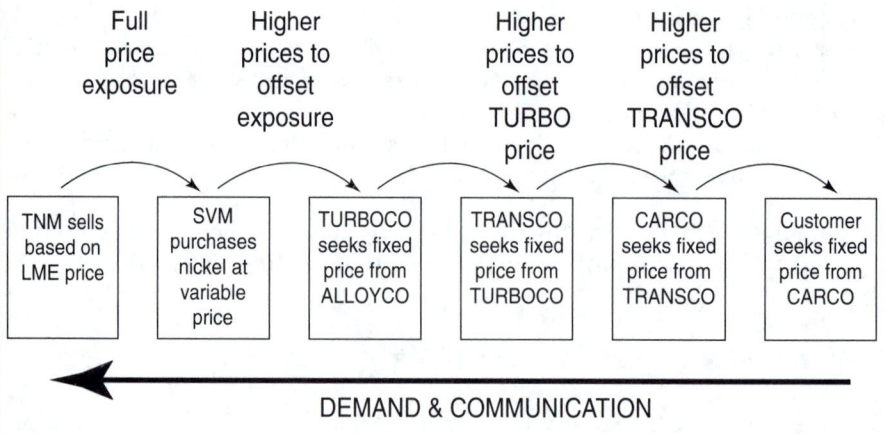

Figure 3.3.1   Value chain – nickel to auto.

Dr Bearing explained that if SVM's customers could understand the forces that affect nickel prices, these customers would be more willing to work towards the creation of contractual agreements that allow improved nickel price management.

> My customers now demand that I give them fixed prices that are valid for one year but they will not commit to any fixed quantity of material. Further, since my customers have 'just in time' (JIT) manufacturing systems, they feel that giving SVM one week's notice of an order is more than enough lead time. I currently deal with this situation by adding an extra markup to my prices to cover the risk associated with my nickel price exposure. If I could get my customers to give me greater lead times specifying both quantities and the timing of deliveries, I could use the hedging options that TNM has developed. If my nickel purchases are hedged* at a fixed price that matches the fixed price commitments given to my customers I can eliminate the safety margin that I now build into my pricing. With more competitive pricing I'm sure that SVM can get more business; increasing our operation's utilization rate from 70% to 95%; driving down fixed costs per unit of production and increasing profits substantially. Everything depends on educating my direct customers and all the various parties in the supply/value chain. By working together with TNM and our customers we can eliminate the nickel price risk and allow ourselves to focus on the aspects of the operation that we can actually control.

SVM's input

Dr Bearing explained that next month senior purchasing executives from the world's major turbocharger manufacturers would gather in Geneva for their annual conference. He wondered if perhaps Keegan could put on a presentation at the conference and include TNM's ideas on how to bring more efficiency to the value chain while managing nickel price exposures.

Keegan responded quickly that she certainly could. After some thought, she proposed that her presentation would cover the following topics:

1 Nickel Supply, Demand and Prices
2 How Nickel Prices are Set: The LME
3 Metal Trading and Hedges
4 How TNM Can Make SVM and its Customers More Competitive Through Value Chain Rationalization.

Keegan then laid out the presentation format (Figure 3.3.2).

---

*Hedging in the nickel market allows a buyer of nickel to obtain fixed quantities for future delivery at a fixed price.

Comments on the presentation

Dr Bearing liked the content of the presentation but added the caveat:

> Don't expect an immediate response from my customer group. It will take a fair amount of education before they will accept the concept of hedging and the improved communications required for effective hedging. To my knowledge, no one in our industry has ever looked at aligning the needs of the whole supply/value chain. Also, remember that during the presentation the audience will be composed of representatives from competing companies – none of them will want to tip their hand in front of each other.

Keegan then had to convince her boss, Hans Volker, on the merits of the approach.

Volker's initial response was: 'I can see how SVM and its customers will benefit, but how will we? TNM is not a nonprofit educational organization. You educate all those people and they are still free to buy their nickel from whomever they wish.' Keegan replied:

> That's true, but the Russians don't have an LME trading operation like we do – they will be cut out of the business. Only ourselves and one other major nickel producer have organizations capable of providing hedging services on an ongoing basis. Further, we have the inside track as far as establishing the required relationships. Once the required communication pathways are set up to facilitate hedging activities it is very inconvenient for a customer to switch suppliers on a whim. Basically the customer's switching costs increase – that becomes our competitive advantage!

Volker's response was: 'How long will it take for the Russians to set up LME trading and hedging capabilities? What will you do then?' Keegan shrugged and responded: 'We will innovate again.'

Volker's final comment was: 'It sounds like it will be a lot of work for little gain, but go ahead.'

With that OK from Volker to proceed with the relationship marketing project she quickly realized that the ball was clearly in her court – just how should she proceed with her initiative so that success for all those involved would be realized? As this was her first solo initiative since joining the Sales and Marketing Department in London, she wanted to do well with it.

Questions

1 Is this a viable relationship marketing initiative for TNM and Jane Keegan? Why or why not?
2 What are the principal criteria which should be present for a successful relationship marketing endeavour? Which ones are present here? Which ones are lacking?
3 Critique Jane Keegan's approach to relationship marketing for TNM. How would you do things differently?

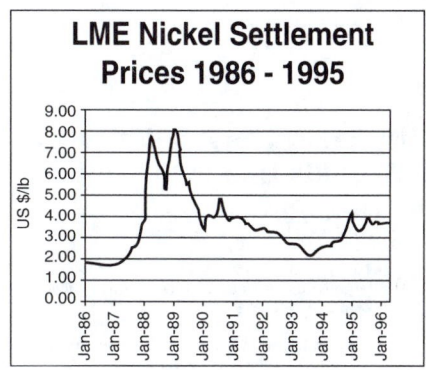

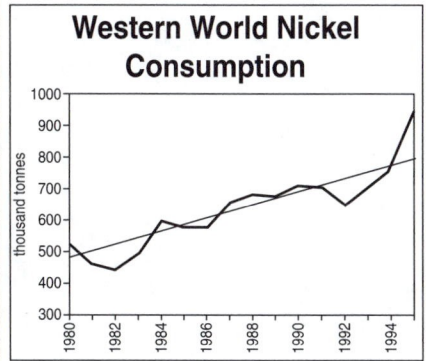

THE NICKEL MARKET	
• SUPPLY AND DEMAND	

Western World Nickel Market Base Case Scenario
('000 tonnes)

	1995	1996	1997	1998
Western Production	652	700	725	750
East-West Trade	164	173	170	162
Sales from DLA	9	9	7	2
TOTAL SUPPLY	825	881	902	915
Demand Growth	20%	0%	1%	3.5%
DEMAND	905	905	914	946
BALANCE	-80	-24	-12	-31
LME Stocks	45	33	21	-10
Producer Stocks	92	80	80	80
Total Stocks	137	113	101	70
As Weeks of Consumption	7.9	6.5	5.7	3.8

Figure 3.3.2 The Presentation

Nickel Stocks and Prices

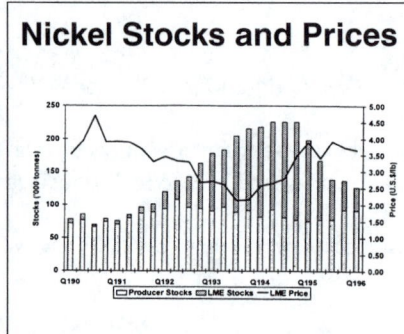

Current and Constant Dollar Nickel Prices, 1960-1995

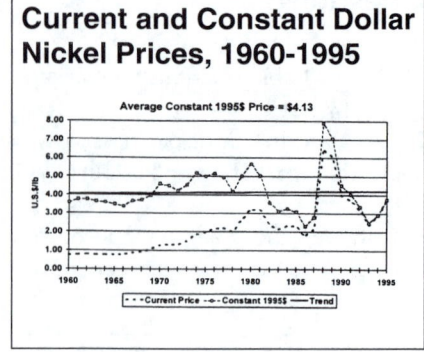

Nickel Prices Act With High Volatility Due To:

- illiquidity / low volume / high unit price compared to other base metals

- funds tend to trade infrequently and in large volumes – not matched to physically-based business

- fund and speculative interest is concentrated into shorter time frames than fundamental business

- price is pushed by many factors (Cu, world money flows, economies, perceptions, ...)

Price Movements
Weekly LME Nickel Price – Lows & Highs

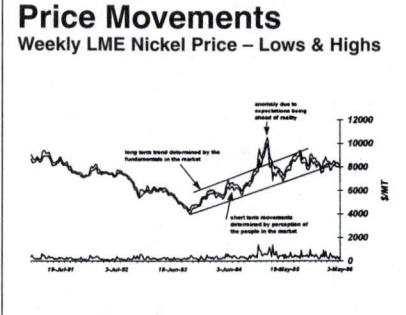

Figure 3.3.2   (*continued*)

Pricing Strategies

- 'Formula' based on LME prices in the future

- 'Fixed' based on a current or future LME price

- 'Ceiling' with a maximum price to Transvaal Nickel Mines (may be expensive)

- 'Ceiling/Floor' with maximum and minimum prices to Transvaal Nickel Mines (cost effective, but risky if market prices fall below minimum)

Metal Trading

Uncertainty feeds the volatility in the metal market

Metal Trading

Even producers and brokers with massive research budgets cannot accurately forecast short-term trends in metal prices

Metal Trading

Does your organization want to commit the resources in order to make metal trading/buying one of your core competences?

Figure 3.3.2   (*continued*)

Metal Trading

Transvaal Nickel Mines has invested millions of dollars to establish a world class LME trading and research group. We can spread our investment over a large group of customers.

Metal Trading

Trading and particularly hedging is a core competence of Transvaal Nickel Mines. We offer this as a service to our customers – it is not a profit centre.

Why Hedge?

WOULD YOU ACCEPT 25% VARIATIONS IN YOUR LABOUR, ENERGY OR OVERHEAD COSTS?

Why Hedge?

WHY ACCEPT IT WITH YOUR NICKEL RAW MATERIAL COSTS?

Figure 3.3.2   (*continued*)

Greater Certainty

- **Greater certainty can allow you to set your prices to your customers more aggressively**

- **Utilize your capacity better**

- **Drive down your fixed costs per unit**

- **Make you more competitive**

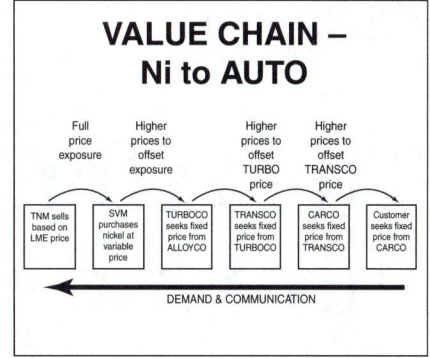

VALUE CHAIN – Ni to AUTO

SUPPLY CHAIN CRITICAL ISSUES

- Fixed prices can only be provided on commitments to purchase fixed volumes

- The longer the advanced notice of future demand the greater the opportunity to provide attractive pricing solutions

- Uncertainties force the inclusion of safety cushions in all the pricing throughout the value chain

- Better communication reduces the need for safety cushions in prices and makes everyone's product more competitive

Commercial Needs

- Transvaal Nickel Mines' LME expertise
  - strategies to benefit from Ni volatility
  - facilitate fixed price quotations
  - minimize financial exposure flr & clg

- Transvaal Nickel Mines' size allows it to receive preferential treatment from LME brokers

- Access to greater market knowledge than any individual industrial user

Figure 3.3.2   (*continued*)

Chapter 4

The referral and influence market domains*

The referral and influence market domains

*This chapter is a summary version of a Cranfield School of Management working paper by Frow, P. and Payne, A. (1998). *Relationship marketing: Developing the potential of referral and influence markets*. © Copyright Cranfield School of Management. Reproduced with permission.

Introduction

This chapter addresses two key market domains within the Six Markets model – referral markets and influence markets. In Chapter 1 of the book, the roles of referral markets and influence markets within the six markets framework were introduced. These two market domains can be described as follows.

- *Referral markets.* The 'referral' market domain consists of two main categories – customer and non-customer referral sources. Frequently, the best marketing is that done by an organization's existing customers; which is why the creation of positive word-of-mouth referral, through delivery of outstanding service quality, is critical. Non-customer referral sources, which recommend an organization to prospective customers, are described by a number of terms. These include networks, multipliers, connectors, third party introducers, agencies and so on. It should be noted that these referral sources are also sometimes referred to as intermediaries. However, this can be a confusing term when applied to referral markets as it is more appropriately used in describing the role of channel members (see Chapter 2 of this book) in the delivery of a product or service to the final consumer.
- *Influence markets.* The nature of the 'influence' market domain is such that it usually has the most diverse range of constituent groups. Illustrative of the wide range of constituent groups which comprise the influence market domain are: shareholders, financial analysts, stockbrokers, the business press and other media, user and consumer groups, environmentalists and unions. Each of these constituent groups has the potential to exert significant influence over the organization and the relationships an organization has with them can be managed through the application of a strategic marketing approach.

The distinction between these two market domains is that referrals (when they are successful by generating a sale) have a direct impact on customers purchasing goods or services from an organization. For example, an existing satisfied customer from a motor car dealership may recommend an acquaintance to this dealership. Here the referral market directly impacts on the customer market. On the other hand, influence markets have a more indirect impact on a diverse range of groups who are usually not customers. For example, a company may develop a marketing programme aimed at groups in the influence market domain, such as environmentalists or City analysts. Here the emphasis may be on avoiding negative actions on the part of the environmentalists or seeking to increase loyalty amongst investors in the company. Thus the influence market typically has greatest impact on groups who are not customers.

It should be restated here that within the Six Markets model one given group may be involved in a number of these markets. For example, in the case of customers they may play a role within the customer market where the interaction is between a firm and its customers and in the referral market where the interaction is between an existing customer and a prospective customer. Thus customers (and other groups) may have a number of differing roles within the six market domains.

The remainder of this chapter is divided as follows: in the first part, a brief introduction outlines the content of this section. In the second and third parts, the nature of the referral market domain and the influence market domain are discussed. As there has been little previous research in these two domains in the context of relationship marketing, the groups within them are examined in some detail. Following a brief summary of the chapter, an overview is then provided of the three case studies which are contained within this chapter of the book.

The referral market domain

As previously noted, there are two broad categories of referral market: existing customers and other non-customer referral sources. The first category, existing (and former) customers can have a potentially very strong impact on most organizations. The importance of the second category will depend upon the organization concerned. The role of the different constituent groups will vary not only between companies in different industry sectors but also within different business units, divisions or product/service areas of a single company. Using the principles of relationship marketing to systematically manage relationships with both these two broad groups can represent a major source of increased revenue and profits for an organization. The referral market domain is illustrated in Figure 4.1.

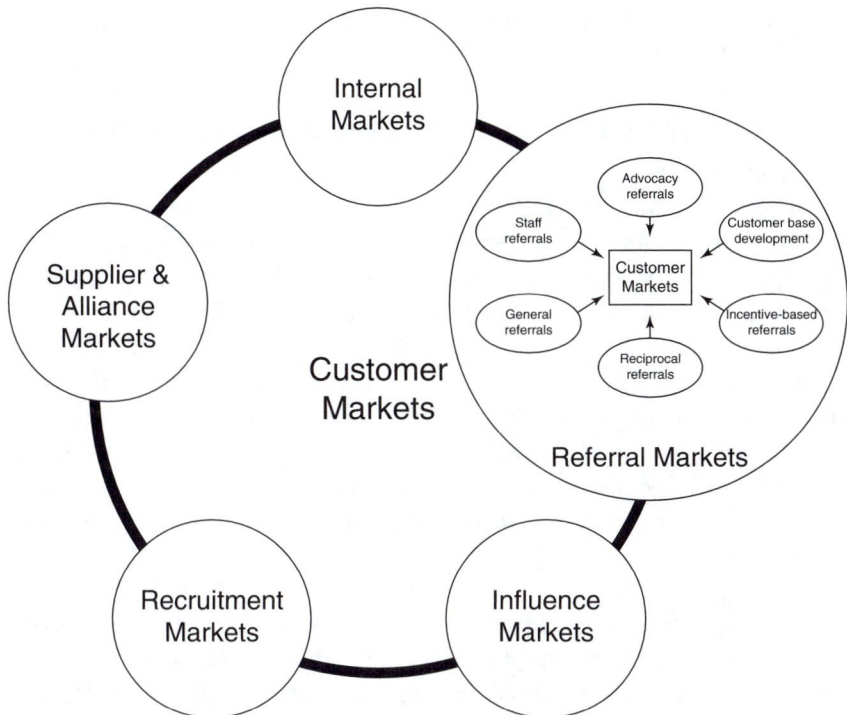

Figure 4.1　The referral market domain.
Source: Payne (1998).[1]

Existing and former customer referrals

Within existing and former customers we can distinguish between two subcategories: advocacy referrals (or advocate–initiated customer referrals) and customer-base development (or company-initiated customer referrals). We discussed the first subcategory in Chapter 2 of this book. We introduced the relationship marketing ladder of customer loyalty (see Figure 2.6) and briefly described the role of advocates in assisting a company's marketing efforts. An advocate, a special classification of customer, was defined as someone who actively recommends your organization to others, who does your marketing for you. An organization's existing or former customers can play a key role in acting as a referral source. The second subcategory of customer-base development relates to an organization's explicit attempts to use its existing customers, as part of its marketing activities, to gain new customers. Each of these will now be addressed.

Advocacy referrals (advocate–initiated customer referrals)

Customers become advocates when they are totally satisfied with a company's products or services. Research by Jones and Sasser[2] has found that, except in rare cases, total or complete customer satisfaction is the key element in securing customer loyalty and that there is a tremendous difference between the loyalty of merely satisfied and completely satisfied customers. They cite Xerox research which found that totally satisfied customers were six times more likely to repurchase Xerox products and services than its satisfied customers, over the next 18 months. In discussing measurement of loyalty, Jones and Sasser discuss a range of measures, including secondary behaviour:

> Customer referrals, endorsements, and spreading the word are extremely important forms of consumer behaviour for a company. In many product and service categories, word of mouth is one of the most important factors in acquiring new customers. Frequently, it is easier for a customer to respond honestly to a question about whether he or she would recommend the product or service to others than to a question about whether he or she intended to repurchase the product or service. Such indications of loyalty, obtained through customer surveys, are frequently ignored because they are soft measures of behaviour that are difficult to link to eventual purchasing behaviour.

Which companies have a high proportion of customers who make such referrals and endorsements and exhibit advocacy? Discussions with many consumers and marketing executives suggest the following as examples:

- Airlines: British Airways, Virgin Airlines
- Banking: First Direct
- Computers: Dell Computers
- Healthcare: Shouldice – a Canadian hospital
- Industrial services: Service Master – a US cleaning services company
- Motor cars: Lexus, Mercedes

- Retailing: Marks & Spencer, The Body Shop, Nordstrom
- Trucks: Scania, Mercedes
- Watches: Rolex

The area of advocacy and referral marketing is significantly under-emphasized from the perspective of both corporate practice as well as academic research. Few organizations have any formal processes which focus on seeking to maximize the use of referrals from existing customers. Whilst many organizations recognize that customers can be the most legitimate source of referrals to their prospective customers, there is a strong tendency by most companies to simply let referrals happen rather than to develop marketing activities seeking to leverage the power of advocacy.

Virgin Atlantic is an interesting example of this. In a recent executive seminar in the UK 120 managers were asked to indicate whether they were advocates of various commercial organizations. When asked about Virgin Atlantic, about 20 per cent claimed that they were advocates of the airline. This group was asked which of them had ever flown on a Virgin flight. Five of them even claimed to be advocates of Virgin despite *never* having flown with them. When asked about this, one of these members of the audience spoke at some length about how he recommended Virgin Airlines despite having never travelled with them. On a later occasion, one of the authors asked Richard Branson to what extent he had sought to leverage this advocacy. 'It just happens and we enjoy the benefits of it,' Branson said. He believes a more proactive approach can be taken to leverage the power of advocates and to attempt to get benefits beyond customers' ad hoc offers of advice and recommendations.

A relatively small number of organizations benefit greatly from proactive efforts in creating advocates of their customers and obtaining advocate-initiated customer referrals from them. First Direct, a UK bank, who became the world's first all-telephone bank, has been a great success story through customer referrals. By 1996, seven years after it commenced operations, it has over 640 000 customers and has been highly successful in attracting high net worth individuals. Despite less-than-successful TV advertising campaigns, this bank has grown – largely through referral from its very satisfied existing customers. Research conducted by First Direct shows that around 85 per cent of its customers have referred new customers to the bank, whilst the average for the other five major banks network is around 15 per cent.

Other organizations have been able to grow to a strong market position relying heavily on the process of referral from existing customers whilst undertaking no advertising. Shouldice Hospital in Canada, a highly focused hospital which specializes in hernia operations, undertakes little advertising and is a good illustration of the power of such advocacy. Other organizations such as The Body Shop (see Case Study 4.3) have achieved considerable success despite spending no money on advertising. It has achieved this through a combination of gaining referrals from advocates and by building relationships with the media and other groups within the influence market domain.

One of the best examples of advocacy referrals is Nordstrom, a US chain of superior quality fashion department stores. This chain is regarded by many as the best retailer in the USA. Despite its heavy geographic emphasis, in terms of store location, on the west coast of America, it has a strong following right across the USA. Many people living in overseas countries are also advocates of this organization and travel to the USA specifically to visit a Nordstrom store. Nordstrom has a high degree of advocacy amongst its customers and formal organizational processes aimed at leveraging referrals. It is also renowned for its superior recruitment and internal marketing (see Case Study 5.3) which leads to outstanding customer satisfaction and advocacy.

The practice of unconditional reimbursement for any merchandise, whether used or unused, without a Nordstrom sales receipt can be used to demonstrate how Nordstrom views referrals and how they think about investment in word-of-mouth referral marketing compared with advertising. Pascale[3] quotes an executive vice-president who explains:

> Nordstrom literally grinds up truckloads of shoes each year. These are returned shoes that have been worn and cannot be sent back to vendors. At face value, it seems nuts to have a policy like this. But if we run one full-page promotional ad in the Seattle newspaper, it costs us the equivalent of 500 pairs of shoes. And we don't know if the ad works. But give a customer a new pair of shoes with no hassle and it's a story that gets told and retold at parties and at the bridge table. Word-of-mouth endorsement really works. We know it.

Heskett, Sasser and Schlesinger[4] argue that such stories have become more important to Nordstrom than the relatively low advertising which the company undertakes: 'It does not take many of these "service encounters" to encourage the development of a relationship either with Nordstrom or with the individual sales person. Nordstrom does whatever it can to encourage these relationships.'

Customer base development (company-initiated customer referrals)

Advocacy referrals (the previous subcategory) involve the customers initiating the referral. In this second subcategory, the company undertakes a set of activities or a programme, aimed at their existing customers, leading to customer referrals. This may involve a simple request for referrals from customers or offering some form of inducement to the customer. Whilst advocates on the relationship marketing ladder, described in Chapter 2 of this book, will frequently be active in initiating referrals, 'supporters' – the next rung down the ladder – whilst positive to the organization, tend to be more passive. Asking for a referral can be a very good way of generating business from this group of supporters. For example, a study by File, Judd and Prince[5] of referrals by lawyers' clients found that only 49 per cent of the clients reported they were asked by law firms for a referral. Of those who were asked, 95 per cent provided at least one referral, whilst only 8 per cent of those who were not asked provided a referral. In most organizations we have had discussions with, few of them have any formal processes for requesting referrals.

Membership organizations represent one sector where considerable effort is made to use the membership base to reach further members. American Express, The Institute of Directors, wine clubs and many similar organizations have regular promotions aimed at generating new customers by using their existing customer base. Frequently these 'member get a member' marketing efforts are accompanied by some form of incentive, inducement or reward. The rewards may be offered at several different levels based on the number of new members recruited. They typically include reduced or free annual subscription, special privileges, Air Miles, gifts of champagne and bottles of wine, Mont Blanc pens, or entry in a special prize draw.

Referrals from customers are amongst the most relevant, effective and believable sources of information for customers. Kotler and

Armstrong[6] have argued that the most effective information sources for consumers are personal ones; commercial sources normally inform the buyer but personal sources legitimize or evaluate products for the customer. The legitimacy aspect of referrals makes the step of converting prospects into customers on the relationship marketing ladder of loyalty much easier.

Non-customer referrals (third party and staff referrals)

In addition to an organization's customers, many other parties have the potential to act as a referral source for a business. Referrals may be made informally when an individual's experiences of an organization and its general reputation cause them to recommend the use of that organization to others. In some circumstances a more formal system of referrals may be set up. For example, an accounting firm will develop formal (and informal) relationships with law firms, banks and other accounting firms with the aim of receiving referrals from these organizations. On other occasions referrals may be generated from an organization's current or former staff.

A consideration of the wide range of non-customer referrals suggests they can be divided into a number of groups. These are now examined in greater detail and include:

1 general referrals
2 reciprocal referrals
3 incentive-based referrals
4 staff referrals.

General referrals

A broad range of referrals that result in the generation of business for an organization can be termed general referrals. These can be further divided into four subgroups.

1 *Professional referrals.* This type of referral includes those where one particular professional may recommend the services of another. Often these are inter-industry, or at least within the same broad sector, such as health care. For example, a general medical practitioner may refer a patient to a specialist consultant; or a solicitor may refer a client to a barrister. In the case of the general practitioner and solicitor, the referrals are typically one-way. The specialist consultant and barrister usually

do not refer work back to the general practitioner or solicitor. Conflict of interest, where a professional cannot work for two competitors at the same time, may be a further reason for a professional referral.

Referrals made by a manufacturer to specific customer groups may be considered within this category; for example, a computer hardware or software manufacturer may refer a customer to a value-added reseller (VAR), who packages hardware, software, training and service for customers who need special help.

Where an opportunity exists to systematically refer work back to the referrer or to offer some form of inducement to them to refer further work these become reciprocal referrals or incentive referrals, which are discussed shortly.

2 *Expertise referrals.* Referrals may be sought by a customer because of the referrer's specialist expertise or knowledge. For example, a business school academic may be asked to recommend a management consulting firm, a market research firm or advertising agency to a company; or a technical expert may recommend the use of a particular testing laboratory. Often such referrals are made on an irregular or ad hoc basis.

In such cases, professionals from a range of different industry sectors receive acknowledgement of their professional competence and knowledge of a market in particular or business in general by making suitable referrals. As a result, the person making the referral receives thanks and on-going goodwill from their customer or contact. Further, their professional status and their reputation as an informed and knowledgeable professional may be enhanced. It is in every professional's best interest to be willing to help and offer advice, either solicited or unsolicited, to their clients.

3 *Specification referrals.* These referrals cover the situation where an organization or person specifies or strongly recommends the use of a particular product or service. For example, architects building a number of superior homes may mandate, within their specification, that Miele electrical appliances such as dishwashers and ovens are used within every kitchen. A further example would be where a real estate agent insists that a particular removal firm undertakes the transportation of his or her client's household contents.

4 *Substitute and complementary referrals.* These groups have been identified by Herriott[7] and could be viewed as specific cases of professional referrals. Substitute referrals occur in circumstances where organizations which are at over-capacity, have long lead times to undertake work, or cannot fulfill a specific need may refer a customer to one of its competitors. Such referrals may not be made voluntarily. Another example

would be a referral from a regional bus line who does not travel on a particular route, to another regional carrier who does. Complementary referrals occur in circumstances such as when a taxi driver is asked for a referral to a local hotel or a 'bring your own' restaurant recommends a nearby wine merchant.

Reciprocal referrals

Historically, the existence of professional ethics in many of the professions, such as law and accounting, have precluded advertising and aggressive competition. Up until the 1980s most of the professions were highly restricted by their professional bodies with respect to their marketing activities and had to rely extensively upon referrals from third parties. As a consequence, referrals have often constituted the principal source of work for professional service firms. Thus the professions represent an interesting sector to consider from the perspective of referral markets.

Some referrals, especially those between professional firms, are mutually dependent and referrals may be made backwards and forwards between firms. Under these circumstances a referral system exists, where referrals are made between one organization in one profession such as an accounting firm and other organizations such as law firms and banks, and vice versa.

The nature of a referral system varies considerably across different industries with respect to the members of groups, their relative importance and the degree of potential symmetry between referrals given and referrals received. These are some of the referral system members for a law firm and a non-retail bank:

- **A law firm referral system:**
  - accounting firms
  - banks and building societies
  - social security agencies
  - consulting firms
  - other law firms.
- **A non-retail bank referral system:**
  - insurance companies
  - property brokers
  - accountancy firms
  - surveyors/valuers
  - other banks.

Where members of a referral system are mutually dependent on each other for referrals, there is usually strong potential for reciprocation between them. Where the potential for a high degree of mutual dependency exists, this can be managed by the use of a 'balance sheet' in which referrals received and referrals made are documented. This serves two purposes. First, it enables the firm to monitor how well it is doing in encouraging its referrals sources to generate referrals for it. Second, it enables the firm to determine how proactive it is in generating referrals back to its referrers.

It also provides a management tool to help ensure that mutuality is preserved. If the number of referrals given compared with the number of referrals received becomes significantly unbalanced, it provides an opportunity to rectify this. For example, an accounting firm could go to its bank and say: 'Did you know we referred six clients to your M&A section, four clients elsewhere in your corporate finance division and nine clients to your business banking division? In the same period we received only three referrals back from you. Two of these were not relevant to our expertise and type of business. Could we sit down and talk about how we can work together more closely in generating mutually beneficial referrals.'

Incentive-based referrals

There are a number of circumstances under which incentive-based referrals are appropriate. First, where members of the referral channel are mutually dependent they may find some advantage in creating a formal arrangement that helps reinforce this dependence. In certain circumstances financial incentives may be useful to augment the goodwill and professional courtesy that act as the source of referrals, especially where the referee seeks to gain exclusive referrals or preferential referrals from the referrer. Second, where the referral system is one such that the firm is receiving many referrals but giving back relatively few and there is not the potential to change this, it may seek to address this imbalance by providing some incentive-based method of compensating its source of referrals.

Incentive-based referral systems may be financially or non-financially based. Herriott[8] points to North American examples where financial incentives are either banned or are part of normal business marketing: estate agents are prohibited from taking finders' fees from banks and doctors are prohibited from taking a direct financial pay-back from hospitals; on the other hand, accountants may receive commission on business they refer to others and car dealers are permitted to take commissions from loans they send to banks.

The introduction of financial incentives raises the question as to whether someone who is financially compensated for referring business is a referral source or an intermediary (and thus a member of the customer market domain). It is suggested the criterion that should be used in deciding this is the primary motive of the person or organization who is receiving an incentive for its referrals. If its primary motive is driven by professional needs to help its clients and to develop a good relationship with referral sources, and it receives a financial incentive, it is a referral. On the other hand, if the company receives a significant fee for referring the work and views this as an important part of revenue generation for its business, it is better characterized as an agent or intermediary within the customer market domain. Here we are concerned only with the referrals that fall within the first category.

Incentive-based referrals do not necessarily involve direct financial remuneration by way of a commission or finder's fee. In some cases a written letter of thanks, a dozen bottles of wine, or another attractive gift may be used. Invitations to major sporting events, the opera or ballet are typical ways in which financial service institutions and professional service firms acknowledge the assistance of referral sources. One company organized a high level management development event with several leading business school professors and invited the directors from the business referral sources to attend this event free of charge.

The potential for use of incentives varies considerably across industries. In a referral system which incorporates financial incentives, an important issue is to ensure that it is managed in an ethical manner. In some industry sectors incentives are considered unethical or may even be prohibited under the rules of a professional or regulatory body. As a consequence, ethical considerations, regulatory practices, commonly accepted norms within the business sector and commonsense should all be taken into account before a decision is made regarding the use of incentives.

Staff referrals (from existing and former staff)

Internal staff referrals represent an important source of referral within a number of industry sectors. The most common examples of these are probably within professional services and financial services but there are also many other examples of organizations which have a number of different divisions or products aimed at similar customer segments where referrals can be generated between them.

For example, many organizations within the banking and insurance sectors have a relatively low level of product cross-sales. One large insurance company has an average of 1.2 product sales per customer. This company is hierarchically structured and the products are sold within different divisions. These divisions hold customer information on different databases, using different software and hardware and different legacy computer systems. There has been no attempt to create a data warehouse so that information on customers can be readily shared. Creation of a data warehouse provides an opportunity to develop a platform of customer-based information systems rather than product-based information systems. By having customer-based information residing in a data warehouse, there is often a great opportunity to initiate internal referrals of existing customers within one division who have the potential to be sold products or services by another division.

Within professional service firms, the opportunity for internal referrals is enormous. Whilst some organizations are starting to develop reasonably sophisticated internal referral systems, others have a long way to go. One of the problems is that separate practice areas such as audit, taxation and consulting in an accounting firm tend to operate within 'silos' and often there is not a cross-functional or cross-departmental view of the customer. In many accounting firms the audit departments carefully guard their major audit customers. They may be suspicious about the quality of work done by their management consulting practice area and may be concerned that a referral of business to their consulting colleagues will result in the basic audit relationship being put under threat if the assignment does not work out.

A final group of referrals are those made by former staff members who have left the organization. Former staff members may be a useful source of business referral for certain sorts of organizations. Again, professional service firms provide a good example of where their former staff are used to generate referred business. When the staff of major management consulting firms such as McKinsey & Co and Andersen Consulting and accounting firms such as Price Waterhouse and KPMG leave these firms, they often take up senior appointments in industry. Firms such as McKinsey & Co and some of the 'big six' chartered accounting firms place considerable emphasis on these 'alumni' and run a number of regular activities to involve their staff members on an on-going basis with their former firm. Although this approach is typically used by the professions, other suitable organizations may be able to benefit from utilizing their former staff to generate referrals.

Developing relationships with referral markets

Most organizations fail to exploit the opportunity of maximizing referrals from their own customers, from third parties, and, where appropriate, from their own staff. With respect to its customers, many organizations still do not realize the power of customer delight and the benefits that accrue from significantly exceeding customer expectations. However, there is a small, but increasing, number of companies which do significantly exceed customer expectations. These organizations have been able to grow through word-of-mouth referrals from highly satisfied customers (and other groups) and include The Body Shop, Shouldice Hospital, Nordstrom and First Direct. In the case of the first two they have achieved a strong market position without advertising and the latter two have achieved their market position with either relatively low levels of advertising (Nordstrom) or television advertising that has not been considered particularly effective (First Direct). Similarly, referrals by third parties have the potential to play a major role in many industries, especially in professional and financial services. In the context of professional firms Wilson[9] has pointed out the benefits, which include extremely low cost and high yield in acquiring customers:

> There can be few business activities wherein this overwhelming value of business is acquired through the effectiveness of a single marketing tool and yet so little knowledge exists of that tool or so few attempts have been made to establish whether, despite its undoubted success, it is nevertheless capable of further exploitation.

Our work in this area suggests a number of steps need to be followed to ensure successful implementation of referral marketing. They include:

- Review sources of past business.
- Undertake a relationship audit with all major referral categories, identify-

ing current and potential referred work from each major referral group.

- Use Pareto analysis to identify the 20 per cent of referral sources with the potential to generate 80 per cent of the business.
- For the most important referral groups, identify the potential of individual organizations (or individuals within them).
- Complete a relationship marketing network diagram (spidergram) to identify current and desired emphasis.
- For those organizations where there is a mutual dependency, create a referral balance sheet.
- For those organizations where networks are heavily asymmetrical, or one-way, explore the use of incentives to develop and sustain that referral network.
- Develop a marketing plan aimed at referral markets.
- Develop an educational programme to educate all staff members in the objectives of their plan and their role in it (internal marketing).
- Ensure a regular feedback and monitoring system of an appropriate nature.
- Create a high visibility mechanism for publicizing success.

The influence market domain

It was noted in the introduction to this chapter that the influence market domain has the broadest and most diverse range of constituent groups. Further, influence markets vary considerably according to the industry sector being addressed. For example, companies involved in selling infrastructure services such as telecommunications or utilities will place governments and regulatory bodies high on their list of important constituents within their influence market domain. Highly visible public listed companies may place much of their attention on shareholders, financial analysts and the financial and business press. Manufacturing companies and the petrochemical sector may be especially concerned with environmentalists and government. Figure 4.2 illustrates several of the major groups within the influence market domain.

The relative importance of different groups within the influence market domain will also vary at different points in time. For example, a bank faced with fraud or illegal dealing may suddenly have the press, regulatory bodies and the central bank brought to the top of their influence market agenda. Recent examples of this

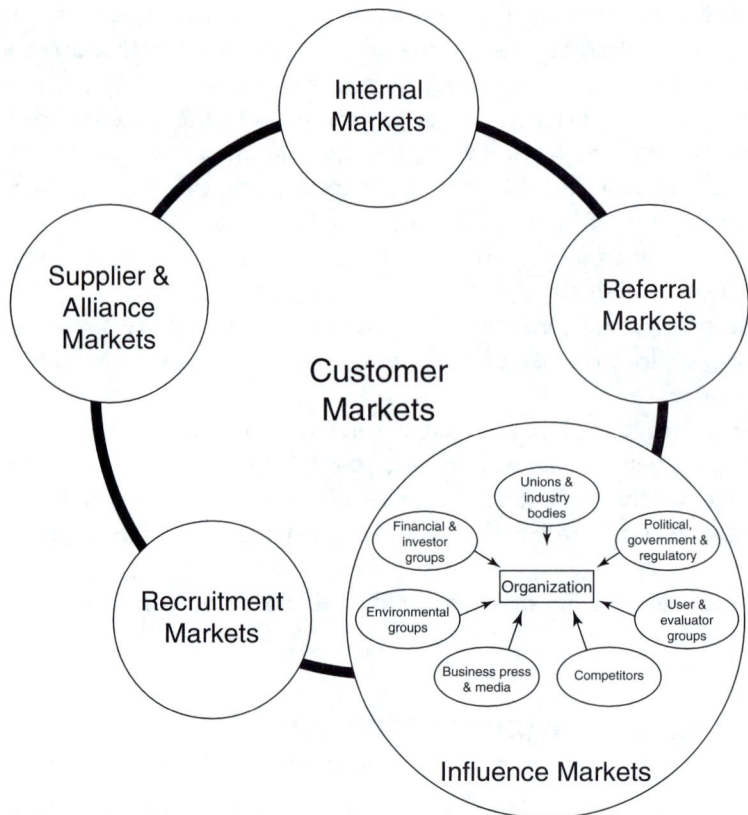

Figure 4.2   The influence market domain.
Source: Payne (1998).[1]

include the situation faced by the Hong Kong & Shanghai Bank following the Nick Leeson affair in Singapore and the 'scam' Standard
Chartered Bank had to address in India in the mid-1990s. Similarly,
actions by Greenpeace and other environmentalists with respect to
Shell's Brent Spar platform brought environmental issues to the top
of their agenda.

Some comment should be made on the distinction between the
constituent groups within the influence market and the terms
'stakeholders', used in strategy and corporate affairs, and 'publics',
used in public relations. Over the last 15 years stakeholder theory[10]
has developed as an important area. Also, within the public relations literature, it is common to focus on a wide range of 'publics'
which may include investor relations, community relations and
media relations.[11]

The principal distinction between the groups in the influence market domain and the concept of stakeholders or publics is in the marked difference in relational emphasis that is placed on them. Addressing these groups as markets involves the rigorous use of marketing techniques such as segmentation, positioning and the development of marketing strategies and plans to achieve specific strategic market objectives. We view this market-based approach as fundamentally different in the manner in which relationships are managed. This can be contrasted with the concept of influencing or communicating with these groups, as emphasized in public relations. Further, the concept of stakeholders implies a 'stake' in the business. However, some constituent groups within the influence market domain may have no formal 'stake' in the organization itself but nevertheless may have an important influence on it. We also consider that the analytical tools and disciplines of relationship marketing can overcome some of the difficulties (raised by authors such as Argenti[12]) of determining trade-offs between different stakeholders.

Campbell[13] has provided some compelling arguments for the need to consider the constituent groups in what we term influence markets. He argues that companies compete for 'space' in an 'economic jungle' and that to survive in this jungle a company must win the loyalty of important groups and that it must maintain loyalty from these groups; without this loyalty the company's space will gradually erode until it disappears. He further points out that this approach helps managers think of the full range of communities from whom they need loyalty, as well as where other major constituent groups should be considered.

Major categories within the influence market domain

Who then are the constituent groups an organization should consider? There is considerable debate regarding this in the stakeholder literature. Argenti[14] suggests an infinite number of potential groups. On the other hand, Freeman[15] points to excessive breadth in identification of stakeholders.

The number of potential groups that may constitute the influence market domain is considerable. However, the important ones for a firm to address at a given point in time may be relatively few. Figure 4.3 outlines many of the key groups within the influence

FINANCIAL & INVESTOR GROUPS
- broker analysts
- institutional analysts
- portfolio managers
- individual investors
- institutional investors
- credit rating agencies
- the Stock Exchange

UNIONS
- union groups
- unofficial groups

INDUSTRY BODIES
- Chamber of Commerce
- Bankers' Association
- other trade bodies

REGULATORY BODIES
- central bank (e.g. Bank of England)
- securities regulators
- environmental agencies
- health and safety agencies
- ombudsman

BUSINESS PRESS & MEDIA
- trade press
- national press
- business press
- consumer media & evaluator

USER & EVALUATOR GROUPS
- customer-initiated user groups
- company-organized user groups
- industry surveys
- testing authorities

ENVIRONMENTAL GROUPS
- environmentalists
- consumer groups

POLITICAL & GOVERNMENT
- central government
- local authorities
- specific government department, e.g.
  - Department of Trade & Industry
  - Federal Drug Administration
- MPs
- lobbyists

COMPETITORS
- existing competitors
- potential competitors

Figure 4.3 Constituent groups in the influence market domain.

market domain. Most of these can be conveniently grouped as categories or under subheadings, as shown in this figure. This list has been developed from experiences in analysing the influence markets of over 20 organizations. It represents many, but by no means all, of the possible groups; in fact, each time we examine several further organizations, one or two additional groups are often identified. In the future we will develop more detailed listings of groups and create more developed classifications of them.

We have suggested above that the influence market domain is a particularly diverse one. However, several of the categories shown in Figure 4.3 are of special interest because they are common to so many organizations. These include:

1 financial and investor influence markets
2 environmental influence markets
3 competitor influence markets
4 political and regulatory influence markets.

Financial and investor influence markets

Gaining the support and loyalty of a wide range of financial markets is an issue that has to be addressed by many organizations. Financial and investor influence markets are especially crucial for organizations who are listed on a stock market or who plan to achieve a listing. However, they are also important to other organizations such as 'mutuals' (organizations owned by subscribers, depositors or customers, which include non-listed insurance companies and building societies).

For example, a credit rating agency recently decided to credit rate certain mutuals, including non-listed insurance companies. One mutual chose not to go through the formal credit rating process. The rating agency subsequently gave them a provisional rating which was arguably much lower than was justified. This impacted negatively on the organization in a number of ways, not the least of which was to make the cost of their capital higher. Relationships between this mutual and a number of its influence markets were adversely affected by this.

Another further illustration will serve to illustrate the importance of managing financial influence markets. In a large British manufacturer the chairman and deputy chairman of the company developed a coordinated marketing programme aimed at all the important members of the financial influence market. It created an integrated marketing programme directed at improving its market perception and share price. The chairman and deputy chairman spent a high proportion of their time making presentations to key groups, including stockbrokers, bankers, investment associations and financial journalists. They marketed their vision for the strategy and future of the company, which included plans for a series of ambitious acquisitions. The company's acquisition plans were very well thought through and details of them, in broad terms, were made available to these groups as part of the presentations. After the period of a year of such activity, the financial community became very interested in this particular company. There was favourable commentary in stockbrokers' circulars and in the financial and business press. As a result, share prices improved by over 20 per cent and it is considered that this was a direct result of the marketing activity aimed at these financial markets. The company subsequently acquired a number of companies. Although some criticism was made of certain of the acquisitions, in terms of them paying too high a premium for some of the companies that were acquired, they were purchased mainly through share placement. Because of the increased share price the total cost to the company was considerably less than it might have been. In effect, through management of their financial influence markets, the company ended up successfully acquiring these companies at a reasonable price.

The financial and investor influence market, because of its overall importance, is one where academic attention is now being directed in applying some of the principles and approaches of relationship marketing that have been used to address the customer market domain. Tuominen[16] has focused on several key markets within this market, including investors, stockbrokers, financial advisers and analysts. She has characterized this as investor relationship marketing and defines it as 'the continuous, planned, purposeful, and sustained management activity which identifies, establishes, maintains and enhances mutually beneficial long term relationships between the companies and their current and potential investors, and the investment experts serving them'. Tuominen uses the relationship marketing ladder of customer loyalty (discussed in Chapter 2) to develop a ladder of investor loyalty which illustrates five levels of investor relationships, including advocacy investors, supporting investors, regular investors, new investors and potential investors. She then uses an interactive approach to examine the short-term investor episodes which form the basis of the long-term investor relations which lead to the creation of relational bonds.

A key issue of importance within the financial and investor influence market is the loyalty of investors. The importance of this group has been highlighted in recent work by Reichheld[17] which shows that investor churn in the average public company in the United States is more than 50 per cent per annum. He concludes that managers find it nearly impossible to pursue long-term value-creating strategies without the support of loyal, knowledgeable investors. He further points out that many of the world's leading companies, in terms of customer loyalty and retention, are privately owned, and are mutuals or are public companies where there is high investor loyalty. This is illustrative of research starting to focus on financial influence markets as they are increasingly recognized as an important part of relationship marketing.

Environmental influence markets

Environmental influence markets represent a key group for many organizations involved in industries such as petrochemicals, mining and manufacturing. Many environmental bodies and pressure groups are now becoming highly active and sometimes militant in their behaviour and have the potential to make a very significant negative impact on organizations whom they target as being environmentally unfriendly. Environmentalists not only focus on organizations whom they view as polluting the environment but also those who abuse scarce resources, hunt or use animals in testing procedures, or manufacture products where satisfactory natural substitutes could be used instead. For example, the impact of animal activists has led a number of leading department stores around the world to stop retailing products made of fur. As environmental influence markets become increasingly vocal, organizations need to develop relational strategies for dealing with these groups.

A further example relates to the events which followed Shell's decision to dispose of its Brent Spar platform in the North Sea. Activities which included the occupation of the Brent Spar platform by Greenpeace, highly militant action by activists in countries like Germany where Shell petrol stations were fire-bombed, and other events, including probable behind-the-scenes involvement by the British Government, led Shell to reverse its decision regarding the disposal of the Brent Spar platform. It is clear, from a review of the events surrounding the Brent Spar affair, that Shell failed to develop appropriate relational strategies with these key environmental groups and failed to communicate effectively with them.

By contrast, The Body Shop (see Case Study 4.3) is an excellent example of an organization which has managed its relationships with environmentalists and other influence market groups very well. For example, it has formed various alliances with Greenpeace and Friends of the Earth and developed very close relationships with other influence market groups, such as local communities, by ensuring that every one of their retail shops develops local community projects.

Competitor influence markets

Large organizations, especially those which have high visibility or are dominant within their industry sector, need to consider carefully the relationships they have with their competitors. An appropriate response for such firms is often to adopt a profile of an industry statesman.

The competition between British Airways and Virgin Atlantic in the mid-1990s is an example of how BA's industry statesmanship was eroded for a period of time. British Airways received considerable negative publicity over the so-called 'dirty tricks campaign' and as a consequence Virgin Atlantic's position in the marketplace was enhanced. Virgin received much favourable free publicity and its favourability and familiarity amongst the public at large was significantly increased.

Further examples of inappropriate relationship management between competitors are illustrated by price wars. Intense rivalry and price wars between firms may result in the erosion of profitability within industry sectors which are often potentially very profitable.

Today, the nature of complex alliances and supply chains may result in an organization viewing one particular company in a number of roles, including that of a competitor, a customer and a supplier. Under these circumstances it is essential that these different forms of relationships are understood and managed appropriately.

Political and regulatory influence markets

The political category within the influence market domain includes a number of groups, including members of parliament, government ministers, central and local government departments and other government and semi-government bodies. These may impact on organizations within a given country, within an economic region such as within the European Community or on a global basis.

Marketing activity may need to be directed at government and regulatory bodies. Gummesson[18] has pointed out how this is of particular relevance for companies who sell equipment of an infrastructural character, such as nuclear reactors, telephone systems and defence products. These are products which may impact on the country's economic performance, employment levels or financial status or may be important from a political point of view.

An important area for a wide range of manufacturing companies is to seek to ensure that the standards adopted by a regulatory body, for example a standards setting authority within the European Community, favour their production capabilities and competencies as well as being ones which are consistent with the materials they use in the manufacture of their products. For example, certain German companies have the reputation of being highly skilled at influencing standards setting bodies within the European Community to favour their businesses. This might involve persuading the standards bodies to adopt a specification for a particular type of fire-retardant or to insist on the use of a specific material. Further examples of this approach are provided by Stigler[19] and Kangun and Polonsky.[20]

Companies involved in sensitive areas such as defence or drilling programmes in the North Sea may recognize the importance of these influence groups, but may not have formulated detailed and coherent relationship marketing strategies and plans to gain maximum advantage from managing these relationships. As discussed earlier, a marketing approach that is based on a closely defined set of specific objectives with a detailed market plan and an appropriate monitoring system which measures results is likely to improve the chance of success in addressing a programme aimed at a specific influence market.

Strategic credibility in influence markets

The dangers of ignoring the constituents of the influence market domain are illustrated by Fisons, a UK mini-conglomerate which is discussed in Case Study 4.2. Fisons failed to build appropriate relationships with a number of key members of the influence market domain, including shareholders, the financial press, environmentalists, competitors and the US Federal Drug Administration. Failure to address these key influence markets damaged the firm significantly, as the case study illustrates.

A further example of poor management of influence markets is provided by Ratners, a large jewellery retailer which is the subject of Case Study 4.1. In 1992, Gerald Ratner made a speech at the Institute of Directors in which he described his jewellery products as 'total crap'. This story was picked up in the general press and reported widely. A failure to manage strategic credibility before and after this event, together with an inappropriate and incomplete recovery programme, led to a massive fall in the fortunes of the Ratner Group.

Both these examples illustrate a failure to manage strategic credibility effectively, especially with respect to the influence market domain. Work by Higgins and Diffenbach[21] has identified strategic credibility as a critical issue impacting on the influence market domain. They see strategic credibility as the result of the interplay of four key factors: a firm's strategic capability, past corporate performance, communication of corporate strategy to key stakeholders, and the credibility of the firm's top management team – especially the firm's chief executive officer. They argue that management of these four key areas enables a firm to improve its strategic credibility. This strategic credibility is especially important as adverse events occur from time to time. The relationships that an organization builds over time, together with the strategic credibility it establishes in the marketplace, will have a very significant impact on how an adverse situation is perceived by both influence and customer markets. In particular, the business press and the general media may respond very differently, depending on whether the organization has high or low strategic credibility.

The response of Johnson & Johnson following the Tylenol disaster is a good example of how an organization established strategic credibility amongst key members of the influence market domain (as well as with other of the six markets). When someone tampered with the contents of Tylenol in drug stores in the USA, leading to the death of several people, the manufacturer, Johnson & Johnson, was able to benefit from the strategic credibility which it had already established with influence markets before the event. This, together with the adoption of appropriate responses following the event, caused observers to conclude that Johnson & Johnson dealt extremely well with their influence markets and by doing so the company successfully avoided a potentially disastrous outcome for their business.

Developing relationships with influence markets

The previous discussion has outlined many of the potential groups in the influence market domain who may be important to an organization. Identification of relevant groups within the influence market can usually be done fairly quickly by holding a series of brief meetings with senior executives. The key steps are:

- Identify the relevant groups within the influence market.
- Break down these groups into appropriate subgroups.
- Determine their relative importance and potential to have a positive or negative impact on the organization. (Some of the frameworks used in stakeholder theory[22] can be used to assist in this process. For example, constituent groups in the influence market domain can be categorized as ones of primary and secondary importance. They can also be categorized in matrices based on their likely positive or negative impact upon the organization.)
- A relationship marketing network diagram (spidergram) can then be used to identify the present and desired emphasis that should be placed on each constituent group. The level of detail in and structure of the spidergram will depend on the number of constituent groups within these market domains. If an organization wishes to address a large number of markets it can first construct a spidergram based on the broad categories described above. A second level of analysis can then be undertaken for each broad category. For example, the financial market category can then be broken down into constituent groups, such as City analysts, institutional investors, etc. and then a further spidergram can be used to consider them.
- Develop a written market audit for each group. (Usually this can be done in a maximum of two pages. The structure of a situation review can follow closely that used in the marketing planning process.)
- The other steps in the marketing planning process,[23] with appropriate modification, can then be used to focus on achieving the specific market objectives that are set for each group within the influence market domain.
- An appropriate measurement and monitoring system is put in place.

The use of a rigorous marketing planning process has the potential to develop a relational approach with the groups within the influence market domain that is superior and different in emphasis and impact when compared with the tasks of communicating and influencing suggested by much of the public relations literature.

Summary

This chapter has examined two key market domains – referral markets and influence markets. We have described how both these market domains can impact significantly on the organization.

Referral markets are a critical means of generating referrals of business and increasing customer revenues. Further, customer referrals represent one of the most legitimate sources of information – one which is frequently viewed as being more relevant than information provided by either the companies or their employees. Influence markets are a diverse group that needs to be managed in order to maximize their positive value or minimize the negative impact they can have on an organization's activities. The approach that needs to be followed for both domains is to first identify all the constituent groups within each market domain and then develop appropriate marketing plans for them.

In the next chapter of this book we examine the internal market and recruitment market domains. These domains are concerned with acquiring the best quality people into the organization through the recruitment process and maximizing their productivity after they join the organization. These market domains play a major role in the delivery of customer value and help ensure that the marketing activities directed at the other market domains continually focus on sustained delivery of customer value.

References

1   Payne, A.F. (1998). *Relationship marketing – the six markets framework: A review and extension*, Draft Working Paper, Cranfield School of Management, Cranfield University.
2   Jones, T.O. and Sasser, W.E. (1995). Why satisfied customers defect. *Harvard Business Review*, November–December, 88–89.
3   Pascale, R. (1993). *Nordstrom*. (This case study is reproduced in Chapter 5 of this book.)
4   Heskett, J.L., Sasser Jr, W.E. and Schlesinger, L.A. (1997). *The Service Profit Chain*, The Free Press, New York.
5   File, H.M., Judd, B.B. and Prince, C.A. (1992). Interactive marketing: The influence of participation on positive word-of-mouth and referrals. *Journal of Services Marketing*, **6**, No. 4, 5–14.

6 Kotler, P. and Armstrong, G. (1993). *Principles of Marketing*, Prentice Hall, Englewood Cliffs.

7 Herriott, S.R. (1992). Identifying and developing referral channels. *Management Decision*, **30**, No. 1, 4–9.

8 *op. cit.*

9 Wilson, A. (1994). *Emancipating the professions*, John Wiley & Sons, Chichester.

10 Freeman, R.E. (1984). *Strategic Management: A stakeholder approach*, Pitman/Ballinger, Boston.

11 Center, A.H. and Jackson, P. (1990). *Public Relations Practices*, Prentice Hall, Englewood Cliffs.

12 Argenti, J. (1997). Stakeholders: The case against. *Long Range Planning*, **30**, No. 3, 442–445.

13 Campbell, A. (1997). Stakeholders: The case in favour. *Long Range Planning*, **30**, No. 3, 446–449.

14 Argenti, *op. cit.*

15 Freeman, *op. cit.*

16 Tuominen, K.M. (1990). *Investor relationship marketing – a theoretical framework and empirical evidence*, paper presented at 26th EMAC Conference, Warwick Business School, May.

17 Reichheld, F.F. (1990). *The Loyalty Effect*, Harvard Business School Press, Boston.

18 Gummesson, E. (1987). The new marketing – developing long term interactive relationships. *Long Range Planning*, **20**, No. 4, 10–20.

19 Stigler, G.J. (1971). The economics of regulation. *Bell Journal of Economics and Management Science*, **2**, 3–21.

20 Kangun, N. and Polonsky, M.J. (1995). Regulation of environmental marketing claims: A comparative perspective. *International Journal of Advertising*, **14**, No. 1, 1–24.

21 Higgins, R.B. and Diffenbach, J. (1989). Strategic credibility – the basis of a strong share price. *Long Range Planning*, **22**, No. 6, 10–18.

22 Freeman, *op. cit.*

23 McDonald, M.H.B. (1995). *Marketing Plans: How to prepare them, how to use them* (3rd edn), Butterworth-Heinemann, Oxford.

Chapter 4 case studies

The preceding discussion in this chapter has described a number of different forms of referral markets and influence markets. Three case studies are included in this section which focus on a number of aspects relating to these two market domains.

Case 4.1: Ratners: A case of corporate reputation

This case is written by Neil Botton, a Principal Lecturer in Strategic Management, University of Westminster, and Dr Chris Carr, a Senior Lecturer in Strategic Management at the Manchester Business School, University of Manchester.

Abstract

During the 1980s the Ratner Group grew to become one of the world's leading jewellers. It developed a highly distinctive, volume-oriented approach to the marketing of jewellery and embarked on an ambitious growth programme through acquisition. In 1991 Gerald Ratner, Chairman of Ratner, achieved notoriety by having part of the content of a speech he made at the Institute of Directors printed on the front page of the *Sun* newspaper, in which he described one of his products as 'total crap'. He was reported further as saying that his products 'have very little to do with quality' and that one of his products – an imitation book – 'was in the worst possible taste'. This led to a downturn in the Ratner Group's fortunes. The case traces the misfortunes of the Group, leading to the resignation of Gerald Ratner and the steps taken by the new Chairman, James McAdam, in attempting to turn around the company in the face of hostile shareholders and former customers.

Learning points

This case study documents one of the most visible mistakes a company chairman has made in giving a speech in front of an external public audience in recent years. By describing his product as 'total crap' he alienated many constituents within the referral and influence market domains. This case covers strategy and marketing issues as well as relationship marketing issues.

Broad marketing and strategic issues:

- dealing with structural changes in an industry
- managing rapid growth and acquisitions
- repositioning a major retail business.

Relationship marketing issues:

- recovery after offending customers
- managing referral markets
- developing responses to hostile influence markets, including:
  - the press
  - the financial community
  - shareholders.

The coverage of Gerald Ratner's negative comments about his own company were widely reported on all the media. Inevitably this created negative word-of-mouth amongst loyal customers. Influence markets also had a strong negative impact on the business. Participants in a case study discussion can engage in a good debate on how important negative customer word-of-mouth and upset influence market members were in causing a decline in profitability and share price.

Case 4.2: Fisons: The fall from grace

This case is written by Helen Peck, a Research Fellow at Cranfield School of Management, Cranfield University.

Abstract

This case study explores the problems experienced by Fisons Plc in the early 1990s. Fisons, a mini-conglomerate, operating in a range of sectors including agrochemicals, pharmaceuticals, scientific equipment and horticulture, was once regarded as a superbly managed company.

The case commences with the problems encountered by Fisons in the early 1990s in its horticultural business and in particular with its behaviour with respect to large peat deposits. The case documents the formation of the 'Peatlands Campaign', which involved 10 highly influential conservation groups and was led by the Prince of Wales, who called on gardeners to boycott some of Fisons' products. As the campaign mounted against Fisons, 34 local authorities publicly supported a boycott of peat products. Further, B&Q, Britain's largest DIY and gardening chain, announced that, on conservation grounds, it had decided to ban all peat cut from Sites of Special Scientific Interest. Further problems then occurred within

Fisons' pharmaceuticals division, including delays for approval of drugs from the US Federal Drug Administration. By 1992 institutions were questioning the board's failure to manage effectively both its business and its investor relations and share prices plummeted.

Learning points

This case study charts the rise and fall of Fisons Plc. It provides a good insight into how a company failed to develop appropriate relationships with a wide range of influencer markets. These include shareholders, the business press, the financial press, the popular press, City analysts, environmentalists, local government and central government. It also failed to manage its internal affairs in terms of product quality with respect to new product development. As a consequence, it disenfranchized the Federal Drug Administration – a critical influence market who had to approve Fisons' drugs.

Broad marketing and strategic issues:

- managing a diverse range of businesses in a 'mini-conglomerate'
- adding value to commodity and generic products
- marketing products under limited patent protection
- the role of governance and external stakeholders
- failure to develop appropriate quality standards.

Relationship marketing issues:

- managing influence markets, including:
  - environmentalists
  - the financial community
  - the media at large
  - shareholders
- failing to develop a relationship marketing approach
- managing customer markets.

Fisons is a unique case in that it shows how *not* to manage relationships with a myriad of markets. The case focuses primarily on influence markets. However, its failure to manage relationships with key customers, such as B&Q, is also explored. Although not addressed specifically in the case, participants involved in a case study discussion may wish to discuss their perceptions of the company's internal marketing activities.

Case 4.3: The Body Shop International: The most honest cosmetic company in the world

This case is written by Andrew Campbell, one of the Directors of the Ashridge Strategic Management Centre in London.

Abstract

The Body Shop International originates, produces and sells naturally based skin and hair products and related items through its own shops and through franchised outlets. It opened its first shop in Brighton in 1976 and, despite a recession in retailing, the company has grown with great speed. This case study examines The Body Shop's development from the late 1970s to the early 1990s, by which stage it had a turnover of over £85 million and 457 shops.

The Body Shop is an unusual company which relies very much on referrals from its satisfied customers and which undertakes no advertising. It has been highly successful in developing relationships with a wide range of influence markets, including the press, local communities and environmentalists. It has also benefited from referrals from its existing customers.

Learning points

Unlike the two previous case studies in this section, which show how two companies failed to manage their relationships with referral and influence markets and suffered dramatically as a result, The Body Shop International case study is an excellent example of how to manage relationships with diverse groups of constituents in the influence market domain. There are many learning points in this case study which relate to strategic and marketing issues as well as relationship marketing issues. These include:

Broad marketing and strategic issues:

- the development of a sense of mission
- the illustration of ideological strategy
- the importance of the characteristics of the chief executive.

Relationship marketing issues:

- personality based marketing
- the ability to get extensive media coverage through developing relationships with a wide range of media, within the influence market domain, without having to use costly formal advertising
- the development of relationships with local communities by having each shop involved in a local community project
- the development of relationships with a range of environmentalists, including Greenpeace and Friends of the Earth.

The case may also provoke some discussion about The Body Shop's ability to interact successfully with the City, especially with the investment community. There are also obvious linkages with other of the relationship markets, including recruitment and internal markets.

Case 4.1   Ratners: A case of corporate reputation

This case was written by Neil Botton, a Principal Lecturer in Strategic Management, University of Westminster, and Dr Chris Carr, a Senior Lecturer in Strategic Management at the Manchester Business School, University of Manchester.

Introduction

On 11 January 1992 it was announced that James McAdam would take over from Gerald Ratner as Chairman of the troubled Ratners Group and that Gerald would concentrate on the day-to-day running of the business. Mr McAdam had apparently been hired because he was well known in the banking community and the City. Gerald Ratner had grown the company from relatively small beginnings after a boardroom coup when he took over from his father, Leslie, in 1984. At its peak, under Gerald's leadership, as both Chairman and Chief Executive, the company had well over 1000 stores worldwide and a ten-fold increase in share price to 383p. The company had been grown by a combination of aggressive acquisition on both sides of the Atlantic, the introduction of a formula of volume-oriented retailing and aggressive management of the supply chain. Gerald Ratner was forced to resign from the company on 25 November 1992 because of the 'continued negative press' that had been associated with his speech at the Institute of Directors in April 1991.

The early years

Leslie Ratner, Gerald's father, founded the company in 1949 on his son's birthday, and ran it as a traditional company in a conservative, seemingly highly constrained industry. By 1984, when Gerald ousted him, he had built up 130 branches but, with losses of £350 000, the company showed signs of struggling.

At the start of the 1980s most jewellers traditionally made the majority of their sales in December, converted their cash receipts to gold stocks in the New Year and waited out the coming year until the following December to repeat the cycle. Without a good Christmas they had a very thin time. The value of stock constrained any modern trend to self-service and promotional display designs, despite opportunities afforded by newly introduced 'deep tunnel' store entrances.

There was a distinction between real jewellery and costume jewellery. Real jewellery was, at that time, made from solid, but not plated, gold, silver or platinum. There might be precious or semi-precious stones set in the pieces. Real hollow-gold jewellery was to appear later, introduced by the Ratner Group. Costume jewellery consisted of pieces made from all other materials, including gilt and gold plate. In the early 1980s the manufacturing industry was very fragmented, with around 2000 manufacturers in the UK and globally tens of thousands.

Jewellery was still considered a luxury item, infrequently purchased and designed to last for life; but given so many retailers and manufacturers, branding was considered impossible because of the high advertising spend required. Traditional independent jewellers were very product oriented, an effect reinforced by the training of staff, who were more likely to be recruited for their ability to repair or value a wide range of stock, rather than any ability to sell. The image was of engagement rings bought in a sedately furnished store, from a sombrely dressed assistant, who whispered the price to the purchaser so that the intended wearer could not hear. Despite the threat of increasing competition, many considered this traditional approach a recipe for continuing success. Mark-ups were still 100 per cent, with an annual price rise of 10–15 per cent regardless of inflation. British Home Stores´ David Cassidy was subsequently quoted[1] as saying that:

> You in the jewellery trade, as customer surveys indicate, are in a very, very happy position. Price is not high in the purchaser's decision. You have the opportunity to observe the change in lifestyles and to build an environment and an attractiveness which will ensure your profitable survival. But I believe you have a bigger opportunity, you are predominantly independent. You are small, and in retailing small is beautiful … you can motivate staff in ways that we [British Home Stores] would find impossible. If you lose out to the major chains who can only offer a price advantage you will only have yourselves to blame, and I do not believe you will.

1982, when Gerald became joint managing director beside his father, was, however, a particularly poor year, with the whole of the jewellery industry suffering bad results. The recession had reduced consumer demand, bad weather had affected the all-important Christmas sales period the year before, and an increase in VAT, compounded with a sharp increase in the price of gold, had all added to the retailers' problems. Stock replacement had resulted in higher cash costs, as stock was traditionally paid for in advance. Caution was spreading through the industry, with rumours of significant de-stocking by some companies. Some commentators believed that many firms were only surviving by the generosity of their bankers and of tolerant landlords. It was also felt that the recession had changed the nature of the purchase for many consumers, opening the way for the department stores and mail order catalogues to offer mass-produced, imported items at the lower end of the market. This affected the perceived value of the craftsman-manufactured products of the old school. Others offered a contrasting opinion, stating that the discounters and catalogue companies, such as

Argos, were actually expanding the market by attracting a whole new segment of customers. Prospective customers were able to carry the attractively laid out catalogues around with them to make price comparisons. However, no one believed that the large independent sector would expand, particularly by organic growth, as the traditional methods of trading were under threat. The market was becoming more clearly segmented, with a polarization between the luxury end of the market and the products offered by the multiples. A number of industry commentators thought that it would be difficult for consumers to distinguish between one multiple and another.

The performance of Arthur Conley & Sons indicated to many the attractions of moving to the 'better' end of the market. Part of the Sears Group, they had been in the low to middle price range. The company changed its name to Walker & Hall and proceeded to refurbish its 60 stores. The first store converted was the in-store branch at Selfridges, where they raised the quality, improved the range and increased many prices between two- and five-fold. The turnover increased by an order of magnitude in the space of a year. Brian Franklin, the managing director, criticized jewellers who suicidally reduced their prices, and other jewellers were similarly experimenting with modern re-fits.

Gerald Ratner's growth strategy

Taking over as Chairman after a boardroom coup, Gerald Ratner's response to this fragmented, depressed and confused industry was characteristically iconoclastic. He had started work in the family business at the age of 16, having been expelled from his grammar school after telling his headmaster at a teacher's funeral that he shouldn't bother leaving the cemetery. Working as a salesman, he lacked patience with the sedate traditional jewellery industry and had little impact until 1982, when he became joint managing director beside his father. The working relationship was not particularly successful; his restlessness and openness led to many boardroom rows when he felt that his ideas were not given due attention. When his father was away in America, Gerald abruptly moved the business downmarket for the first time, putting notices in shops advertising reductions on watches. The business picked up. When his father returned he told him that the other directors did not want him to run the business, and he told the other directors that his father wanted him to run the business.[2]

At the time Ratners was losing money and share prices were generally low: Ratners was quoted at 39p and Walkers at 43p. The concern was for 'further steps to improve efficiency and stock rationalization'. Gerald nevertheless set about the expansion of the business by buying up sites and acquiring other smaller chains, whilst squeezing margins to increase turnover. As he was quoted:[3]

Ratners are not interested in special offers or gimmicks. We don't feel it is the correct way forward for jewellers. We don't feel it does them any credit and it does not improve credibility as far as the general public is concerned. I think they expect more than that from jewellers. We will operate under more than one name, but as a force we will be unbeatable.

The first significant acquisition, in December 1984, was Terry´s, a chain of 26 stores selling fashion jewellery in the South East, with a reputation for being highly competitive. The Terry's format was adopted by the Ratners Group, surviving until almost 1989, when Terry Jordon, the previous owner, retired. Using the Terry's formula the company started to address the problems they perceived in the industry. Outlets were made more inviting to make the customer feel more comfortable and the product was offered cheaply with no frills. The Ratners Group started to move downmarket; the range was made more fashionable and keener priced to attract the younger generation. The range still consisted solely of real jewellery but it was recognized that this section of the population would increasingly have the highest spending power as the decade progressed. Correspondingly, the average age of staff was reduced, particularly in Ratners and Terry's branches, and a fashionable image was considered very important. Incentive payments became the norm: bonuses and merit payments were made every eight weeks to top branch managers, who were encouraged to spend more time selling rather than administering.

Ratners solved the problem of store branding by giving the impression of a permanent sale. Shop fronts were continuously covered with posters, emphasizing price, money-back guarantees, price matching and credit terms:[4] 'Price is crucial. Image isn't. You can't sell a new logo or colour scheme.' New store designs were, however, taken very seriously. Detailed full-size mock-ups were made in a warehouse, approved, photographed and passed to every store manager for implementation.

Costume jewellery increased in importance throughout the decade, with increasing awareness of dress coordination down the age range. Initially the effect came in the lower price ranges but later higher prices were also being realized. The tastes of the Princess of Wales and Joan Collins helped to create an additional upmarket fashion jewellery sector in which, by 1988, prices of £300 were not uncommon. This new sector attracted department stores and successful new players such as Next, Miss Selfridge and Top Shop, together with specialist accessory stores such as Salisbury's, Torq and Cio, who recognized brand opportunities.

A new segment of the market was being created and at the lower end of the range impulse purchasing became a reality, with the product being promoted as an everyday, disposable fashion accessory. The 'flash for cash' image was born, as cheap imports were used to keep prices down, particularly in the 9ct gold ranges. Significant cost savings were achieved by using their increased volume in their dealings with suppliers, trading 27 tonnes of 9ct gold and a quarter of a million carats of diamonds in their most successful year. The closer relationship with suppliers allowed them to dictate design, which was vital to a group selling a 'disposable' fashion item. Most prices were held year on year and in some cases reduced. By April 1990 Victor Ratner, deputy managing director, was able to claim[5] that 'a particular bracelet sold for £38.95 at Christmas 1988 was sold at £29.95 the following Christmas with only a slight loss of weight, and would be cheaper again at Christmas 1990'.

Innovation also involved rapid copying of expensive items. Princess Diana's engagement ring, originally priced at £30 000, was copied and available in Ratners stores in four days, priced at £28.50, selling 10 000 in one day. The group were the first to offer hollow jewellery, representing a significant cost, and price, reduction. The reduced margins were also made possible because of the increased efficiency that Ratners were able to bring to their stores. Stocks were kept to a minimum in the stores, with increased holdings in the central warehouses connected by EPOS. Restocking usually occurred twice a week rather than the previously fortnightly cycle. This also helped to address the cash-flow problems caused by seasonality and the required payment in advance.

By the beginning of 1986 the group had grown to 173 stores and announced its intention to become the largest jewellery store in the UK. Although plans were in place to add a further 40 stores organically, funded by a 1 for 4 rights issue at 112p, expansion eventually came by the usual method. In May, Ratners paid £4.3 million to take a 27.7 per cent stake in H. Samuel's, having acquired the shares from family members who were frustrated by the recent performance of their shareholdings. By the end of the month Mr Edgar, Samuel's Chairman, had been persuaded to accept £149 million and the chairmanship of the combined group for his share-holding, with Ratner as group chief executive. In August he resigned and Gerald Ratner assumed both roles. Andrew Coppel became Group Finance Director, joining from Morgan Grenfell Finance Ltd, where he had been the director advising Ratners over the previous 18 months. Sale and lease-back of the stores started to recoup some of the costs of the acquisition, and the merchandising and marketing methods used in Ratners and Terry's were quickly introduced to the H. Samuel's stores, which continued to operate under their own name. The buoyant property market had helped the group as one portfolio of properties realized £8 million over book value and brought lease costs of £1.1 million per annum on a 25-year lease, with five-year rent reviews. By the end of the year the group were able to report a 40 per cent increase in turnover at the H. Samuel's stores, bettered only by the improvement in the Ratners branches, where 50 per cent increases had been achieved. Additionally, the group opened 40 new stores.

The acquisition of Ernest Jones for £25 million, funded by a 3 for 10 rights issue, filled a portfolio gap in the middle to upper end of the market. Once again, the family members of the board of Ernest Jones stayed in the combined company for a few months but then left after 'their ability to perform their duties' was undermined by the Ratners' interference.[6] The 61 stores which were in prime sites and had recently been refurbished filled in gaps where the group was geographically underrepresented. Operating under their own name, the stores were quickly subjected to the marketing and logistics which had proved so successful elsewhere in the company. The bid was not referred to the Mergers and Monopolies Commission (MMC),[7] a decision criticized by the Goldsmiths group, the UK's fourth largest jeweller with a market share just over 1 per cent higher than Ernest Jones; but it was felt that this would be Ratners' last major acquisition in the UK.[8]

By the end of the decade, jewellery had been transformed from an occasional luxury purchase to the third most popular gift and Gerald Ratner had achieved almost total segmentation of this market on price. Watches of Switzerland was aimed at the highest level, offering brands such as Rolex, Patek Philippe and Audemars Piaget; then came Ernest Jones; then H. Samuel and its subsidiary James Walker, catering to the middle range; then Ratners and Terry's at the lower end.

Almost simultaneously with the Ernest Jones takeover, Ratners announced its acquisition of Sterling, the fourth largest jewellery chain in the US. Sterling operated in the market segment that matched the Ratners and Terry's format, and with which Gerald was so familiar. He was reported as saying that the US market was 'ripe for a shakeout and consolidation'. The US market, worth $20 billion per annum, was highly fragmented, in the hands of small chains and independents, and very similar to the UK market of 1982. The market had been buoyant, enjoying nearly 8 per cent per annum growth for the previous five years. In the States, Ratners then acquired Westhall and Ostermans, whilst in the UK the loss-making chain of Stephens Jewellery was acquired for a nominal cash sum and added 13 stores in the South East. Shortly after, they added a further 16 jewellery stores and 4 accessory stores in the UK with the purchase of Time, a Jersey-based retailer, for £5.3 million.

The market share of the group had now grown to 19 per cent from the 2.5 per cent that Gerald Ratner had inherited when he took over as chairman. Turnover at H. Samuel had doubled since acquisition, 18 months earlier, to £62 per square foot, and Ratners had risen to £95. The buying power of the group allowed them to offer for £94.50 a diamond ring for which other multiples had to charge £200. Turnover was still pursued at the expense of margin, as the company continually searched for cheaper products which sold quickly, or what Gerald Ratner was already describing as 'crap'[9] in his conversations with journalists.

Concerns in the City as Ratners over-extend?

Towards the end of 1988[10] concern was expressed in the City that the Group was being too ambitious in its latest acquisition.

Undeterred, Ratners' acquisition strategy continued unabated. From Next they acquired 130 Zales stores (consolidating these under the Ernest Jones name); 73 Collingwood and J. Weir stores; and 235 Salisbury fashion accessory stores that the group had previously stated it didn't want.[11] Fortunately the yearly results were excellent: turnover up 76 per cent, profits up 60 per cent, respectively 20 and 40 times levels in 1983.

In October 1989 Weisfields in the US was purchased, which made a total of 470 outlets in the US and increased geographical coverage. With the US now 30 per cent of operating profits, a short-term target of 1500 US stores was mentioned. By August 1990 US sales had grown to $402 million, with three distinct merchandising divisions but with integrated EPOS and delivery systems. Heavy advertising, promotion and discounting continued improving volumes and brought in profits growth again of about 50 per cent in April, a growth achievement surpassing all other retailers; but their acquisitions-oriented finance director was replaced with Mr O'Brien, signalling a shift towards more detailed financial controls. The City remained equivocal:[12] 'the phenomenal success at taking jewellery downmarket might eventually backfire', but this 'reckoned without the growth prospects in the US or Ratners' unique position in the UK market, with the 20% growth in the company the share price would take care of itself', and the market had it in for Ratners, with the shares receiving a lowly rating from the City at 227p.

This was shortly followed by an agreed bid for Kays, a 500-store chain in the US. The acquisition would make the company second only to Zales, who had 1700 stores in the US. Although there was geographical logic in the acquisition, substantial investment and integrations was required and a number of analysts were concerned as to the financial wisdom of the deal. Kays had been rumoured[13] to be heading for voluntary bankruptcy and a Chapter 11 filing to stave off creditors in the face of heavy losses. Ratners claimed that Kays' previous performance was irrelevant, all they were buying was real estate and staff. In the event, much of the stock acquired with Kays was not up to scratch and had to be shifted through alternative outlets. Ratners also misjudged the quality of the staff; by September 1991 fully 25 per cent of the managers had been replaced. A bid too far perhaps?

The results in January 1991 showed a slowdown in performance. Despite improved stock control, and just-in-time delivery systems which had allowed stock to remain at constant levels whilst turnover grew to over £1 billion, profit had shrunk, for the first time in 10 years. Generally the retail sector was not looking very good, with the suspicion that things would get worse.[14] The company started to pursue a rationalization programme, cutting costs in the pursuit of margins as well as volume, but still recorded a first half loss.

The speech at the Institute of Directors

Gerald Ratner's media flair and popularity proved invaluable in early acquisitions, most crucially in the case of H. Samuel's. On Tuesday 23 April 1991, however, he achieved what may have been a unique feat by getting the Institute of Directors conference reported on the front page of the *Sun* newspaper, by describing one of his products as 'total crap'. It was to culminate, ultimately, in his downfall.

He proudly told the audience that the way to survive the recession was to give the people what they wanted, 'never mind the quality, buy in quantity and have some cheap fun'. Cheap fun came in the form of silver earrings at 99p which 'have very little to do with quality, costing less than a Marks & Spencer prawn sandwich and most probably not lasting as long'. Another winner was an imitation open book with curled up corners and 'genuine antique dust'. Gerald declared, cheerfully, that it was in 'the worst possible taste' but it had already sold 250 000. He added:

> We also do cut-glass sherry decanters complete with six glasses on a silver tray your butler can serve you drinks on, all for £4.95. People say how can you sell this for such a low price. I say because it is total crap ... Our Ratners shops will never win any awards for design and they are not in the best possible taste. In fact, some people say they can't even see the jewellery for all the posters smothering the shop windows. But these shops, that everyone has a good laugh about, take more money per square foot than any other retailer in Europe, because we give the customer what they want.

Reaction to the speech was mixed. The jewellery trade was in no doubt, however, a scornful East London diamond trader insisted that Ratner wasn't a jeweller but a marketing man who didn't understand the business. 'He had ripped the guts out of it, selling a commodity which might just as well be onions.' One customer, having just bought £50 of earrings, was quoted[15] as saying, 'It's cheap and convenient. Knock him if you like, but he's made a bloody fortune.' The City, initially, gave the same reaction.

The marketing director, Simon De Mille, hastily insisted that the comment had been made with tongue firmly in cheek and that the media had misinterpreted what were only gags. By the end of the month Gerald Ratner, having received hundreds of letters (half in protest, half in support), issued a formal apology for his comments: 'My remarks were meant as a joke and I would like to apologise for any offence they have caused … I would like to put on record that I am very proud of my shops, my products, my staff and my customers.'

Even as the apology was issued new posters were appearing in some of the store windows plying the merits of 'Cheap, Reliable and Affordable Prices' – eating humble pie did not come easy. Gerald followed up with an apologetic appearance on a TV chat show, but nevertheless found himself labelled by some as the 'businessman the great British public loves to hate'. After an initial rise of 6p on the day of the speech the shares had shed 7p, to settle at 180p. The fall continued until a recommendation[16] from analysts Hoare Govett brought a rise to 169p. 'The worries about the recent comments about the quality of his goods were overdone. Consumers will soon have forgotten about them.' But by the end of the month other analysts had downgraded their forecasts on the company, based not on company performance but on the strength of the economy. The shares closed the month at 150p.

By the AGM Gerald was insisting that he had been mis-reported, saying that the term 'crap' had been solely applied to the sherry decanter, not the jewellery products, which he said had always represented good value for money and high quality. However, he did plead guilty to naivity. He had tried to make what was meant as a light-hearted comment to brighten up an otherwise gloomy discussion of recession at a conference. It was reported that his contriteness was well received by the shareholders.[17] Roger Cowe also defended Ratner:[18]

> He should be in a Batman cartoon as 'The Salesman Who Told The Truth About His Products'. He made the mistake of repeating a well-worn joke, that he had used many times before, once too often and found it splashed over the front pages. The outrage was misdirected. He has never pretended the Ratners chain sells high quality products. The point of his speech to the stuffy directors was that Ratners provides customers with what they want – something shiny that doesn't cost very much.

It wasn't enough. The media were now alert for any piece of news about the group: Exeter magistrates fining Zales and Ernest Jones for wrongful claims about the durability of Laser Rope gold bracelets made front-page news.

The losses continue

The interim results 1991 reported a loss of £17.7 million compared to £9.3 million for the previous half year. This was attributed to the recession rather than the earlier comments, since 8 per cent more people were coming into the Ratners stores; but with the average transaction value coming down 15 per cent, cost increases of 10 per cent seemingly could not be passed on and the share price continued downwards. In the US the group's preference stock was downgraded by Moody's Investors Service, adding $2 million to the annual interest bill and making further debt auctions more expensive. Shares fell further to 83p as the market anticipated poor results and a cut in the dividend. One analyst commented:[19]

> Gerald Ratner has, in the past two to three years, thrown a lot of money at the Christmas sales and the expectation is that this year he will throw anything he has at the sales. The question is whether the returns justify the spending.

Although the New Year brought a rally for some retail shares, Ratners continued to suffer heavy selling pressure, falling to 27p by 2 January 1992, and a further 5p in the next two days, after 5.3 million shares were traded in a day. Ratners would face breaching banking covenants, believed to require interest cover of at least 2×, and having to pass on the preference dividend. The group's own principal bankers, Barclays and Midland, along with some analysts and newspapers, played down any risk of any actual financial meltdown, allowing some slight share rally; but all perceived the need for radical restructuring. One banker commented:[20]

> Ratners may break a banking covenant. But that in itself is no reason to panic. What we have to assess is how much cash the business is generating and is capable of generating. Only then will we need to discuss whether its debt has to be restructured. There has to be a serious possibility that Gerald Ratner will give up either his role of chairman or chief executive.

Mr Fuller, chairman of the British Jewellery Association, reportedly highlighted the ill feelings of suppliers following Ratners' aggressive buying policies, and unfirmed-up orders resulted in infuriating stock returns: 'He got away with it once, but a lot of the bigger companies got together and said they were not prepared to deal with him again on that basis, so he had to change.' Gerald was respected for what he had achieved but not particularly liked.

His City credibility was likewise waning. Finally on 11 January 1992, after further downgradings from Moody's and the American credit rating agency Standard & Poors, Gerald Ratner abdicated his role as Chairman, in favour of the 61-year-old, tough-speaking Scot Mr James McAdam, the former Deputy Chairman of Coats Viyella and Chairman of two major British clothing associations. An unusually subdued Mr Ratner was quoted as saying:[21]

> I am very happy about the decision. I shall continue to have responsibility for the running of the business on a day-to-day basis as any chief executive would

do, but there will be a load of things that Jim will be able to deal with. It is more stable.

He added[22] that Mr McAdam would be a 'tremendous asset' to the company: 'He has got a lot of experience – in retailing as well – and is well known by the banking community.' Mr McAdam's immediate role was to keep the bankers happy and his appointment elicited an unusual statement of their support. Although Ratners had already cut the headcount by 10 per cent, Mr McAdam believed his experience of rationalization in the clothing industry would complement Gerald:[23]

> I am battle-hardened from the textile industry in just this sort of exercise. We have been doing it for a hell of a long period of time. There is going to be no quick fix. It is going to be a long, hard slog ... I will be tackling the financial issues and looking at the organisation and the structure. Gerald will be running the commercial side of the business. I think we complement each other. We both feel very comfortable with the chemistry. I have a real respect for him and I think we'll make a good team.

How the two men would work together seemed to be a source of endless fascination. By temperament, background and outlook they could hardly be more different. Where Ratner was spontaneous, wry and softly spoken, James McAdam was described as single-minded, loud and blunt: having run the Country Casuals clothing chain, he had also had experience of the retail side but was more of a big company man.

Despite huge price-cutting promotions there was a 15 per cent fall in UK jewellery sales in the vital six weeks before Christmas and Ratners announced that it was likely only to break even in the year to 1 February. The results were worse than the City had feared. The company reported a total loss of £72 million and Ratner shares closed 2p down at 21p.

Restructuring and departures

Following a management reorganization, Victor Ratner, deputy managing director and cousin of Gerald, left the company, though James McAdam stated[24] that both he and Gerald were 'together on this'. All the group's UK jewellery operations were reorganized under a single management board, chaired by Gerald Ratner. Masarrat Hussein, the administrative director, also left the company. The final board now comprised four executive and two non-executive directors, though a third was added later to strengthen the legal side. Mr McAdam stressed that he was not stamping his authority on the company for the sake of it: 'We have created a sensible operating structure. Gerald is driving the business forward at the sharp end. That is what he likes doing and what he does best.'

Ratners' offer document[25] and the sale of Watches of Switzerland to Asprey in June pleased bankers but were less well received by analysts. This 25-branch chain, dealing particularly with quality Swiss brands such as Rolex, Patek Philipp, Audemars Piaget and Cartier, had generated £22 million of turnover the previous year and profit of £0.9 million before exceptionals and tax. Its price of £23 million compared with £50 million sought only five months earlier, though it was thought to have remained relatively immune to the IOD speech. Ratners said that the stores had limited synergy with the core business and it was going to 'the right sort of home'.

Elsewhere it was rumoured[26] that the Bank of England was putting pressure on Ratners' 25 bankers to take an enlightened attitude and extend their £500 million lending facilities to the struggling jewellery group to avoid a high-profile collapse. By 22 August 1992 agreement had been reached, with more stringent covenants and higher interest rates. Some of the debt was now secured. The shares fell back to 11.5p.

Published annual results showed group sales up at £1.13 billion, mainly due to the inclusion of a full year of Kays' turnover. Apart from slight increases at Kays and Salisbury, all other chains suffered falls in turnover, with Ratners down 24 per cent on a comparable basis. Pre-Christmas price cutting had cost an estimated £36 million, whilst operating costs had risen by an extra £113 million. Pre-tax losses were £122 million. Total debt had risen to £480 million, offset by cash of £257 million and shareholder funds of £303 million. James McAdam commented that 'adverse publicity' had affected sales in Ratners' stores, though this was offset by shoppers avoiding Ratners' shops 'only to go into one of the group's other stores'.

He reiterated his comment[27] that the future of Gerald Ratner was not in doubt, insisting that the four-strong executive committee was united in its mission. 'People say, "My God, is there a lot of blood on the walls in relation to that?" The answer is that everybody has understood what has to be done and the whole thing has the commitment of the key management. We have not had problems.' David Wellings, managing director of the Cadbury Schweppes confectionery business, replaced the retiring Sir Victor Garland as a non-executive director, contributing retailing experience from sweets, which shared similar seasonality and impulse buying characteristics. He had known James McAdam since the mid-1980s.

Gerald Ratner, himself, announced further details of the rationalization programme. Blaming the recession, he said that they could no longer be a sales-led organization, and would start to concentrate on profits. Nick Bubb, retail analyst at Morgan Stanley, was quoted[28] as being amazed: 'I never thought I'd see the day when Gerald Ratner would admit that the old approach was dead.' 180 stores in the UK and 150 stores in the US were to be closed with the probable loss of 1000 jobs, and they would reduce their over-dependence on low prices and promotional activity. Gross margins were to be pushed up and the marketing strategy would be more tightly focused, placing more emphasis on customer service and making greater distinctions between the different outlets. H. Samuel would be aimed at the middle market, the 'powerhouse' moving clearly above Ratners, whilst Ernest Jones and Leslie Davies would move upmarket to take sales from the independent jewellers. Gerald Ratner would spend time visiting stores to build morale and re-build bridges with the increasingly nervous suppliers. 'Suppliers are one of the most important elements, especially now that we are not selling on price,' he said.

Excalibur Group, the jewellery manufacturer and engineer, announced[29] that it was happy to supply Ratners 'now that banking support is in place'. Suppliers had been wary of selling to Ratners, fearing bad debts. About 13 per cent of Excalibur's jewellery sales went to Ratners, a proportion that had declined from more than 20 per cent. But Mr Griffiths, the managing director, said that he would now consider doing more business with Ratners if asked. Other suppliers continued to be nervous.[30] Abbeycrest, which refused to comment on customers' business, was known to be a leading supplier to Ratners who took nearly 60 per cent of their turnover in 1989. However, the group had since built up its sales to other customers and said their largest customer would account for only around 15 per cent of that year's turnover. They would not confirm that this was Ratners.

Ratners was fined £1000 and ordered to pay £350 costs after stud earrings, priced at £3.50, described as opal when bought in May 1991,[31] were found to be plastic and melted when tested. The company pleaded guilty and said it was an administrative error from which no profit had been made. To counter the negative publicity Ratners announced[32] a few months later that it now employed a trading standards officer. Gary Cullimore, formerly a government trading standards inspector, was responsible for due diligence at Ratners to ensure that £325 'diamond' rings were what they purported to be.

Despite higher operating margins, described by some analysts as 'amazing in the circumstances', pre-tax losses declared for the first half of 1992 rose to £30.6 million. Operating profits had risen slightly, but UK results were hit by sales down (on a comparable basis) by 24 per cent at Ratners, 18 per cent at H. Samuel's and 9 per cent at Ernest Jones. The experiment of renaming some Ratners stores as James Walker had 'failed materially to improve trading' and had been halted. At the meeting to announce the results, Gerald Ratner and his salary came in for heavy criticism, particularly from representatives of the American preference shareholders, and some analysts considered the results represented an 'appalling indictment of past management practices'.

On the night of 25 November 1992, Gerald Ratner resigned as chief executive. He was quoted[33] as saying:

> I am obviously saddened to be leaving a business of which I am so proud. However, the continuing negative press I have attracted leads me to believe that this is in the interests of the group and the people working in it.

James McAdam, chairman, acknowledged that press comment had been a major factor in the resignation.

> It's his decision and it would be wrong for me to add to it. It's sad for everybody. It's the end of an era. He's been in the group for 26 years and built it up from small beginnings ... The bad press has not died down. It was something that happened on an almost daily basis.

Associates of Mr Ratner insisted,[34] however, that he had been the victim of a boardroom coup. 'They have been waiting in the long grass for him for some time,' said one. He offered his resignation to a full board meeting, and it was accepted – normal City parlance for a forced resignation. Mr McAdam took direct responsibility for the UK jewellery business and the US operation would report to him through Nathan Light.

References

1 *Retail Jeweller*, 1 July 1992.
2 *Guardian*, 24 April 1991.
3 *Retail Jeweller*, 10 December 1992.
4 *Independent*, 29 November 1992.
5 *Financial Times*, 24 April 1990.
6 *Daily Telegraph*, 9 September 1987.
7 *Financial Times*, 31 July 1987.
8 *Financial Times*, 4 July 1987.
9 *Financial Times*, 28 January 1988.
10 *Financial Times*, 12 October 1988.
11 *Financial Times*, 7 May 1987.
12 *Financial Times*, 24 April 1990.
13 *Financial Times*, 3 July 1990.
14 *Financial Times*, 4 January 1991.
15 *Guardian*, 25 April 1991.
16 *Financial Times*, 4 May 1991.
17 *Financial Times*, 9 July 1991.
18 *Guardian*, 16 September 1991.
19 *Guardian*, 4 October 1991.
20 *Financial Times*, 7 February 1992.
21 *Financial Times*, 11 February 1992.
22 *Financial Times*, 11 January 1992.

23 *Guardian*, 11 January 1992.
24 *The Times*, 11 February 1992.
25 *Sunday Times*, 10 May 1992.
26 *Sunday Times*, 2 August 1992.
27 *Guardian*, 22 August 1992.
28 *Sunday Times*, 23 August 1992.
29 *Financial Times*, 4 September 1992.
30 *Financial Times*, 16 October 1992.
31 *The Times*, 11 September 1992.
32 *The Times*, 1 October 1992.
33 *Independent*, 26 November 1992.
34 *Guardian*, 26 November 1992.

Case 4.2   Fisons: The fall from grace

This case was prepared from published sources by Helen Peck, Cranfield School of Management, as a basis for class discussion rather than to illustrate effective or ineffective handling of an administrative situation.
© *Copyright Cranfield School of Management, June 1994. All rights reserved.*

On 14 January 1992, the board of Fisons Plc announced that it had 'with considerable regret, accepted the resignation of Mr John Kerridge as chairman and chief executive, as a consequence of his retirement on the grounds of ill-health'.[1] The announcement was made by non-executive director, Patrick Egan, who would oversee the pharmaceutical, horticulture and scientific equipment group until a new chief executive could be found. Egan stressed that 'The board wishes it to be clearly understood that there has been no pressure on Mr Kerridge, either from his fellow board members or from our principal institutional shareholders, for him to take this course of action.' Egan did, however, concede that his former colleague's health had not been improved by recent well-publicized difficulties between the company and US drugs industry regulators, the Food and Drug Administration (FDA).

Kerridge's resignation came at the end of a traumatic four months for Fisons. Supply problems with two of its drugs had meant that, for the first time in many years, financial performance had failed to live up to expectations. With Kerridge's departure, the remaining management were confident that they could now draw a line under the unfortunate episode, and look forward to a speedy revival of Fisons' reputation as one of Britain's best managed companies. The group's shares gained 10p on the news of Kerridge's resignation. It was an inglorious end to what had been a remarkable career.

From ailing dinosaur to rising star

Kerridge joined Fisons as a marketing controller in 1967, after spells in consumer marketing at Cadbury Fry, AEI-Hotpoint, Lyons and Rothmans. At Fisons he had risen steadily through the ranks of the company's core Fertilizer Division, eventually emerging as chief executive following a boardroom fracas in June 1980. The mini-conglomerate he inherited looked anything but promising. The company's origins lay with its 130-year-old fertilizer business, which by 1980 appeared to be in terminal decline. For 20 years it had fought a losing battle to maintain its grip on the UK fertilizer market in the face of stiffening price competition from multinational giants, Shell and ICI. There were early indications that Fisons' profitable agrochemicals business would suffer a similar fate. Together the Fertilizer and Agrochemicals Divisions accounted for 60 per cent of the company's turnover; the remainder being generated by its Pharmaceutical, Horticulture, and Scientific Equipment Divisions.

On his accession to power Kerridge immediately set about a programme of financial and business restructuring. Cost-cutting measures were introduced right across the business. The company's headquarters were relocated – with half the number of staff – from London's Mayfair to more modest premises back in the Suffolk coastal town of Ipswich. In the divisions, finance directors were appointed and tighter financial controls introduced, to underline the fact that senior managers' objectives were tied to profit. Return on capital employed became the definitive performance measure.[2] Within three years the fertilizer and agrochemicals businesses had been revamped and – to the amazement of onlookers – promptly divested. The cost cutting continued, but alongside it a programme of reinvestment and growth through acquisitions to build the remaining businesses.

The Pharmaceutical Division was profitable, but was generally regarded as being over-dependent on one product, the anti-asthma drug, Intal. Intal was a truly innovative anti-asthma treatment, first launched in the UK in the late 1960s. Existing treatments simply attempted to ameliorate the asthma attack; Intal, on the other hand, was a preventative used to treat background inflammation in the patients' respiratory systems. By the mid-1970s Intal accounted for 60 per cent of Fisons' pharmaceutical sales of £4 million a year, becoming one of the top 15 selling drugs in the UK by the end of the decade.[3] In 1982, the web of UK patents protecting Intal's base compound (sodium cromoglycate) began to expire. Overseas patents had longer to run, but they too would start to unravel before the end of the decade. Fortunately, a much-needed reprieve was secured with the development of an aerosol-administered version of the drug in the mid-1980s. The breakthrough bought an extension of Intal's patent protection. Moreover, it made the drug more accessible for children and the elderly, opening up a whole new sector of the respiratory market. During this time new applications were found for other derivatives of sodium cromoglycate, resulting in a batch of new specialist anti-allergy treatments; Norcrom for food allergies, Opticrom for eyes, with Lomusol and Rynacrom for nasal allergies. Around the world sales of anti-allergy drugs were on the increase, and a renewed effort from the sales force produced a marked improvement in Fisons' market share both at home and abroad.

Meanwhile, a severe pruning of the Scientific Equipment Division resulted in the closure of some of its manufacturing operations and the shedding of a third of its labour force. It was decided that from now on the division would concentrate on its strengths: distribution expertise, marketing, and aftersales service. To reduce its dependence on the UK, it expanded overseas, particularly into developing countries where new laboratories were being kitted out. By 1982, Fisons was the largest non-American supplier of scientific equipment anywhere in the world. Two years later the purchase of Curtin Matheson Scientific Inc., a distributor of laboratory products, firmly established the division in the important US market – home to around 40 per cent of the world's research activity.

The tiny Horticulture Division (a recent spin-off from fertilizers) was also searching for complementary acquisitions both at home and abroad. Fisons already dominated the peat and lawn fertilizer market in the UK, and had acquired 30 per cent of the North American peat market with the acquisition of Western Peat Canada. The division was now working hard to develop added-value peat products to launch into the US.

The best managed company in its sector

By mid-1984 the turnaround was complete and, in recognition of his achievements, Kerridge was appointed to the newly merged position of chairman and chief executive. He had by now established a reputation as a strong and determined leader, whose aggressive 'hands-on' style was respected and admired (from a distance) by the investment community of the City of London. The City's approval was underlined in January 1986, when a poll conducted by stockbrokers James Capel voted Fisons the best managed company in its sector. Throughout the rest of the 1980s, Fisons turned in impressive year on year increases in profits, while making significant acquisitions in all areas of its business.

As the 1990s dawned, Fisons was generally regarded as a superbly managed company; highly skilled in the art of mergers and acquisitions, and very efficiently run. The company's director of finance, Roy Thomas, was widely acknowledged as the best in the sector, if not within the entire FTSE 100 list of companies. Under his guidance, the company's financial function had been transformed from a pedestrian book-keeping operation staffed by accountants into one of the slickest finance divisions in British business. Thomas's financial wizardry had consistently kept the company's tax charges at an enviably low level, and had made substantial profits from currency hedging and treasury management. Thomas was a popular figure with the City, and while many commentators struggled to follow his financial manoeuvrings, he was usually forthcoming when asked about the financial management of the company. The same could not be said for Kerridge. At his rare forays into the public arena, Kerridge tended to concentrate on an overview of the business, showing a marked reluctance to be drawn into detailed discussions. Though widely respected for his management abilities, he was known to be extraordinarily sensitive to criticism. Kerridge's personal distaste for high profile management was in sharp contrast to the regime he had displaced. In the early days following his appointment as chief executive, the popular magazine *Management Today* had made the following observations about the company:

> The overblown publicity that tends to surround every move made by Fisons belies its actual size. In the past, public interest in Britain's first fertiliser producer may have been simply a function of an active in-house PR department; but it may also have had something to do with how the company saw itself. One analyst suggests that one of Fisons' historic problems has been that it has always tried to be a big company in its chosen activities, sometimes when there was call for a lower-key approach.[4]

A lower-key approach was exactly what Kerridge adopted, with the internal corporate PR department becoming an early casualty of the cost cuts. In a subsequent interview for the *Financial Times*, Kerridge explained the rationale behind its closure.[5] He believed that while the expense of divisional public relations departments – dealing with divisional product publicity – could be justified because of their 'expertise, the experience and the repetition of work', a corporate public relations department could not. Corporate PR could be contracted in when needed. He therefore looked to an agency to develop a clear corporate image of Fisons.

The agency would advise on the best approach to adopt when announcing financial results, takeovers and other matters, thus preventing what he described as 'inaccurate slants' from emerging. An important facet of PR, he felt, was that it should help to create an image of the company in which personalities were unimportant, 'I want to reach a stage where Fisons' image is such that chief executives can come and go and you don't notice the change,' he explained.

By the mid-1980s, memories of Fisons the failed fertilizer business were fading, and a small, low-key, in-house Corporate Affairs Department was re-established. Kerridge and Thomas met with the company's most important institutional investors once or twice a year, but the appointment of Peter Woods to the post of Director of Corporate Affairs in September 1989 improved the flow of information to other sections of the City. Woods, himself a City insider (formerly director of Health Care Research at Warburg Securities), cultivated links with a select handful of key analysts, with whom he met on a regular basis. Efforts were also made to strengthen what had become an arms'-length relationship with the financial press.

Nevertheless, Fisons remained noticeably less communicative than most of its pharmaceutical competitors. Unlike Glaxo, Smithkline Beecham and Wellcome – who communicated assiduously with the City – the senior management of Fisons did not give regular presentations to analysts. Field trips to the sites were not permitted, and there was no access to divisional or operations management. The company consistently declined to release details of its research and development activities, refusing to disclose any information on the products in the R&D pipeline, or even the extent of its R&D expenditure. Nor would it provide breakdowns of divisional sales by region or by product, as was usual with other drugs companies.[6] Wherever and whenever possible, Fisons preferred to maintain a 'golden silence'. Despite this tendency towards secrecy, the majority of analysts slept soundly in their beds at night, confident that Fisons' remarkable financial performance could be relied upon for the foreseeable future.

In September 1990, the UK economy – like most of the industrialized world – was sinking into recession, but as Fisons announced its half-year profits, the company showed no signs of succumbing to economic pressures. A 35 per cent rise in pre-tax profits for the half year reinforced its reputation as a safe and predictable performer, 'a safe haven for investors in troubled economic times'.[7] The company's achievements were summed up by *The Times* in the following way:

> Any company that has net borrowings of £150 million but can still include an interest receivable item of £1m in its interim accounts deserves to be taken seriously. That company is Fisons, the pharmaceutical and scientific equipment group, which continues to amaze its followers by never putting a foot wrong … With Fisons' five-year record of rising sales and profits pretty much second to none, the inevitable question is how long can it all continue.[8]

According to the *Financial Times*, Fisons itself was in no doubt, claiming that its 'current 20% earnings growth can be repeated annually for the rest of the millennium'.[9] The results led analysts to raise their estimates of pre-tax profits for the full year to £230 million, up £61 million on 1989. Sure enough, when the full year figures for 1990 were announced in March the following year, Fisons did not disappoint, turning in pre-tax profits of £230.2 million. These were halcyon days for Fisons. The normally publicity-shy Kerridge was feted by the media amidst speculation that a knighthood must be just around the corner. There were, however, occasional dissenters.

Lingering doubts

Globally, the pharmaceutical industry was undergoing a rapid process of restructuring and consolidation. In terms of size, Fisons was a small player in a land of giants, ranking around fiftieth worldwide. Because of its size, some commentators questioned the company's ability to finance the development of new products in the longer term, and to muster the marketing resources required to bring them to market. To achieve greater critical mass, some mid-sized pharmaceutical firms had entered into the cross-company joint marketing agreements; others had gone for all-out mergers. Kerridge was notoriously reluctant to follow either of these routes, believing that Fisons' steady earnings growth and a string of successful acquisitions could achieve the same ends.

By the late 1980s the R&D costs of the average new drug were estimated to be in excess of $125 million[10] from test tube to market; gaining access to the US market could double that figure. The costs incurred in securing US approval for new drugs had been rising steadily since the mid-1980s, as the mighty FDA progressively tightened its already stringent standards, demanding ever more documentation to accompany every application. A scandal in which four of its employees were convicted of receiving 'illegal gratuities'[12] had only served to heighten the regulators' demands for probity. Changes to the rules relating to alterations in the design of medical devices – such as drugs dispensers – meant that they too would require FDA approval. The escalating cost of applications, often exacerbated by inefficiencies within the FDA itself, was disproportionately damaging to smaller pharmaceutical firms. However, a poll of companies in an American magazine had indicated that few were prepared to risk the wrath of the regulators by complaining.

Fisons had frequently demonstrated its adeptness at repackaging elderly drugs, but some commentators feared that its dependence on an ageing drugs portfolio remained an Achilles' heel. The progress of Tilade, a long-awaited successor to Intal and Fisons' first major new drug for many years, was therefore watched with interest by competitors and commentators alike. Tilade was a non-steroid preventative; as such it was seen to have several advantages over its main rival, Glaxo's Becotide, a post-attack steroid. Steroids were disliked – especially in the US – because of their associated side effects. There was also a growing body of medical opinion that questioned the safety of the traditional post-attack asthma treatments. These concerns had created a window of opportunity for Tilade. First launched in the UK in 1986, the drug had since been approved in several other European countries where it had become a steady, though not spectacular earner. In 1990 Fisons was awaiting the outcome of applications to market Tilade in Japan and in the US. Securing approval in the US was crucial if the company was to hold its share of the North American asthma market, once Intal emerged from patent protection. For some reason though, the FDA seemed to be dragging its feet over Tilade. Fisons had hoped to have the drug on sale in the US by late 1988, but had had to wait until June 1990 before the FDA voted 5–1 in favour of its initial approval. The split decision had made some City analysts slightly wary about the company's prospects. But when the decision prompted analysts from stockbrokers BZW to raise doubts about the quality of growth at Fisons, Kerridge was incensed. Feeling that their comments misrepresented the company, he responded by declaring a freeze in relations with the BZW team.[11] Meanwhile assurances were given that the final go-ahead for Tilade would follow shortly.

On the scientific equipment front, Fisons had continued to grow more by acquisition than by organic means throughout the late 1980s. By 1990, the scientific equipment business ranked third or fourth in the world, behind the computer giant Hewlett Packard, and some leading German firms. The cyclical nature of the business meant that it was generally regarded by analysts as vulnerable to economic downturns, despite protestations to the contrary by Fisons.[8]

Meanwhile the small, sound Horticulture Division plodded along happily, providing a modest but steady stream of profits, with only the occasional hiccup resulting from fluctuations in the price of peat. As a 'third leg' to the business it was too small to be capable of offsetting losses in the larger divisions, so was largely ignored by analysts. Latterly, however, the division's peat-cutting activities had developed into something of a *cause celebre* among other sections of the population.

Problems with peat

Fisons had first become involved in the peat business with the acquisition of large peat deposits in the 1960s. The tracts of land brought with them long-standing planning permission for commercial peat extraction. To exploit the resource, Fisons had invested heavily over many years in peat-working and processing equipment. Traditionally peat had been used for fuel or animal litter, but a sustained marketing effort by the company succeeded in establishing it as an important horticultural product. Value-added peat products, such as peat plant pots and growbags, were subsequently developed. By 1990, consumption of peat in the UK had risen to some 2.5 million cubic metres per year. In terms of volume, between 70 and 80 per cent of the market was in commercial horticulture. Fisons held around 60 per cent of this sector. In monetary terms, the UK market for peat and peat products was worth around £200 million per annum, with annual sales to the lucrative amateur gardening sector estimated to be in the region of £66 million.[13]

Britain has over one and a half million hectares of peatlands, but the fine-quality peat sold to gardeners came almost exclusively from active peat bogs classified by conservationists as 'raised lowland mires'. These mires accounted for only a tiny proportion of the country's total peatlands, but were important wildlife habitats, havens for many rare and endangered wetland species. By 1990 less than 10 000 hectares of the raised lowland mires remained intact. The peat producers owned just over 5200 hectares of them; over two-thirds of these (3432 hectares) had been designated as

Sites of Special Scientific Interest (SSSIs) by the Government conservation body, the Nature Conservancy Council (NCC).[32]

Fisons was by far the largest cutter of peat in the UK. According to the conservationists, 90 per cent of its cutting was on land which had been awarded SSSI status.[14] The company was legally entitled to continue working the sites where planning permission predated the SSSI designation. Despite protests from environmentalists, Fisons resolved to continue cutting the peat, claiming that there were no viable substitutes for its customers. But in March 1990, Fisons overstepped the mark when it illegally drained and severely damaged a 65-acre peat bog on the edge of Thorne Moor, a South Yorkshire nature reserve. The company owned large areas of the surrounding moorlands, and had been cutting peat legally from the adjoining Hatfield Moor for 30 years. Both moors had been designated as conservation areas of international importance since the early 1970s.

Fisons was clearly sensitive about outside interest in the site. In 1989 a *Sunday Times* photographer had to be rescued by police after being held against his will by Fisons' employees, who blocked in his car for 45 minutes, demanding that he hand over his film.[15] The company admitted destroying the Thorne Moor bog, but claimed that it had been a 'genuine mistake'. Environmentalists were not convinced. Similar incursions had been made before, when the company had accidentally cut peat from important conservation sites on neighbouring Snaith and Cowick Moor.[16] This latest 'mistake' had come to light when the local government authority, Doncaster Borough Council, demanded that the work on the site be stopped and threatened legal action. A report by the NCC had revealed that in the year 1988–89 peat extraction destroyed 2250 acres of lowland peat mires with SSSI status and as such it was the largest single cause of damage to SSSIs in Britain. The NCC duly set about gathering evidence to support the first ever prosecution under The Wildlife and Countryside Act, 1981.

The Thorne Moor incident prompted 10 highly influential conservation groups – including Friends of the Earth, The Royal Society for Nature Conservation, The Royal Society for the Protection of Birds, and the Worldwide Fund for Nature – to take up the cause, banding together to form The Peatlands Campaign. The campaigners, led by The Prince of Wales, called for gardeners to boycott peat products and save the remaining peat bogs from total destruction. In a well-publicized statement, Prince Charles announced that:

> As patron of the Society for Nature Conservation, I was most concerned to learn that 96% of the lowland peat bog in this country has already disappeared. The campaign by a number of respected conservation bodies, with over two million supporters, to attempt to secure the continued existence of the remaining 4%, has my full support.[17]

The Prince disputed Fisons' claims that there were no viable alternatives to the use of peat in forestry and horticulture, citing research by the respected Henry Doubleday Research Association showing that substitutes were available. The Prince went on to announce that peat would not be used on his own estates in the Duchy of Cornwall, and urged other landowners to follow his lead.

The Peatlands Campaign did not stop there. In May 1990, it turned to Pensions Investment and Research Consultants (PIRC) – a consultancy which advises clients on social and environmental issues – for expert help. PIRC duly contacted 50 investment managers controlling over £870 million worth of the company's shares (around 42 per cent of the total) in advance of the company's forthcoming annual meeting.[18] Institutional shareholder, The Borough of Lewisham, supported by the South Yorkshire Pensions Authority (SYPA) and the Pearl Assurance Pension Fund, agreed to press the company to end its peat-cutting operations, urging it to invest instead in the development of peat substitutes. The campaigners received sympathetic hearings from other large institutional shareholders, including the British Telecom Pension Fund, Eagle Star and the Cooperative Insurance Society, but failed to secure sufficient support ahead of the company's annual meeting to raise a motion on the matter. Fisons' response to the campaign had been to distribute half a million leaflets promoting its case to garden centres. In the press it was pointed out that Fisons had already given 10 per cent of its total peat holdings for conservation purposes, including 106 acres for a habitat for an endangered species of bird – the nightjar.[19] As for the demand that it should develop peat substitutes, the company retorted with claims that it was already spending large sums in the search for peat substitutes, and had considered alternatives such as coconut scrapings but these were too expensive. It went on to complain that it was unfair to single out Fisons for such attention, because other companies had bigger environmental problems.[20] An advertising campaign was later launched, claiming that 'Buying British peat-based compost in no way endangers our remaining wetlands of conservation value'. The campaign subsequently earned Fisons the less-than-coveted Friends of the Earth 'Green Con of the Year Award'.

At the annual meeting itself, Fisons produced a video and detailed handouts for shareholders explaining 'The Case for Peat'. PIRC's Stuart Bell questioned Kerridge on the matter, and again urged the company to cease cutting peat. Kerridge politely declined the request, informing the meeting that peat was not a scarce resource globally – 3 per cent of the world's landmass was peat – and restated the company's position that no practical alternative to peat was available.[19]

Pressure on Fisons continued. The issue was debated in the House of Lords, and in the House of Commons, where an Early Day Motion in support of The Peatlands Campaign was signed by 25 Members of

Parliament. Fisons remained firm. As far as the company was concerned, abandoning peat cutting was simply not an option. In the words of Robert Stockdale, a director of the Horticulture Division, 'Any dialogue with the company has to ask how we can work together – recognising that peat harvesting has to continue'.[21]

The row rumbled on, keeping Fisons in the headlines throughout 1990, and into the following year, with a repeat performance by PIRC at the company's 1991 annual meeting. The Peatlands Campaign was a thorn in Fisons' side, and while its reputation among certain sections of the press, the public and the powers that be was looking distinctly tarnished, the overall impact on the group's bottom line was negligible. The entire Horticulture Division only accounted for 7 per cent of the company's turnover and 5 per cent of its profits. But by August 1991 the campaigners had managed to up the stakes. Surrey County Council became just one of 34 local authorities to publicly support a boycott of peat products. More worryingly for Fisons' shareholders, B&Q – Britain's largest DIY and gardening chain – announced that, on conservation grounds, it had decided to ban all peat cut from SSSIs from its stores. Its leading competitors quickly followed suit.

Meanwhile, the South Yorkshire Pensions Authority – holder of £1 million worth of Fisons' shares – continued to demonstrate concern for its local bogs, requesting that it be allowed to visit the SSSIs in its own area where peat cutting was taking place. The SYPA received a letter from head of corporate affairs, Peter Woods, informing it that 'Relations between Fisons and major customers are operational and confidential and as such we would not comment on these to shareholders or any other third party.' Permission for the visit was refused. The matter was reported in the *Sunday Times*,[22] which explained that Woods and two other directors had spent the day discussing the bog with the institutional investor. According to the press report, the company felt that 'the SYPA had had preferential treatment, a fair share of its time, and is so intransigent it deserves no more'.

It would be some time before the impact of the peat boycott could be judged. In the meantime investors' attention focused on the forthcoming announcement of Fisons' results for the half year to June 1991.

Disappointing results

On 6 September, the eve of the meeting, analysts were still gleefully predicting an 11 per cent rise in interim profits. Little growth was expected from the scientific instrument or horticulture businesses, but hopes were high for the seemingly recession-proof Pharmaceuticals Division. Recent reports had highlighted the fact that the death rate from asthma – the most common chronic disease in the industrialized world – was climbing steadily. At the same time, misgivings over the safety of some forms of inhaled post-attack steroids were boosting sales of anti-inflammatory asthma drugs like Intal and Tilade. The company was still waiting for final approval for Tilade from the FDA, but this was widely believed to be only months away. However, the results announced by Kerridge the following day were not quite as everyone had expected. For the first time in over a decade, Fisons' profits – while up by 6 per cent – had failed to meet expectations. The reason, explained Kerridge, was that £10 million had been wiped off profits following supply problems with two of the company's older drugs – Imferon (a blood product), and Opticrom (an anti-allergy eye treatment). Kerridge went on to explain that the disruption to supplies was caused by an 'extremely pedantic' US Food and Drugs Administration, which, having recently been hit by drug approval scandals, was clamping down on the minutiae of drug manufacturing.[23] The FDA had demanded changes to the way Fisons produced ultra-pure water for Imferon, and changes to the quality control methodology on Opticrom. Kerridge assured the assembled interested parties that there was nothing wrong with any of the company's products, and that sales of Imferon in the US had already restarted. Opticrom would also be back on the market in the near future. The City was stunned. Analysts' forecasts for the year were duly cut by £20 million to £255 million. Kerridge remained upbeat about the company's growth prospects. Sales of Intal had risen by a further 15 per cent, providing most of the group's profits for the year. As for US approval on Tilade, Kerridge was confident enough to declare that 'We have ticked all the boxes and are awaiting the final letter'.[24] Nevertheless the share price took a 35p downward leap, closing at 464p on the day.

Following the announcement, the company stepped up its routine of visits to key brokers and financial institutions. The visits did nothing to reassure the analysts who, critical of the company's reluctance to discuss key areas of the business, marked the shares down further on suspicions that the US problems may be more serious than the interim statement had implied. Their worst fears were confirmed – and exceeded – on 11 December 1991, when Fisons proclaimed that:

> The Interim results announced in September identified the adverse impact on Fisons' results of the voluntary withdrawal in the USA of the iron dextran product Imferon and Opticrom, an anti-allergic eye preparation. At that time there was good reason to believe that the production and quality control issues, which were the only reason for these withdrawals, would be resolved quickly. Unfortunately, this has not been the case, and it is now clear that supply of these products will not be resumed in 1991. This further delay has a significant impact upon the Company's profits in 1991, which are heavily geared to the final quarter ... The total impact on profits, including the resultant increased finance charge, is approximately £65m. This will have the effect

of reducing the Group's 1991 pretax profit to a level materially below that achieved last year ... It must be emphasised that this reduction in profits is essentially short term in nature as it arises specifically from the production issues outlined above, and relates exclusively to one factory, the Holmes Chapel site in the UK.

... It is expected that once full supply is resumed next year, it will be possible to recapture the sales levels and market share enjoyed in the US for these two products before the disruption occurred. It is likely that the first six months of 1992 will suffer to some extent, and that it will be towards the end of that year before full normality is restored ... Because of the Board's confidence in the future growth of the business it sees no reason to alter the policy of sustained growth of the Company's dividend.[25]

Disagreements with the FDA

Fisons was reportedly furious with the FDA, blaming the delay in clearing Imferon and Opticrom on a dispute between the FDA's Washington and Buffalo offices.[26, 27] Within minutes of the announcement, a further £340 million was sliced off the value of the company. The affair had by now reduced the price of the company's stock by a third, wiping over £1 billion off its stock market value.

Fisons' longer-term growth prospects were now in doubt. As one analyst observed, 'They've been growing through acquisitions based on their shares, and with the shares so weak, they won't be able to keep it up.'[28] There were suggestions too by some media commentators that, in its determination to deliver above-average earnings growth, Fisons had neglected to invest in its production facilities. The *Financial Times* raised this and similar questions:

Even in the context of Fisons' accident-prone history, yesterday's gruesome announcement leaves the company with much to explain. Exasperated shareholders will want to know why the loss of £40m worth of US drug sales is costing the company £65m in profits this year. They may also wonder why Fisons is taking 18 months to raise manufacturing standards on two of its main products to the levels required by the world's largest pharmaceutical market. Fisons' share price has been weakening for six months as the scale of the disaster became gradually apparent. But there has been more to it than that. Fisons' management style, marked by occasional secrecy and apparent self-satisfaction, is more respected than trusted in the market. The company now has to convince investors that the problem is indeed isolated and that profits, having fallen to perhaps £185m this year, can rebound to £250m in 1992.[29]

The City took a deep breath, crossed its fingers and consoled itself that Fisons' management had at last come clean on the extent of the problems with the US regulators. Two weeks later, as the stock markets reopened from the Christmas break, Fisons' share price nosedived again as further revelations about its difficulties with the FDA swept the City.

According to an article published in *FDC Reports* – an American technical magazine with close links to the FDA – there were long-standing leakage problems with Intal inhalers. The cause had been found to be a fault in the inhalers' aerosol valves. The company had then switched to an alternative design of valve without FDA approval. The article claimed that the regulators had found that the new valve was 'cheaper' but 'not an improvement on the one it replaced',[30] and that the company had allegedly produced deceptive test results on the new equipment. It quoted FDA sources saying that despite 'violations', Intal had been allowed to remain on the market only because of 'the medical necessity of the product' (i.e. because no alternative product was available).[31] The report went on to reveal that the FDA inspectors had also uncovered significant deviations from best practice at the Holmes Chapel manufacturing plant, including the practice of storing drug solutions in what the inspector identified as 'beer kegs'. Most damaging of all though, it alleged that supplies of Imferon and Opticrom were unlikely to return to normal before the end of 1992.

The company's immediate reaction was to dismiss the allegations as 'scaremongering' and 'nonsense', adding that 'the reports the magazine is talking about are negotiations that took place between 1987 and 1990'.[30, 31] The City was unimpressed. Three days later, on 30 December, a detailed press statement was issued by the company. It stated that 'At no time did Fisons use "beer kegs" in the manufacture of Imferon' and refuted or explained in detail the background to each of the allegations made by the *FDC Reports*. The statement steadied the stock market, but for at least one institutional investor – the SYPA – this was one mishap too many. On 6 January 1992, in a letter to the *Financial Times* (Figure 4.2.1), the institution's clerk and financial officer expressed his concern over 'the board's failure to manage effectively both its business and its investor relations'. He went on to question their ability to extricate the company from its current difficulties before the next annual general meeting, and urged every other Fisons shareholder to do the same. The SYPA was looking for institutional support for a motion to challenge the board at the forthcoming annual meeting. Fisons' shares were widely held, with the six largest shareholders controlling around 15 per cent of the equity. The following day, an article in London's *Evening Standard* revealed that the SYPA's letter had drawn a 'positive response'. Kerridge's resignation – with immediate effect – was announced one week later.[33]

From Mr R. C. Johnston.

Sir, Fisons' recent revelations about its difficulties in complying with the new US Food and Drug Administration manufacturing requirements seems to demonstrate once again the board's failure to manage effectively both its business and its investor relations.

The City's reaction to the bombshells has been clearly reflected in the dramatic fall in share price.

As an institutional shareholder in Fisons, my concern at the mishandling of the group's corporate relations is not just confined to the failures of the last few weeks. For some considerable time I have been anxious over Fisons' attitude towards peat extraction from sites of scientific interest and have been involved in lengthy and protracted correspondence with the company over the issue.

Despite leading retailers such as Do-It-All, Texas Homecare, B & Q, and Sainsbury's Homebase agreeing not to sell products dug from environmentally sensitive sites, the company's director of corporate affairs stated that relations between Fisons and its big customers were an operational matter and not a matter for comment to shareholders.

Earlier in the year, you will recall, Fisons had been represented on the Confederation of British Industry working party looking at communications with small shareholders.

Presumably, the board of Fisons considers that its appalling track record over Opticrom and Imferon and the FDA is an operational matter as well and, therefore, not of importance to shareholders either.

One leading stockbroker said recently that the company is clearly like a patient on a stretcher at the moment. I prefer to think of it as a board whose members are sinking slowly up to their necks in a deep peat bog.

Whether they can dig themselves out of the mire in time for the annual general meeting is another matter and one that I would urge every single shareholder in Fisons to give very serious thought to.

R. C. Johnston, P.O. Box 37, Regent Street, Barnsley, South Yorkshire S70 2PQ

Figure 4.2.1 Institutional shareholder takes issue with Fisons' investor relations.
Source: Financial Times, 6 December 1992. Reproduced with permission.

The post-Kerridge era

The man who stepped into Kerridge's shoes was Patrick Egan, a non-executive director of the company since 1985. Egan was immediately appointed executive chairman by his fellow directors. Currently a main board member of Unilever NV, with responsibility for Latin America, South Africa and central Asia, he was due to retire from the post in May 1992. He would juggle both posts in the meantime. The search for a new chief executive was already underway, and Egan stressed that both internal and external candidates would be considered for the post. Once a new chief executive had been appointed, Egan would take over as non-executive chairman, concentrating his energies on the company's strategic direction, and on improving relations with the City and the company's shareholders.

Promising a more open style of management, Egan acknowledged that Fisons' relationships with institutional investors and City opinion-formers had not always been well handled in the past, 'But if you think there are any further skeletons to come out of the cupboard, then you are wrong,' he said.[34] Egan at once set about rebuilding bridges with the investment community. His easy, open style contrasted reassuringly with Kerridge's guarded manner, lifting hopes that Fisons would recover quickly under new leadership. To prove the point, it was announced that Peter Fothergill, Director of the Pharmaceutical Division, would be giving the company's first R&D briefing to analysts for five years, later that week. It seemed that *glasnost* had finally reached Suffolk. The news that an FDA inspection of the Holmes Chapel plant was due within the quarter also provided a glimmer of hope that progress was being made regarding Imferon and Opticrom's restoration to the market.

The respite was short-lived. On 16 January, the full report of the FDA's latest visit to the Holmes Chapel plant became available under US freedom of information legislation. The report revealed that the FDA had in fact inspected the site back in 1990, when a catalogue of faults had been identified. Several months later a reinspection by the regulators had found that some of its recommendations had still not been put into practice. A spokesman for Fisons had dismissed the FDA's findings as 'not significant', adding that 'All drug companies get these reports. The FDA is ever increasing its quality standards.'[35] The following day, Peter Smith, Associate Director of the FDA's International Programmes and Technical Support Branch – the department responsible for overseas inspections – made plain his views on the matter. In a statement in the *Guardian* newspaper, Smith insisted that 'The points in the inspection report were very significant from the FDA standpoint ... We have invested more time and money in Fisons than any other company ever and [they] keep failing inspections.' Smith went on to explain that an inspection usually takes two to five days, but visits to Fisons could take up to three weeks. 'And this is all at the expense of the FDA. There are always problems there.'[36]

Meanwhile the *Financial Times* had also been reading the FDA's documentation. Fisons had maintained for the last 18 months that final approval of Tilade was imminent, but the documentation revealed that its production too had failed to meet regulations.[37] It had since passed a second inspection. The share price yo-yoed as each setback fuelled speculation that Fisons was about to become the target of a takeover. On 3 March 1992, Egan announced the final pre-tax profits for 1991. They were very much in line with expectations. He expressed his regret that the results were not better, but said that 'we explained there would be a shortfall. I believe that it is of a comparatively temporary nature and that by the second half of this year we will be back on our growth pattern'.[38]

A month later the company was able to announce that Cedric Scroggs, chairman of the Scientific Equipment Division, had been appointed to the post of chief executive. Other board members were reshuffled. Bob Lankester, head of the horticulture business, replaced Scroggs in scientific equipment. Scroggs would also take over the post of chairman of the Pharmaceutical Division, following the resignation of Peter Fothergill. The City had made no secret of the fact that it would have preferred an outsider, but Egan insisted that having built the Scientific Equipment Division up from nothing to a £68 million profit in 1991, Scroggs was 'far and away the best man for the job'.[37] The announcement was accompanied by the news that a date had been set for the long-awaited FDA inspection of Opticrom's facilities. The visit would take place on 20 April, meaning that, subject to a clean bill of health, approval should follow almost immediately, putting the drug back on the US market in time for the autumn hayfever season.

On his appointment Scroggs acted at once, slaying two of Kerridge's sacred cows to demonstrate that changes were afoot. The horticulture business – together with its troublesome peat bogs – was up for sale; so was the over-the-counter medicines and vitamin units of the Pharmaceutical Division. Their sale would provide cash to reinvest in the core drugs business. Fisons would also be seeking joint marketing agreements to improve the marketing of Tilade in some parts of Europe.

At the company's annual meeting on 12 May 1992, it became apparent that, although the FDA had yet to formally report on the Opticrom inspection, all was not well. There were real doubts as to whether the drug would make it back onto the US market this autumn. Worse still, it came to light that the US patent on Opticrom would expire early the following year. Rival versions of the relatively simple drug would by then be on the market. Long-standing rumours about the company's intention to abandon the manufacture of Imferon were also given credence, when the company confirmed that American drugs group, Starius, had just received FDA approval to market a similar product. The shares took their now customary skip downwards.

On 12 June a profits warning was issued, indicating that profits for the half year to June 1992 would be only around half those for the same period last year. It became clear that the most gloomy analysts' forecasts had been grossly overoptimistic. The shares crashed 77p on the day, wiping almost a quarter off the company's value. Only speculation that a bid must now be imminent prevented them from falling further. Investors were infuriated by the imprecise nature of the statement and bailed out of the stock as quickly as possible. Criticisms that Fisons should have warned investors of the extent of the problems a month earlier at the annual meeting were rejected by Scroggs, who felt that they showed 'a fundamental misunderstanding of the nature of our pharmaceutical business'.[39]

Fisons struggled through the next year. The consumer healthcare interests were duly sold, but the UK horticulture business was proving more difficult to unload. The company continued its negotiations with the FDA, finally securing approval for Tilade on 31 December 1992. Discussions regarding the relicensing of Opticrom continued, but Imferon was formally abandoned. The forced abandonment of the company's next generation asthma drug, Tiperdane, in a late stage of development (an occupational hazard in the drugs business), added to the company's woes. In late 1993, analysts were alerted to the fact that the scientific equipment business was in trouble. Customers had been reluctant to invest in capital equipment in a depressed economic climate. A price war in the US had also taken its toll.

On Tuesday 13 December 1993, Egan announced that Fisons was once more seeking to appoint a new chief executive, following the dismissal of Cedric Scroggs. Finance director, Roy Thomas, would also retire immediately. The news was accompanied by yet another profits warning. Difficulties in the Scientific Equipment Division had wiped out all of the company's profits for the entire year (see Tables 4.2.1 and 4.2.2 and Figures 4.2.2 and 4.2.3 for details of financial performance). The collapse, it transpired, had highlighted some accounting practices in the Pharmaceutical Division that were 'unwise, though not illegal'.[39] The news sent the company's shares plunging to 102p. 'It looks like a basket case' was the

Table 4.2.1 Fisons group sales, 1980–93

Year	Sales (£m)	Profit before tax (£m)
1980	454.7	3.8
1981	494.4	9.2
1982	350.5	21.1
1983	365.4	31.2
1984	552.6	48.3
1985	646.7	72.3
1986	702.6	85.1
1987	760.3	109.1
1988	823.7	132.1
1989	1019.8	169.0
1990	1205.1	230.2
1991	1225.3	190.5
1992	1284.2	117.4
1993	1324.4	1.0

Source: Fisons Annual Reports, 1980–93.

Table 4.2.2 Fisons sales and trading profit by activity, 1984–93 (£m)

Year	Pharmaceuticals		Scientific equipment		Horticulture	
	Sales	Profit	Sales	Profit	Sales	Profit
1984	198.5	31.2	291.1	15.8	63.0	5.8
1985	220.8	39.0	358.2	19.2	67.7	8.7
1986	249.8	49.8	380.6	23.2	72.2	8.0
1987	281.9	62.8	410.0	27.0	68.4	8.6
1988	327.6	91.5	419.5	27.0	76.6	5.2
1989	473.0	127.7	467.7	31.2	79.1	8.1
1990	500.7	151.7	620.0	67.2	84.4	10.4
1991	484.1	120.8	644.5	68.4	96.7	11.5
1992	417.6	71.4	662.9	34.6	108.0	11.0
1993*	445.8	42.2	749.0	(12.0)	67.3	4.2

Source: Fisons Annual Reports, 1980–93.

*1993 figures for Horticulture are for continuing operations only.

reaction of one industry analyst, adding, 'It's hard to know what to make of it'.[40]

In a press interview, Egan explained that 'Cedric was offered the opportunity to resign, which he refused. He was, in effect, sacked.'[41] In a special edition of the company's newspaper, Egan confidently announced that 'The radical action taken by Fisons this week has wiped the slate clean. Now the company can make a fresh start to secure its future.'[42] Wiping the

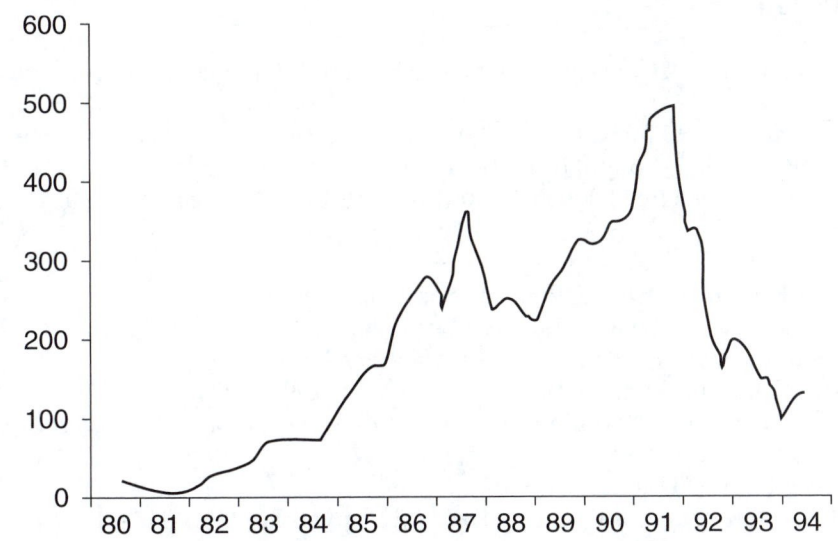

Figure 4.2.2 Fisons share price, 1980–94.

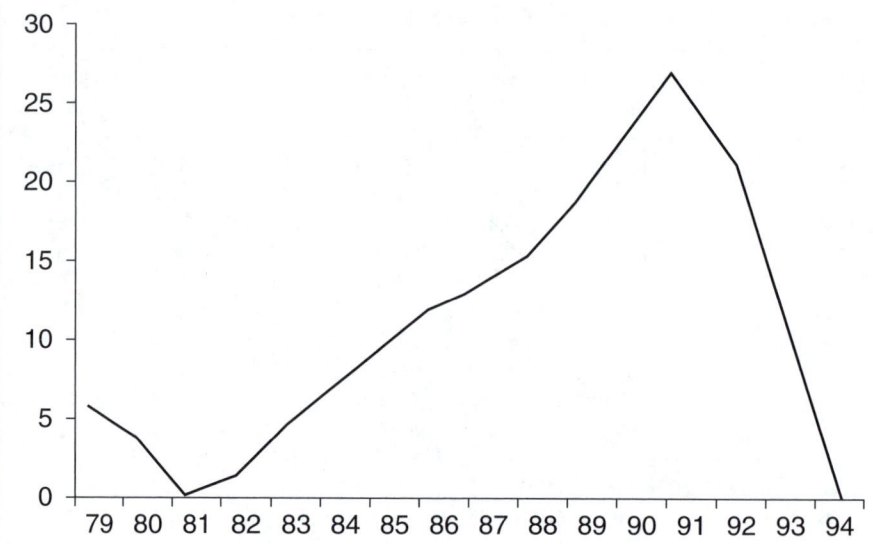

Figure 4.2.3   Fisons earnings per share, 1979–94.

slate clean may not be so easy as he hoped. In the words of one analyst from stockbrokers Salomon Brothers International, 'The credibility of anybody involved in that company is minus seven on the Richter scale.'

References

1   Fisons Plc, 14 January 1992, as reported by the Regulatory News Service.
2   Austin, J. and Finlay, P.N. (1988). *Fisons (C): The years of major restructuring 1980–1983*, Loughborough University.
3   Newman, N. (1982). Fisons' unfertilised future. *Management Today*, May.
4   *Ibid*.
5   *Financial Times*, 22 July 1982.
6   Lex Column, *Financial Times*, 12 September 1990.
7   Reuter News Service, 11 September 1990.
8   Tempus Column, *The Times*, 12 September 1990.
9   Lex Column, *Financial Times*, 12 September 1990.
10 The Economist Intelligence Unit (1990). *The Global Pharmaceutical Industry in the 1990s*.
11 *Observer*, 16 September 1990.
12 Food and drugs and politics. *Forbes*, 22 November 1993, 115–119.
13 ENDS Report 183, April 1990, 18.

14 *Daily Telegraph*, 21 May 1990, 8.
15 *Hansard*, House of Lords, 9 May 1990.
16 *Daily Telegraph*, 27 March 1990.
17 Economist Intelligence Unit (1991). *The Greening of Global Investment.*
18 *Daily Telegraph*, 21 May 1990.
19 *Financial Times*, 23 May 1990, 8.
20 *Independent*, 21 May 1990, 25.
21 *DIY Week*, May 1991, 2.
22 *Sunday Times*, 17 November 1991.
23 Reuter News Service, 17 September 1991.
24 Quester Column, *Daily Telegraph*, 18 September 1991.
25 Fisons Plc. Press Release. 11 December 1991.
26 *The Times*, 12 December 1991.
27 *The Times*, 31 December 1991.
28 Dow Jones News Service, 11 December 1991.
29 Lex Column, *Financial Times*, 12 December 1991.
30 *Daily Telegraph*, 28 December 1991, 16.
31 *Daily Mail*, 28 December 1991, 47.
32 *Guardian*, 27 March 1991.
33 *Evening Standard*, 7 January 1992.
34 *Guardian*, 15 January 1992, 11.
35 *Guardian*, 16 January 1992, 1.
36 *Guardian*, 17 January 1992, 22.
37 *Financial Times*, 17 January 1992.
38 Reuter News Service, 3 March 1994.
39 *Financial Times*, 14 December 1993, 1.
40 *Guardian*, 14 December 1993.
41 *Financial Times*, 14 December 1993.
42 *Connect*, 16 December 1993.

Case 4.3   The Body Shop International: The most honest cosmetic company in the world

This case was written by Andrew Campbell, one of the Directors of the Ashridge Strategic Management Centre in London.

Every year the cynics wait for The Body Shop to trip over its ideologically pure feet and every year they are disappointed. Although imitators have inevitably arisen, The Body Shop benefits from being clearly identified as the leader of the pack in the growing market for toiletries and cosmetics aimed at the environmentally oriented, health-conscious consumer.

The Body Shop 'originates, produces and sells naturally based skin and hair products and related items through its own shops and through franchised outlets'. The business has grown rapidly since it opened its first shop in Brighton in 1976, and in 1990 it had a turnover of £84.5 million and 457 outlets in the UK and overseas. In the seven years since its flotation, the company has increased both turnover and profits by a factor of nine, and has been described as 'the share that defies gravity'. This is despite the onset of a recession in retailing. Financial performance for the last five years is detailed in Table 4.3.1. Although it is not one of the largest retail operators, The Body Shop has been particularly influential because of its phenomenal success, its strong underlying philosophy and the press coverage it has received.

Much of the press coverage has centred around managing director Anita Roddick, a charismatic, outspoken and determined figure who has a simple formula to explain the secret of her legendary success: 'I look at what the cosmetics trade is doing and walk in the opposite direction.'

The extent of this success is such that she claims that The Body Shop is Britain's most international store. 'We produce over 300 products sold in well over 300 shops from the Arctic Circle to Adelaide, covering 31 countries and 13 languages, without once diluting our image.' Actually by February 1990 The Body Shop was operating with 457 shops, 139 in Britain and 318 in 37 other countries. A further 25 UK outlets and 180 overseas were due to open in the following 12 months, including a shop in Japan in October 1990. Preliminary research has been conducted into the feasibility of opening in Moscow.

Anita Roddick has won many accolades from the business community, including the Business Enterprise Award for company of the year, and Business Woman of the Year. In her acceptance speeches she savages corporate approaches to business and in particular traditional ways of doing business in the retail and cosmetics sectors:

> Retailing itself has taught me nothing. I see tired executives in tired systems. These huge corporations are dying of boredom caused by the inertia of

Table 4.3.1   Financial performance, 1986–90 (£'000)

	1990[1]	1989[1]	1989[2]	1987[3]	1986[3]
Turnover: UK & Eire	56 901	41 412	54 754	21 255	13 560
Overseas	27 579	13 997	18 253	7221	3834
Total	84 480	55 409	73 007	28 476	17 394
Pre-tax profit	14 508	11 232	15 243	5998	3451
Earnings per share	10.0p	7.4p	10.2p	4.65p	2.58p
No of outlets: UK	139	112	112	89	77
Overseas	318	255	255	186	155

	1990[1]	1989[1]	1989[2]
Group turnover:			
United Kingdom & Eire	56 901	41 412	54 754
Other EEC countries	6962	4136	5445
Rest of Europe	3910	2717	3966
USA	5839	874	874
Rest of North America	5860	3194	4244
Australasia	3544	2119	2454
Asia	1464	957	1270
	84 480	55 409	73 007
Group trading profits:			
United Kingdom & Eire	13 486	9745	13 015
Other EEC countries	1566	904	1254
Rest of Europe	996	932	1270
USA	(1941)	(1632)	(1820)
Rest of North America	1481	887	1082
Australasia	915	326	422
Asia	389	222	324
	16 892	11 384	15 547

[1]Years ended 28 February, [2]17 months ended 28 February, [3]years ended 30 September.

giantism. All these big retailing companies seem to be led by accountants and they seem to have become just versions of the Post Office or the Department of Motor Vehicles.

Retailing at the moment is a combination of war and sport in designer uniforms, with its obsession with corporate raiding, acquisitions of acres, strategies, niche markets, specialisation and empire building, where their only sense of adventure is in their profit and loss sheet. We have never once been seduced into believing we are anything more than simply traders.

Her belief is that the essential difference between The Body Shop and other retailers is explained in the words of Niemann Marcus: 'Profit is not the objective of my business. It is providing a product and a service.' So how does she do it?

> It is so easy. First, know your differences and exploit them, then know your customers and educate them, then talk about the image of your company as well as the products, and finally be daring, be first and be different.
>
> One of the rules of any successful company is to find out what your original features are and shout them out from the rooftops. We have found that when you take care of your customers really well, and make them the focal point, never once forgetting that your first line of customers are your own staff, profitability flows from that.

The Body Shop has an extraordinary effect on people who come into contact with it. 'It arouses enthusiasm, commitment and loyalty more often found in a political movement than a corporation,' says journalist Bo Burlingham. 'Customers light up when asked about it, and start pitching its products like missionaries selling Bibles.' John Richards, director of retail research at County NatWest Securities, comments: 'I've never seen anything like it. The nearest comparison would be something like flower power in the 1960s.'

The Body Shop story

Anita Roddick was one of four children of Italian immigrants and helped in the family cafe at the Sussex seaside resort of Littlehampton, which is still the base for her retailing empire. She has fond memories of the cafe, a popular meeting place for local children. 'We had the first juke box in the town after the war, the first knickerbocker glories and the first Pepsi-Colas. I didn't know it then but I was receiving subliminal training for business life; I was at the centre of that magical area where buyer and seller come together.'

At an early age she decided that she wanted to see the world and went to work for the International Labour Office of the United Nations in Paris. It was while travelling internationally for the UN that the seeds of her future calling were sown. Visiting such exotic spots as Polynesia, Mauritius and the New Hebrides, she observed the simple but effective way remote communities lived.

'I just lived as they did and watched how they groomed themselves without any cosmetic aids. Their skin was wonderful and their hair was beautifully clean.' She watched the Polynesians scoop up untreated cocoa butter and apply it to their skin with remarkable results. She also observed Sri Lankans using pineapple juice as a skin cleanser, and later discovered that natural enzymes in it help remove dead cells.

Today, Anita spends about two months every year travelling the world picking up tips for natural ingredients to go into Body Shop products. 'Women in other societies know that these well tried and tested ways work and do not need a scientist or advertising agency to sell them.' When Anita gets back from a trip abroad she will regale managers with tales of her adventures. Walls in warehouses and factories are hung with words and images and displays of Third World art.

When she returned from her travels, Anita married and opened a restaurant with her husband, Gordon. He too got the wanderlust and set off on a horse-back ride from South America to New York that was to take him two years. She did not feel that she could cope with the restaurant on her own and decided to open a shop instead, selling skin and hair products made from the natural recipes gleaned on her travels.

'You cannot call this shop The Body Shop'

Starting with a bank loan of £4000, Anita Roddick opened the first Body Shop in Brighton in 1976 with a blaze of publicity. She was jammed between two funeral parlours, who wrote her a letter saying, 'You cannot call this shop The Body Shop', because the coffins would pass twice a day and they were expecting some cute photographer from *She* magazine to take that happy snapshot of the week. Her response was straightforward, as she recalls: 'I have always been petrified of headmasters and solicitors, but I think that the two most talented things that I have ever done in my life were to ignore those letters and then to use that to promote the company. So what I did was quite simple, and I think it should be standard practice for any young company setting up: the anonymous phone call! I rang up the local *Evening Argus* in Brighton and said to them,

> Do you know what is going on in Kensington Gardens? This poor woman on her own, with a new baby, whose husband is trekking across South America on a horse, is being intimidated by two Mafia undertakers. Her little shop is called The Body Shop ... – I mean, I had written the story over the phone and we got our first free editorial. We have never ever paid for an advert since.

Many of the features which made The Body Shop different came about because of lack of funds in those early stages. The company could only afford 20 products to begin with, which was not enough to fill the shop. So making each product in five different sizes gave a wider range straight away. There are now over 300 products, but they can still be bought in five different sizes – customers like to be able to try the small sizes first. The Body Shop still uses the cheapest bottles, referred to in the early days by the press as 'urine sample bottles', and they can still be brought back for refills (a system originally introduced because they were in short supply, but now symbolic of the company's policy of recycling). Similarly symbolic are carrier bags which carry the question: 'Why aren't telephone bills, gas bills, electricity bills, rate demands, income tax forms, public notices, circulars, newspapers, printed on recycled paper? This is a recycled paper bag. The Body Shop introduces changes for the better.'

The success of The Body Shop grew out of Roddick's almost naive belief in herself and the value of sheer hard work. 'We worked hard, therefore we survived,' she says. 'We didn't have any understanding of the commercial methods taught by business colleges. In fact I would suggest that anyone with an ounce of individuality should not go to a business school ... because you are structured by academics who measure you in the science of business. They use a business language that is predictable, and where going out and doing is not part of the course.'

The Body Shop expanded rapidly under the franchise system, which developed a strong camaraderie through the help given to each franchise in setting up and through allowing everyone to do their own labelling. There are currently around 5000 applicants wanting to take up a Body Shop franchise, and it takes three years to succeed. 'Unless you're absolutely obsessed you don't get a look in.'

Applicants undergo strict vetting, including an offbeat questionnaire with unlikely questions such as: 'How would you like to die?' 'What is your favourite flower?' 'Who is your heroine in history or poetry?' Roddick believes that basic business skills can be provided by the company but the right attitude and values cannot. 'We have the back-up to teach almost anyone to run a Body Shop,' she says. 'What we can't control is the soul.'

The Body Shop has managed to achieve a remarkable level of uniformity within its now global network of shops. 'They are all the same – and they all work. I think it is interesting that we are not seen as an English company but as cross-cultural, with a product range with international ingredients.'

Operations

In the early days bottles were filled, labelled and capped by hand and each order picked and filled individually. The process is now fully automated. There is a manufacturing department with a staff of seven, covering manufacturing, quality control, product development and customer complaints.

25 per cent of manufacturing is done in-house, the rest by contract manufacturers. The range grows by 80–90 per cent each year. In 1990 construction is due to be completed on new manufacturing and blow-moulding facilities as well as a research and development and office building in Littlehampton.

Each supplier to The Body Shop is required to sign a declaration guaranteeing that none of the ingredients used has been tested on animals during the previous five years. To show its complete opposition to cruelty in the name of beauty, in 1989 The Body Shop resigned from the Cosmetic Toiletry and Perfume Association amid accusations that the trade body lacked the necessary passion and imagination to eliminate animal testing quickly enough.

Franchise system

At the start of the company, Anita and Gordon Roddick could not afford to open new shops themselves even though business was booming, so they developed the concept of 'self-financing'. If someone else would put up the money to open a new shop, the Roddicks would help with their expertise in running the operation, help re-fit the shop, and grant a licence to use the company name and sell the products. The company now has a franchise manager, who provides a consultancy service and organizes the relocation of older shops to prime sites.

The franchise system also operates overseas, with a head franchisee for a country or group of countries who is granted exclusive rights to use The Body Shop trade mark and to distribute its products. Those who operate their own shops successfully are given the right to subfranchise within their area, and have responsibility for training subfranchisees.

As in the UK, shop designs and graphics are strictly enforced and franchisees have to stock 85 per cent Body Shop products. Unlike the UK, however, no annual operating fees are charged to overseas franchisees and products are sold subject to an overseas distribution discount.

Scandinavia was the first overseas area in which The Body Shop became popular, followed by Canada, which became the first overseas operation to manufacture products itself. Samples are still sent to the UK for quality control before each batch is bottled, and the 'heart ingredients' are provided by Body Shop International and blended in the UK. Other ingredients are approved by the UK before manufacture in Canada.

The Body Shop philosophy

In essence, Anita Roddick promotes an ethical code of behaviour for the global citizen – and that includes multinational companies. She believes in the empowerment of people, through jobs, work, honest earning. 'Our idea of success is the number of people we have employed, how we have educated them and raised their human consciousness, and whether we have enthused them with a breathless enthusiasm. Our solution to third world poverty is trade not aid.'

The philosophy is explained to the customer as follows:

1 The Body Shop continues to trade today on the same principles that have held firm since its beginning in 1976.
2 We use vegetable rather than animal ingredients in our products.
3 We do not test our ingredients or final products on animals.
4 We respect our environment: we offer a refill service in our shops, all our products are biodegradable, we recycle waste and use recycled paper wherever possible, we use biodegradable carrier bags.
5 We use naturally based, close-to-source ingredients as much as we can.
6 We offer a range of sizes and keep packaging to a minimum: our customers pay for the product, not elaborate packaging or for more than they need (and this helps keep the prices down too).

The philosophy is put into practice on many levels, some more visible than others:

1 Our products reflect our philosophy.
2 They are formulated with care and respect:
   respect for other cultures
   respect for the past
   respect for the natural world
   respect for the customer.
3 The Body Shop joined forces with Greenpeace over a two-year period in a campaign to 'Save the Whale' and in the UK we are involved with the Friends of the Earth during 1987–88, in a campaign to raise public awareness of the dangers of acid rain and other environmental hazards, and to encourage others to take positive action to protect their environment, such as recycling household and workplace waste.

Roddick and her employees have real enthusiasm for the company and its products. 'I see business as a renaissance concept,' she says, 'where the human spirit comes into play. How do you ennoble the spirit when you are selling moisture cream? Let me tell you, the spirit soars when you are making products that are life serving, that make people feel better. I can even feel great about a moisture cream because of that.'

There are some very visible manifestations of the company's philosophy in the way the company operates. For example, each employee at the Littlehampton head office has two wastepaper baskets, one for recyclable and one for non-recyclable waste. The company even runs training courses on recycling.

Commitment by top management to such values is vital. 'The people who make the policy decisions ... must lead with integrity, commitment and passion, otherwise a cynicism pervades the whole place,' says Roddick. 'Corporate culture is the most important part of a growing company like The Body Shop – it is the values, the rituals, the goals, the hero's characteristic of a company's style.'

The responsibility of profits

After The Body Shop was launched on the Unlisted Securities Market and the shops were all proving profitable, Roddick set up an environmental and communities department to translate her beliefs and concerns into practical projects. Each franchised outlet is required to take on a community project in its area, 'which is there to give the young women in the organisation additional status and helps them realise that everyone has the ability to change the world for the better'. All projects are taken on within working hours, and franchisees choose what they want to do. There is no coercion.

This determination to use private profit for public good is now reaching out to some of the remotest parts of the Third World, where one of the latest projects is setting up a paper-making plant in a Tibetan refugee camp. The paper is processed from pineapple and banana leaves and will be used for wrapping Body Shop products.

In Southern India, The Body Shop has set up a boys' town for destitute youngsters. The boys are taught rural crafts and to make Christmas cards, and the money from the sales is put into trust funds. When the youngsters leave at the age of 16 they have the means to purchase a herd of sheep or a horse and cart, giving them a vital start. So far, over 3000 jobs have been created and the scheme has made about £100 000 profit, and supplies The Body Shop with soap bags and wooden foot massagers.

The whole process is seen to be self-perpetuating, with ingredients obtained from the Third World, providing work and sustenance for under-privileged societies, making products that are sold to the more fortunate, the profits of which are ploughed back into an educational programme which aims to make people more aware of the critical social issues of our time.

The plight of the inhabitants of the Amazon rain forest has been the subject of a worldwide campaign by the company, which has raised £250 000 for their defence. It has mobilized employees for petition drives and fund-raising campaigns, carried out through the shops and in company time. It has produced window displays, T-shirts, brochures and videotapes to educate people about the issues, and has even printed appeals on the side of its delivery trucks.

The Body Shop was the first company in Britain to use jojoba oil in cosmetics. Jojoba is obtained from a desert plant and is a substitute for sperm whale oil. Apart from helping protect whales, there are other powerful reasons for using oil from the desert plant. Jojoba can be grown on some of the poorest land in the world, which is totally unsuitable for conventional crops, and in regions where people are living in abject poverty.

This approach isn't restricted to far-away places. Chris Elphick of Community Learning Initiatives suggested that perhaps Roddick might care to go and practise some of her philosophies in Easterhouse, an area just to the east of Glasgow which has 56 per cent male unemployment and frequent deaths from solvent abuse. Her response was predictable. Within eight months she had won over all the local councillors, opened a soap factory called Soapworks and dedicated 25 per cent of all profits to the local community. When asked about unions, she told them: 'You only need unions when management are bastards. We will talk to you one to one if there is a problem.' Employees are treated with respect and made to feel that their roles are important. Soapworks is involved in the community and has funded the building of a playground for local children. In the first full trading year Soapworks produced over 4 million bars of soap and expected to produce more than 15 million bars in 1990. A bath-salt filling line has been added, and the workforce now stands at 85.

The commitment to 'profits with principles' is also evident in initiatives for employees, such as the £1 million invested in 1989 in building and equipping a workplace nursery for head office staff.

Journalist Bo Burlingham claims that the campaigning approach is part of a carefully researched and executed business strategy. Roddick wants causes that will generate excitement and enthusiasm in the shops, and says:

> You educate people by their passions, especially young people. You find ways to grab their imagination. They're doing what I'm doing. They're learning. Three years ago I didn't know anything about the rain forest. Five years ago I didn't know anything about the ozone layer. It's a process of learning to be a global citizen. And it produces a sense of passion you won't find in a department store.

No advertising

An important aspect of The Body Shop's product philosophy is that, in keeping with its claim to be the most honest cosmetics company, it does not call its products 'beauty' products, nor does it use idealized images of women to sell them. Roddick explains:

> The cosmetics industry is bizarre because it's run by men who create needs that don't exist, making women feel incredibly dissatisfied with their bodies.

They have this extraordinary belief that all women want is hope and promise. They have this absolute obsession with not telling the truth, which is bizarre because some of the products they make are actually good. But to me it's dishonest to make claims that a cream that is basically oil and water is going to take grief and stress and 50 years of living in the sun off your face. It's bullshit to consistently endorse its main product line which is garbage, waste and packaging.

Salespeople in The Body Shop are expected to be able to answer questions, but are trained not to be forceful.

The Body Shop does not advertise and its point of sale materials concentrate on giving information about the ingredients in the product, and educating the customer about its use. In fact, the information sheets about the products were first introduced because the products in the early days sometimes looked unappetizing: 'We thought we had to explain them because they looked so bizarre. I mean, there were little black things floating in some of them. We had to say these were not worms.' Containers have clear, factual explanations of what is inside and what it is good for. On the shelves are notecards with stories about the products or their ingredients. There are stacks of pamphlets with such titles as *Animal Testing and Cosmetics* and *What is Natural?* There is a huge reference book called *The Product Information Manual* with background on everything The Body Shop sells.

The level of information The Body Shop offers provides a powerful source of competitive advantage. It differentiates the company from its competitors, and it creates obstacles for would-be copycats. Customers feel they *know* its values and business practices, and the effect is to create a loyalty that goes beyond branding. 'I've just taken what every good teacher knows,' says Roddick. 'You try to make your classroom an enthralling place … I'm doing the same thing. There is education in the shops. There are anecdotes right on the products, and anecdotes adhere.'

Although the company does not advertise directly, good public relations is fundamental to their marketing strategy. Roddick quickly learned the same lesson as Marks & Spencer – that product advertising is unnecessary when a company has built up a strong and continuing public image. Says journalist Michel Syrett: 'Roddick courts publicity. She makes herself deliberately available to the Press and is a constant source of good copy. "The Press like us," she commented at the CBI last year. "I'm always available and I'm loudmouthed and quotable." Her views on healthcare, environmental issues and the soullessness of big business are not manufactured and are entirely consistent with the aims and philosophy on which The Body Shop has been founded.'

The money that would normally be spent on marketing is largely invested in the company's employees. 'It takes more or less the same approach that it uses with customers, attacking cynicism with information,' says Bo Burlingham. 'It deluges employees with newsletters, videos, brochures, posters and training programmes, to convince them that while profits may be boring, business does not have to be.'

There is a training centre in London which anyone in the company can attend free of charge. Courses are almost entirely devoted to instruction in the nature and uses of the products, and are so popular that the school cannot keep up with demand.

The company newsletter reads almost like an underground newsletter. Burlingham again: 'More space is devoted to campaigns to save the rain forest and ban ozone-depleting chemicals than the opening of a new branch. Sprinkled throughout are quotes, bits of poetry, environmental facts and anthropological anecdotes.'

The move into colour cosmetics

It might appear to the casual observer that a move into colour cosmetics (make-up as opposed to skin and haircare products) would be inconsistent with this philosophy and approach. After all, the mainstream colour cosmetics industry is characterized by glossy advertising showing the customer an idealized image of the woman she could become if only she would use this or that eyeshadow or lipstick.

But Anita Roddick had the answer to any criticism there might be. She commissioned academic research on the psychology of make-up from Dr Jean Ann Graham in the US to prove that women derive psychological support from painting their faces. The range was launched in collaboration with Barbara Daly, a well-known make-up expert who had designed the Princess of Wales' make-up on her wedding day. The packaging was minimal and the design stylishly simple. Products were coded to guide customers as to which colours go together, and again information leaflets were available, as well as a video showing how to apply make-up using both a young and an older woman as models. The range has proved successful with customers and in 1989 represented around 10–15 per cent of The Body Shop's turnover, a steadily increasing proportion.

The toiletries industry

The market for haircare products is complex and fragmented. In the UK in 1988 the total size of the skincare market alone was over £138 million. Although nearly 50 per cent of the market for skincare creams and lotions is accounted for by the top five companies, a large number of small companies makes up the remaining 56 per cent.

The shampoo market is also highly fragmented, with the top 10–20 brands accounting for half the market and the other 80–90 competing for the rest. The conditioner market is becoming more and more competitive, with many brands competing for a small market share. The top companies in the UK in terms of market share in both shampoos and conditioners are Elida Gibbs, Beecham, Alberto Culver, Johnson & Johnson and Revlon. The fastest growing brand is Timotei, which has a share of just under 10 per cent and caters to two growing trends, that for 'natural' products and that for a shampoo designed for frequent use without damaging the hair.

There is a wide spectrum of products, with new products or reformulations continually being introduced as companies seek to create or imitate new fashion fads. This means that advertising is used extensively, with the premium brands being advertised in upmarket women's glossy magazines. Television advertising is also used for both hair and skincare products, often aimed at educating the consumer about a new type of product. The proliferation of new products has also led to a blurring of product categories, e.g. moisturising cleanser, conditioning shampoo.

The industry can be segmented by a number of different criteria, e.g. price range (premium, middle and budget); target market age group; function (health and hygiene, beauty products), and so on. Growth is mainly in upmarket product ranges and consumers are primarily women, although sectors which have shown high growth recently are products for men and own-label products, with Boots now taking 6–7 per cent of the market for cleansers, moisturisers and astringents. After Boots, the most important in the own-label sector are Marks & Spencer, Sainsbury's, Yves Rocher and Superdrug. Safeway and Woolworths also have their own skincare ranges. In terms of distribution, supermarkets now account for about 30 per cent of the total haircare market; there is a growing tendency to view such products as shampoo, hairspray and conditioners as 'grocery' items, and they are increasingly sold in larger or 'family-size' packs. Another area where there has been growth recently is in anti-ageing products. Women (including younger women) are taking an increasing interest in the health of their skin and the adverse effects of wind, sun and polluted air.

Own-label products are also a threat to established brands in the soap market, such as Imperial Leather, Lux and Shield. Sales of soap reached a peak in 1986 and the market is now thought to be declining, with products such as bath oil or foam overtaking soap for the first time in 1987. The three main trends in the soap market in recent years have been 'fruity' soaps, first made popular by The Body Shop but imitated by others; pure, fragrance-free brands, such as Simple and Pears; and liquid soap, which now accounts for around 5 per cent of the total market.

The Body Shop dominates the UK market for natural make-up mainly because most manufacturers of 'natural' cosmetics, such as Creightons, have not entered the colour cosmetics market. Health-oriented manufacturers of toiletries and make-up include Innoxa, whose slogan is 'pure and beautiful' and which is the only make-up recognized by the British Medical Association.

As a result of the recent growth in environmental awareness throughout Europe, numerous small players have been active in the 'green' cosmetics market. However, only Yves Rocher is a significant competitor for The Body Shop. Monsieur Rocher's passion is 'plants and natural beauty' and this is explained in the *Green Book of Beauty*, a mail order catalogue for toiletries and make-up. Mail order constitutes the heart of this company's activities, and it has, in addition, more than 1200 Beauty Centres worldwide.

Chapter 5

The recruitment and internal market domains*

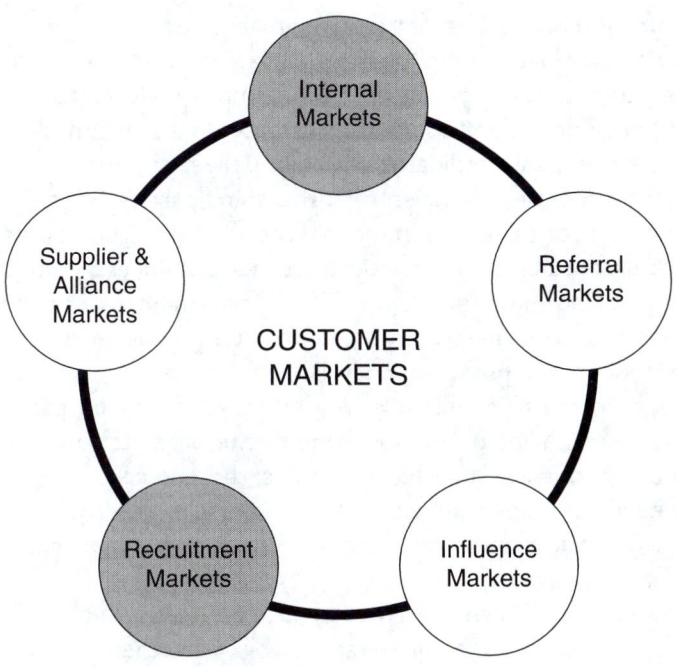

The recruitment and internal market domains

*This chapter is a summary version of a Cranfield School of Management Working Paper by Clark, M. 'Managing Recruitment and Internal Markets: A Relationship Marketing Perspective'.

Introduction

This chapter addresses two key market domains within the six markets framework – the recruitment market and the internal market. It is appropriate to consider these two markets together, as it is the extent to which companies can successfully recruit, develop and train the right calibre of people for their organizations that will determine their future success in the internal market and ultimately with the customer market.

In Chapter 1 of this book, the role of recruitment and internal markets within the Six Markets model was introduced. These two markets can be summarized as follows:

- *Recruitment markets.* The recruitment market domain represents all the potential employees who possess the necessary skills and attributes needed to match the profile that the company wishes to portray to its customers. These core competencies provide a benchmark from which to select potential candidates who can deliver superior service quality. The recruitment market also refers to third parties, for example, executive search consultants, commercial recruitment agencies, management selection consultants, universities or other employers, who have access to pools of potential employees. The recruitment market domain also includes the company's own attempts, at every level from within the organization, to pursue potential employees via advertising and employee referrals, where existing employees may be paid a bonus if they successfully introduce someone to the organization. With so many channel options available to companies, the choice of the recruitment channels is, therefore, an essential ingredient in determining a successful recruitment marketing strategy. The recruitment market domain is illustrated in Figure 5.1.
- *Internal markets.* There are two key aspects to internal marketing. The first is concerned with how staff work together across functional boundaries so that their work is attuned to the company's mission strategy and goals. The second involves the idea of the internal customer. That is, every person working within an organization is both a supplier and a customer. It is this second aspect that this chapter is predominantly concerned with for the purposes of addressing the internal market domain. In attempting to delineate the internal market domain, different levels of staff within the organization can be categorized depending on their degree and type of contact with external cus-

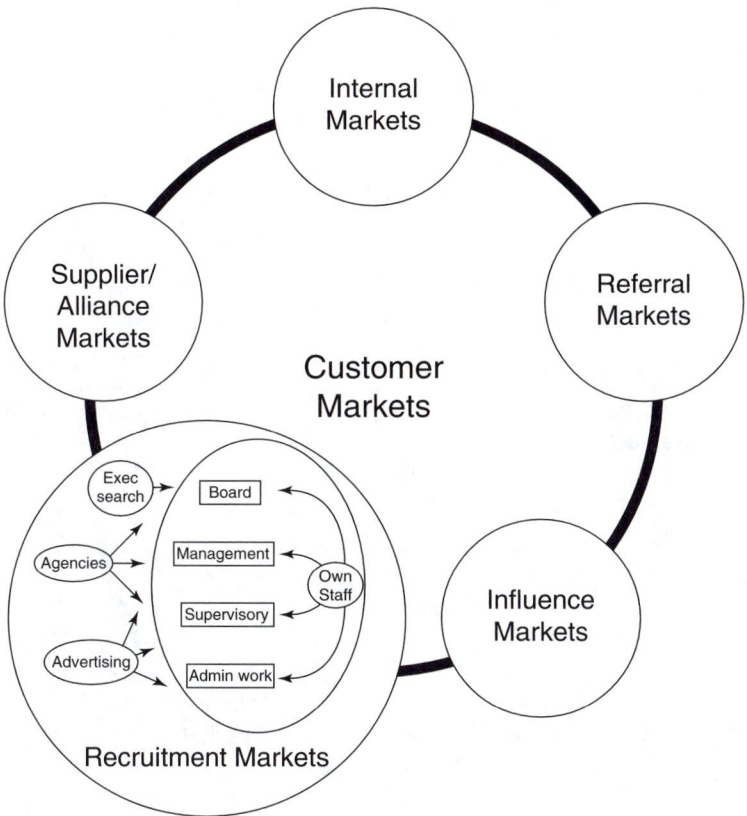

Figure 5.1 The recruitment market domain.

tomers.[1] The scale extends from those who have direct frequent contact ('contactors'), through 'modifiers' and 'influencers' to those who have no contact at all ('isolateds'). These categories make a useful starting point in developing a strategic relationship marketing approach. The internal market domain is illustrated in Figure 5.2.

This chapter is divided into four main sections. The first section provides a brief overview of the recruitment and internal market domains. The second and third sections consider recruitment and internal markets in greater detail. The final section provides an overview of the four case studies, which have been specially chosen to highlight some of the key issues relating to the recruitment and internal markets.

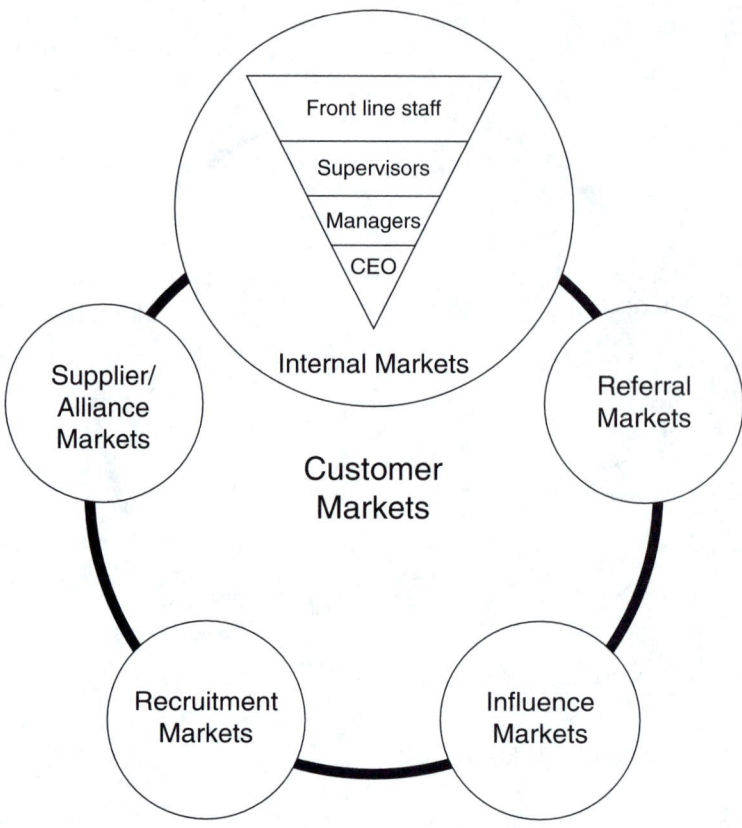

Figure 5.2   The internal market domain.

The recruitment market domain

Despite the economic recovery of recent years, redundancy and downsizing have become a fact of life for British industry. Many highly skilled and capable people have been forced back into the job market and it would seem that employers have even greater sources of potential candidates from which to choose. The reality, however, is that it is becoming increasingly difficult to find good staff. Skills shortages in key areas mean that employers are often facing the dilemma of trying hard to fill vacancies for some jobs, whilst at the same time being swamped by floods of applications for other jobs. In the volatile market of the late 1990s, skill and experience are no longer enough. Employers must also identify those individuals who can contribute to organizational effectiveness and competitive advantage.

In these circumstances good recruitment practices are essential for organizational success, particularly when a company seeks to maintain or develop a certain culture and style within the organization.[2] Organizations such as Disney and Nordstrom in the USA (see Case Studies 5.2 and 5.3) are examples of companies who owe their success to some extent to the care with which they recruit their employees. Yet despite this, the subject of recruitment has received relatively little attention from researchers and academics alike; instead they have tended to want to focus on the more glamorous area of employee selection. However, the best candidates for a job cannot be appointed if they don't apply for it in the first place. It is essential, therefore, for the would-be employer to be able to attract potential applicants and present itself to influential third parties and individuals as an employer of first choice. If it wants to *retain* these employees, it must then prove itself to be able to deliver what is promised to prospective employees. To be a highly attractive potential employer, a clearly defined recruitment strategy that adopts a relationship marketing focus and builds a reputation as a first-class employer is critical if a company is to succeed in this market domain. This section will, therefore, address key issues relating to the recruitment market domain, specifically: the costs of recruitment; finding the best employees; selection techniques; training; recruitment monitoring and evaluation.

The costs of recruitment

Annual employee turnover, particularly in the service industry, can be equal to or greater than 100 per cent.[3] Significant costs are associated with employee turnover and a particularly good example of these costs is highlighted in the Digital case (Case Study 5.4). Costs are typically incurred through expenditure on advertising, staff time taken up in interviewing prospective candidates, administration time in dealing with application forms or CVs. Then there are interview expenses to pay to applicants and possibly relocation expenses if that is required. Finally, when someone is recruited there are the costs of training the person which need to be taken into account. These are all the direct costs incurred when a new applicant is hired;

however, sometimes there are indirect costs, or opportunity costs, which need to be taken into account. Opportunity costs are those costs incurred as a result of the reduction in productivity that may be expected when either a new recruit takes over from an experienced employee or when a situation is left vacant for a period of time. It is estimated that costs of up to 30–50 per cent of an employee's annual salary are accumulated in lost production and expenses incurred in finding a replacement.[3]

Given that the costs of recruitment are so high, it is becoming increasingly important not only to find employees who have the necessary skills and competencies and are able to match the profile that the company endeavours to portray to their customers, but who are also able and willing to be retained by the company. This whole process, therefore, starts with managing the recruitment process and ensuring that the best employees are employed first time.

Finding the best employees

There are a whole range of recruitment channels available to employers who are trying to find the best employees for their organization. A good starting point for looking for potential employees is from within the organization. This method, though not commonly used by employers, is particularly attractive because these recruits already have a relationship with the company and, by being more familiar with the company, are more likely to be retained than those acquired through agencies. Another underutilized channel which is often worth pursuing when a situation cannot be filled from within the company is through referrals provided by existing employees. These potential employees are also more likely to be retained by the company than those obtained via other channels. Both these options, particularly the first one, are very cost effective, as the costs which would normally be incurred through advertising and the administration of numerous application forms can be avoided.

A 1996 IRS Survey into policy and practice in recruitment identified the most successful recruitment channels by job type based on a survey of 165 employers across a number of different industry sectors.[4] In terms of identifying the best recruitment channels by job type the survey asked respondents which recruitment channel they used for each of seven job types. For apprentice/youth training vacancies, the most commonly used recruitment channel is through schools and colleges, with a quarter of respondents considering local newspapers to be the best method of recruitment. Graduates, on the other hand, are usually recruited on a national basis. A range of methods include journals and national newspapers, speculative applications, vacancy directories, recruitment agencies, careers services and contacts with schools and colleges. Clerical/secretarial and manual/craft recruits are generally sourced on a local basis via local newspapers, with less than a third considering agencies. For professional/technical vacancies, the most commonly used channel is specialist or professional journals. Recruitment agencies, however, are also cited as a useful source of recruits. When considering senior managerial positions, recruitment usually takes place in a national labour market. National press advertising is common, as are headhunters and professional journal advertising. Finally, other managerial positions are recruited in both national and local labour markets depending on the specific nature of the vacancy. National and local press, as well as professional journals and agencies, all play a role in recruiting this particular group.

Companies frequently fail to recognize the importance of developing formal relationships with recruitment channel members. These relationships can be a great source of competitive advantage for companies, as third parties are able to act as a referral market, providing word-of-mouth referrals about specific companies to potential recruits. They are also able to act as a gatekeeper to potential employees. When provided with job descriptions and details of the requisite skills and competence which match the profile the company wishes to convey to its customers, they are able to save the company valuable time and effort in sourcing suitable candidates. There is no doubt that during times when there are more jobs available than candidates for those jobs, it is those companies which have already developed long-term relationships with recruitment third parties who are likely to benefit.

Building relationships with individuals who are likely to apply directly for positions is also important in developing a relationship approach for recruitment. All candidates are also potential customers, and, whatever the outcome of their application, a professional approach to recruitment can ensure that applicants are left with a good impression of the company. Potential employees should be given realistic expectations of the job from the very beginning. Press advertisements, brochures or information supplied by third parties should accurately reflect the job requirements and the environment within which the new recruit will be working. A failure to do so can result in disillusioned employees, low employee retention and poor word of mouth, as employees warn other potential recruits not to apply for positions within the company. A good example of how employees' expectations have been poorly managed through the recruitment process can be found in the Club Med case (Case Study 5.1).

One company that is particularly innovative in searching for prospective employees is Cisco Systems, a San Jose networking company, with sales of $6.4 billion in 1997 and profits of $1.4 billion. Because good employees were becoming increasingly difficult to find, they embarked on an innovative strategy to find, lure and keep talented staff. They started by holding focus groups with targeted groups of people to find out how they spend their free time, what websites they visit and how they feel about job hunting. The company then used this information to access potential recruits and devise ways of improving the whole recruitment process. One of the methods that the company uses to maximum effect is the Net. Newspaper job advertisements are used to feature its Internet address and an invitation to apply at Cisco. There are major benefits for Cisco in directing all job seekers to its website address. On the Internet it can inexpensively post hundreds of job openings and information about each one. It can monitor and measure the number of visits to the site, which is important for assessing the effectiveness of the recruitment programme. The company is also able to identify where people work, as most prospective employees visit the website from their place of work. Cisco have also launched a friends programme to help prospective employees make a friend at Cisco. The Cisco employee is able to describe what it is like to work there and if they are successful in helping to recruit a new employee are rewarded with referral fees which start at $500 and a lottery ticket for a free trip to Hawaii. The company, essentially, matches employees to prospective candidates based on their backgrounds and expertise. This strategy was designed to help entice people to come for an interview and smooth the whole recruitment process. To date the programme has been tremendously successful, with 1000 Cisco employees volunteering for the programme and a third of all new recruits coming through the friends programme.

Having identified the main methods for recruitment and the importance of building relationship strategies with recruitment third parties and individuals, the next stage of the process is to select the most suitable candidate for the job. Selection techniques are many and varied and the following section provides an overall summary of some of the main methods used.

Selection techniques

People who staff organizations are the most important single influence in ensuring the future success of the organization. However, creating this level of competence cannot be accomplished overnight. So often poor employee selection begins with poorly trained or untrained interviewers not fully analysing the requirements of the job. Frequently, interviewers do not ask the right questions to determine whether or not the applicants are suitable for the job and they rely too much on 'gut' feel rather than objective evaluation. The first step then in avoiding these pitfalls is to ensure that the interviewers are skilled at employee selection and that they are able to detail a person specification which represents the ideal candidate for the position. One of the most widely used frameworks used for this purpose is Rodger's seven-point plan.[5] This plan includes: physical make-up; attainments; general intelligence; special aptitudes; interests; disposition; and circumstances, as key issues in determining a person specification. Organizational criteria and functional/departmental criteria should also be considered as part of the selection process.[6] Organizational criteria refers to those attributes that an organization considers valuable in its employees and that affect judgements about a candidate's potential to be successful in an organization. For example, an organization may be focused on developing a more customer-oriented culture and wishes to employ people who are warm and friendly and good at communicating with customers. Functional/departmental criteria refers to specific skills required by departments. For example, a finance department may require candidates to have excellent numeracy skills.

Careful selection is critical if companies are to be successful and gain a competitive advantage. Companies should search for individuals whose values and motivations are congruent with the organization's service ethic. Employee suitability, therefore, should not necessarily be based on technical skills, which can be taught later, but on psychographic characteristics. In evaluating applicants there are a whole range of selection techniques available from the more common interview, through to self-assessment, group methods and assessment centres, to the increasingly popular use of psychometric testing techniques. Testing techniques are increasingly being used as a means of identifying the personality profile of people who are likely to be successful in delivering service quality and developing relationships with customers. Traditionally, testing was more likely to be used for management and graduate jobs than for administrative, secretarial or manual jobs.[7] Increasingly, however, companies are using these techniques for a wide range of positions, which emphasizes the importance that companies are placing on the emotional content of front-line positions. The use of tests in employment procedures is still controversial.[8] Those in favour of testing refer to the unreliability of the interview as a predictor of employee performance and the greater accuracy and objectivity of test data. Those against testing often have difficulty in incorporating test evidence into the rest of the information collected about the potential recruit and some employers dislike the objectivity of testing, preferring to rely on their own perceptions of individuals.

When the selection process is complete and a new recruit and company decide they can work together, the next stage is to prepare the individual for the job they must do. The following section considers the role of both informal and formal training in organizations.

Training and development

Research has shown that when there is a lack of clarity about the job role that an employee must undertake, then it is likely to lead to reduced levels of employee motivation and job satisfaction, which in turn can manifest itself in poor customer satisfaction and retention.[9, 10] It is critical, therefore, that new employees are carefully prepared for the work ahead of them. The initial period of employment often lays the foundations of an employee's attitudes and perceptions towards the company. In this regard there are basically two kinds of training. The first, informal training or informal socialization, refers to what new employees are told and what they see happening around them. This determines their view of what is important to the organization and what they must do to survive in that organization. For example, 'if customers are denigrated and/or if supervisors emphasise speed over quality ... newcomers can only draw the conclusion that customers and good service are unimportant'.[11] Great care, therefore, should be taken to ensure that new recruits understand the values and vision of the organization. Many companies use a mentoring system or 'buddy' for this purpose. These people explain the way the company operates, how customers should be treated, where things are and other useful information. They act as a support system during those initial weeks and months of employment.

Formal socialization refers to training. To what extent it is undertaken within the organization, and how good it is, impacts on organizational success. Service skills and technical skills need to be taught and these need to be supported by supervisors, managers and co-workers. Training should also not only be undertaken at induction – as it is in so many organizations – but on an on-going basis as part of a staff development programme.[2] The Disney organization and Nordstrom are both famous for their attention to acculturation and training of their employees (see Case Studies 5.2 and 5.3).

Realistic previews of the job and organization are critical if new employees are to become strongly linked to the job and the organization. Many companies are finding that a particularly effective training approach is through in-house video training, which is able to demonstrate the link between theory and reality. Training workshops are also effective in providing new recruits with the opportunity to role play and become used to the requirements of the job. First Direct provides each new employee with seven weeks of full-time training before they are able to work with the public. Marriott holds an eight-hour initial training seminar for all new employees and for the next 90 days each new employee is then assigned a 'buddy', who serves as a mentor to help and support the new recruit. Corporate universities are also becoming a popular method for training and educating employees. McDonald's Hamburger University offers a range of courses which deal with topics such as management skills, market evaluation and area supervision. These courses enable McDonald's to communicate the style and culture of the company to employees. Unipart also has a university, which provides new and existing employees with the opportunity to develop and enhance their skills whilst the culture and values of the company are reinforced and communicated to the employees.

Recruitment monitoring and evaluation

Recruitment monitoring and evaluation is essential if companies are to determine how effective they are at recruitment. Wright and

Storey suggest that there are four areas to monitor in determining recruitment effectiveness:[12]

- the number of initial enquiries which result in a completed application form
- the number of candidates at various stages in the recruitment and selection process, especially those shortlisted
- the number of candidates recruited
- the number of candidates retained in the company after six months.

This, however, is only part of the picture. Organizations need to have a way of determining the quality of the candidates who applied for the positions. Has the company been swamped by hundreds of applications but only a few of the candidates had the necessary skills? This could possibly point to incorrectly worded advertisements which don't provide enough specific information about the job to be done. The people involved in the recruitment process are often best qualified to determine the success of the recruitment process. They can get informal feedback from potential recruits during the interviewing process and can ask for formal feedback after the process has been completed.

Once an employee has been recruited into the organization and the induction training is completed, the next and probably the most difficult task is how to develop and maintain a customer-oriented culture, where employees are able to excel in assisting and supporting the organization in achieving its goals. The next section will focus on the internal market domain and review different approaches to internal marketing.

The internal market domain

The concept of internal marketing has emerged in the literature in the last two decades as a way of enabling companies to get, motivate and retain customer-conscious employees.[13] The concept was then described more widely in the services management literature[14] and found to be valuable in industrial marketing.[15] An increasing number of companies have recognized the need for internal marketing programmes and the implementation of these programmes has gained momentum in recent years, with perhaps the most famous being Scandinavian Airline System (SAS).[16] In fact, some organizations have started to view internal marketing as a strategic weapon to help retain customers through achieving high quality service delivery and increased customer satisfaction.[13] This section will, therefore, address key aspects of the internal market domain, including: the concept of internal marketing; the scope of internal marketing; and conclude with a consideration of inter-related internal marketing activities which are thought to be essential in achieving sustainable competitive advantage.

The concept of internal marketing

Internal marketing has become a frequently quoted part of marketing vocabulary for practitioners and academics alike. Yet despite the numerous writings on the subject, much of the work to date is descriptive and/or suggests strategies and plans for implementing internal marketing programmes.[17] Internal marketing has not been subjected to extensive research and Foreman and Money conclude 'that nothing specific or substantial has been published on the subject of IM in any of the major journals in the field of marketing'.[18] There is still, therefore, no unified notion of what exactly internal marketing is and how it can be implemented in the organization.

Despite this lack of clarity on the subject of internal marketing, the generally accepted view is that it is concerned with creating, developing and maintaining an internal service culture and orientation, which in turn assists and supports the organization in the achievement of its goals. The internal service culture has a vital impact on how service-oriented and customer-oriented employees are and, thus, how well they perform their tasks. It tells employees how to respond to new, unforeseen and even awkward situations. The development and maintenance of a customer-oriented culture in the organization is, therefore, a critical determinant of long-term success in relationship marketing. It is an organization's culture – its deep-seated, unwritten system of shared values and norms – which has the greatest impact on employees, their behaviour and attitudes. The culture of an organization in turn dictates its climate – the policies and practices which characterize the organization and reflect its cultural beliefs.[11]

The basic premise behind the development of internal marketing is the acknowledgement of the impact of employee behaviour and

attitudes on the relationship between staff and external customers.[19, 20] This is particularly true where employees occupy
boundary-spanning positions in the organization, which can result
in them being as close psychologically and physically to the organization's customers as they are to other employees within the
company, or perhaps even closer. The skills and customer orientation of these employees are, therefore, critical to the customers' perception of the organization and their future loyalty to the
organization.

The link between employee satisfaction and employee performance has been challenged in the literature.[21] However, a meta-analysis conducted by Petty, McGee and Canender concluded that job
satisfaction and performance are indeed correlated.[22] Moreover, job
satisfaction has been shown to relate positively with specific facets
of performance such as organizational citizenship behaviour,[23]
which is employee behaviour that is not formally required in a job
description but is nevertheless critical for organizational success
(e.g. helping co-workers, volunteering for extra work and so forth).
Schneider, Parkington and Buxton and Schlesinger and Heskett
have also found support for linkages between employee satisfaction
and retention and customer satisfaction and retention.[19, 20] The
work undertaken by Bain & Company also suggests a strong link
between these two variables.[24] They maintain that high customer
retention will lead to higher employee satisfaction, as employees
will find their job much easier dealing with satisfied customers
rather than dissatisfied customers. As a result, employees create a
stable and experienced work force that delivers higher service
quality at lower cost. This in turn leads to higher customer retention
and increased profitability.

The advantages of long-term employees are that they are often able
to form personal relationships with customers, understand their
needs and may be able to pre-empt dissatisfied customers leaving the
company. Schneider and Bowen have found that when employees
identify with the norms and values of an organization, they are less
inclined to leave and, furthermore, customers are likely to be more
satisfied with the service.[25] In addition to this, when employee
turnover is reduced, service values and norms are transmitted to new
employees and successive generations of service employees.
Employee satisfaction in internal markets is, therefore, a prerequisite
to customer satisfaction in external markets. The basic philosophy is
that if management wants its employees to do a great job with cus-

tomers, then it must be prepared to do a great job with its employees. Unhappy employees will make for unhappy customers, so unless employees can be successfully taken care of, the success of the organization on its ultimate, external markets will be jeopardized.

According to Gronroos, it is not enough to have customer-conscious employees for effective service delivery, there must also be coordination between front-line staff and back-office staff.[26] Internal marketing is, therefore, seen as a way of integrating various functions to enable staff to work together across functional boundaries and aligning those cross-functional teams with internal and external customer needs and expectations, so that their work is attuned to the company's mission, strategy and goals. This gives rise to 'the notion of the internal customer. That is, every person working within an organisation is both a supplier and a customer.'[27] This largely fits with the TQM approach, where the emphasis is on relationships between employees themselves, who make demands upon each other in their efforts to improve quality. The main concern then becomes one of improving customer service and quality at an individual 'exchange' level. The internal exchanges between the organization and its employee groups must be operating effectively along the entire length of the supply chain before the organization can be successful in achieving its goals regarding its external markets. All employees then, in fact, are part of the process which connects with the customer at the point of sale or 'moment of truth'.[16]

Judd categorizes different levels of staff within the organization depending on the degree and type of contact they have with external customers:[1]

- *Contactors*: employees who have direct frequent or periodic customer contact
- *Modifiers*: employees who have less direct frequent or periodic customer contact which is usually not face to face
- *Influencers*: employees who traditionally have no direct contact with customers although they may make many decisions in relation to customers
- *Isolateds*: employees who have no customer contact at all.

According to Judd, the 'contactors' are not the only people who are involved in service delivery. However, their ability to function effectively depends to a great extent on the support they get from other employees from within the firm. Often there is a large number of support people who do not come into direct contact with the customer, but who nevertheless have a very important role to perform and who directly influence the service ultimately provided to the customers. Gummesson uses the term 'part-time marketer' to describe such employees.[15] Systematic internal marketing, therefore, is a mechanism for developing and maintaining these part-time marketers as service-minded and customer-conscious employees. Through internal marketing all employees can begin to understand how their tasks, and the way they perform them, affect customer satisfaction and contribute to a true marketing orientation.

In determining the linkages between internal service quality and external service quality Heskett has provided a useful model of a service profit chain.[28] This model suggests linkages between internal service quality, employee satisfaction and productivity and external customer satisfaction and the organization's financial performance.

Work by Reynoso and Moores contributes towards identifying internal service quality criteria and linking this to quality dimensions used by customers.[29] Researching employees in health care organizations, they were able to identify and assess quality dimensions that were used to assess internal quality received from other parts of the organization. Dimensions identified were similar to those of the SERVQUAL instrument developed by Parasuraman, Zeithaml and Berry, emphasizing the linkages between internal and external quality management.[30] Their work reinforces the view that organizations wishing to improve customer relationships need to focus not only on external but also internal quality enhancement.

This section, so far, has identified the key aspects of the internal market domain, including: the importance of the internal service culture; the notion of the internal customer; the impact of employee satisfaction and behaviour on customer satisfaction and retention and the value of long-term employees. The next part to this section will now consider the scope of internal marketing.

The scope of internal marketing

There have been many different approaches to internal marketing activities advocated by academics and practitioners alike. The following is a brief discussion of the major themes that have emerged in the literature in recent years and some of the problems in implementing internal marketing.

The application of traditional marketing concepts to the internal market

In order to motivate employees towards customer consciousness and service orientation, internal marketing has been proposed as a way of applying the traditional marketing concepts and marketing mix philosophy, originally developed for the company's external customers, to the internal customers of the organization. It was hoped that by doing this internal employee relationships would improve and hence corporate effectiveness.[15] Piercy and Morgan argue, 'the underlying purpose of what we are calling "internal marketing" here is the development of a marketing programme aimed at the internal marketplace in the company, that parallels and matches the marketing programme for the external marketplace of customers and competitors' (p. 84).[31]

This approach was eventually considered to be far too simplistic. Specifically, it does not take into account all management and employee-related issues which are fundamental to organizational survival. Fundamental to the traditional approach of marketing is the concept of exchange, that is, customers receive benefits by way of products or services in exchange for payment. However, when this concept is applied to employees it causes a number of problems. Firstly, it can be argued that external customers ultimately have a choice and can decide not to buy products or services from a particular supplier if they don't want to; whereas with internal customers the 'product' that they are being 'sold', for example, new ways of providing customer service, may not be wanted or may even be perceived by employees to make their jobs harder or more stressful. In this sense they are obliged to 'accept' the new practice or otherwise they may be forced to do so by the threat of disciplinary action. Also, employees do not have available to them the same open market competition as the external customer, who is able to select products and services from a number of different suppliers. In the internal market, there is usually only one 'offer'.

Considerable attention has also been given to the development of an internal marketing mix based on an analysis of the strengths, weaknesses, opportunities and threats in the internal market and directed towards key target groups in the company. This approach contrives to adapt McCarthy's 4Ps framework into an extended marketing mix for services,[21] originally intended for focusing activity on the external market, to fit with the internal marketing framework. For example, 'product' in the internal marketing mix can refer to marketing strategies and 'price' the psychological cost of adopting new methods of working. It is not, however, readily apparent what 'product' and 'price' refer to, and as such this concept has limited appeal. Whilst this approach can provide a loose framework for guiding internal marketing activities, it is generally considered not sufficiently robust to capture the true essence of internal marketing issues.

Other traditional marketing activities include market segmentation and market research, which have been applied to the internal market.[21, 31] Internal market segmentation refers to the method of grouping employees with similar needs and wants, for example front-line staff,[1] and targeting specific internal marketing plans towards these groups. This practice is already widespread throughout organizations, as evidenced by functional and departmental structures within companies.

In recent years there has been a proliferation of employee surveys as many companies seek ways of identifying the needs and wants of employees and internal employee issues which may hinder improving service quality. Such surveys do, however, need to be handled with much care, as employees are often suspicious of the real intent of such surveys and frequently fear repercussions from their 'honest' responses. The author, whilst working with a particular industrial goods manufacturer, found employees holding their questionnaires up to the light to look for watermarks. The employees were convinced that despite the anonymity of the questionnaire and the fact that the questionnaire was being posted by the individuals themselves to an independent third party, the 'management' would still be able to identify them and enact their retribution. Employees' responses should, therefore, always be interpreted carefully, as many times employees will reply in the way they think the organization wants them to. It is essential to manage employee expectation by telling them precisely why the survey is being conducted and what will be done with the results. Finally, it is important to show that action has been taken over any causes for concern raised in the survey.

Internal marketing and organizational design

In endeavouring to find satisfied customer-conscious employees, internal marketing has expanded its remit into the realms of human resource management. According to Berry and Parasuraman, 'Internal marketing is attracting, developing, motivating and retaining qualified employees through job products that satisfy their needs.'[32] In this context it is quite understandable that some should consider internal marketing as just another name for human resource management. According to Bateson, 'At first sight this appears to be a massive invasion of the prerogative of the personnel function, and indeed, in many ways it is.'[33] There is, therefore, much debate and confusion about whether internal marketing is predominantly part of the personnel function or whether it should be part of the marketing function.

Rafiq and Ahmed, in an attempt to delineate the scope of internal marketing, clearly define the boundary between marketing and human resource management.[21] They suggest that 'the most useful contributions that marketing can make to the human resource management area is in the ideas of generating a customer (or employee) orientation, using a co-ordinated set of promotional or communications techniques, and internally directed application of marketing research techniques' (p. 230). Gilmore and Carson, however, are less concerned with defining a boundary and more concerned with adopting an integrative approach to functional responsibility for internal marketing: 'internal marketing activity can be deemed to be an intra-organisational concern transcending all functional boundaries, largely dependent on the degree of integrative activity within company functions'.[34] They go on to suggest that key managers should take responsibility for the development of internal marketing initiatives and cross-functional communications.

Organizational structures must, therefore, be conducive to internal marketing concepts and philosophies. Traditional vertical organizations, which are hierarchically structured and functionally oriented, often optimize individual functions at the expense of the whole business and the customer. They are not fertile ground for developing internal marketing initiatives. Market-facing organizations, on the other hand, which draw key employees together in multidisciplinary teams, which seek to marshal resources to achieve market-based objectives, provide the ideal environment for developing an internal marketing orientation.

Varey attempts to present a model of internal marketing as a process or mechanism for integrated market-oriented management which does not assume the pre-existence of structures of organization (Figure 5.3).[35] He argues that the effectiveness of the process modelled depends on an holistic application to achieve an 'open system' form of organization. Adoption of individual elements

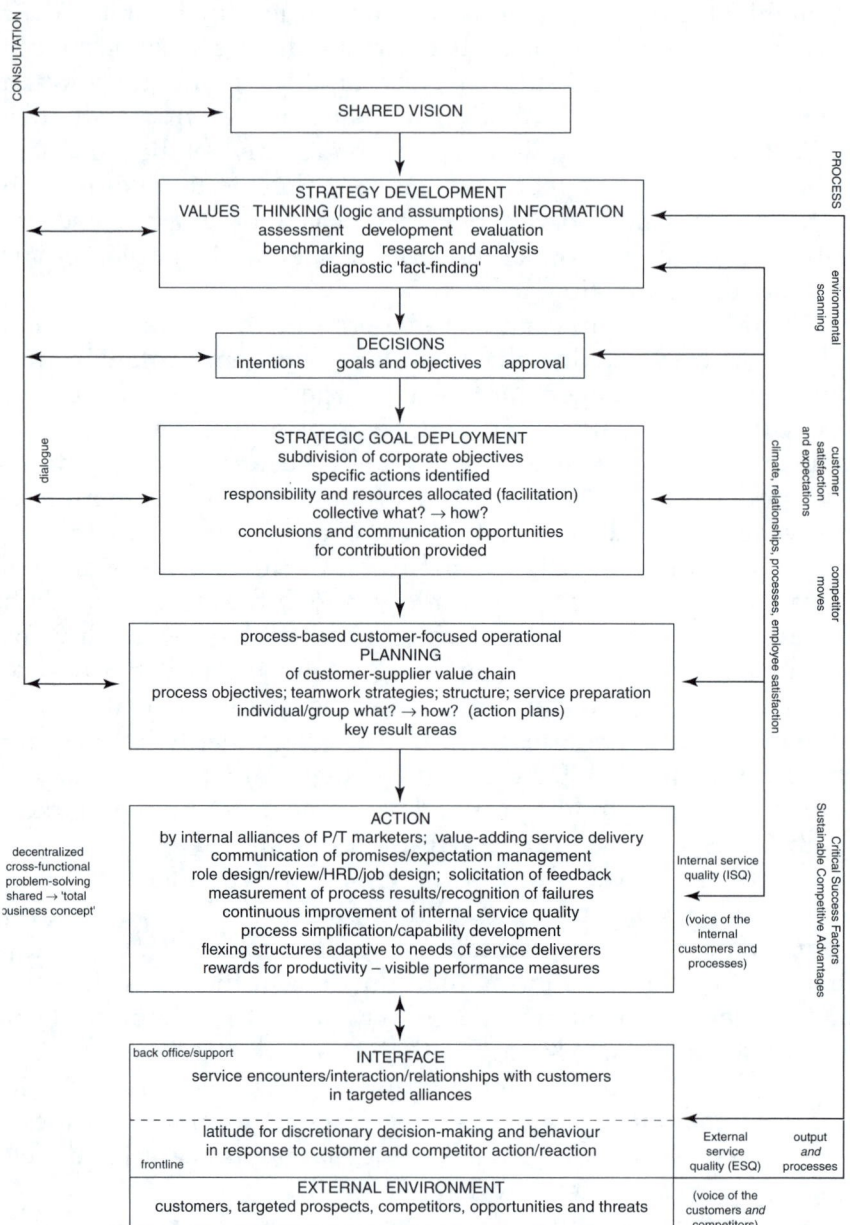

Figure 5.3 Integrated market-oriented management.

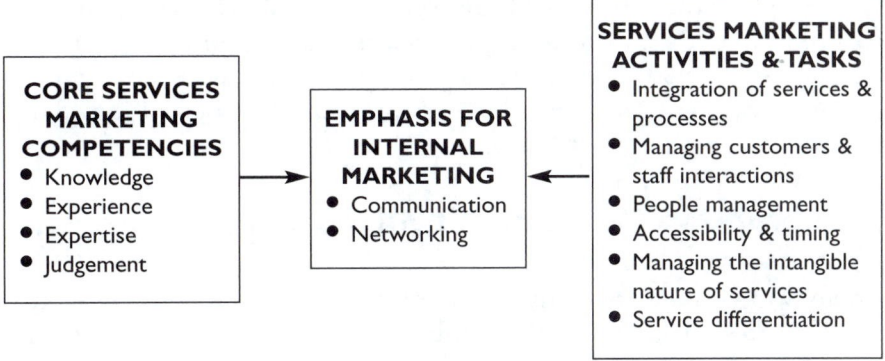

Figure 5.4 Core competencies for internal services marketing.

destroys the emergent properties of the system and would break the complex interconnectedness. He considers it unlikely that this approach would operate well if superimposed on a traditional, rigid, functionally based hierarchical organization structure and that an evolutionary change management programme would be necessary.

Competencies for internal marketing

In recent years there has been increasing recognition that managers should have competencies in their relevant functional area and that the development of these competencies is critical to enhanced management performance. Gilmore and Carson argue that this is equally the case with internal marketing.[34] They argue that, in a services context, the competencies required for the internal management of service tasks and service delivery will revolve around internal communication and networking (Figure 5.4).

An internal communications competency is necessary in order to build on and develop knowledge, experience and expertise. It may manifest itself in a variety of communication media, for example, company newsletters, video conferencing, interactive TV, brochures, etc. It is also a valuable method for achieving coordination and leadership. Other advantages of internal communications include facilitating cross-functional participation and ensuring that everyone knows what to do and their role in the activity.

Gilmore and Carson go on to discuss the importance and significance of using and listening to formal and informal networks between internal customers in the organization in developing internal marketing competencies.[34] They suggest that encouraging networks in organizations will build on managers' external competencies of knowledge, experience, expertise and judgement by concentrating on internal communication, which in turn will contribute to an environment for empowerment, staff responsibility and improved service performance.

The empowerment and involvement of employees

The empowerment and involvement of staff to enable them to use their discretion to deliver a better quality of service to their customers is another key internal marketing activity. This is a radical reversal of the philosophy that declared discretion to be 'the enemy of order, standardisation and quality'.[36] There are various types of empowerment, from all-out empowerment, where employees have absolute power to do whatever is necessary to satisfy the customer, to milder forms of empowerment, which is essentially 'suggestion involvement'. This is where employees can offer suggestions, but the decision-making power rests with the management. There is also 'job empowerment', where jobs are redesigned so that employees can determine how they work.

Empowerment means that the company must create the right culture and climate for employees to operate in. This includes four empowerment criteria: providing employees with information about the organization's performance; providing rewards based on the organization's performance; providing employees with knowledge that enables them to understand and contribute to organizational performance; and giving them power to take decisions that influence organizational direction and performance.[37]

Although there is a growing number of organizations which have adopted an empowerment approach, there are many that have failed in their attempts or have met with barriers in their efforts to create a more empowered culture. Failure is often the result of middle managers feeling threatened by delegating power and authority to subordinates. Sometimes staff are reluctant to take on such responsibility and consider making such empowered decisions to be managers' responsibility.

There are of course advantages of empowerment, which can include a faster and more flexible response to customers' needs by confident and well-informed staff. In fact where service failures do occur, there is evidence that a satisfactorily resolved problem by trained, empowered staff, who take prompt action to resolve a problem, may even raise the customer's perception of service quality.[38] Bowen and Lawler have also argued that empowerment can improve employee motivation and job satisfaction, which can in turn improve customer satisfaction and retention.[37] The benefits of empowerment must, however, be balanced with the increased labour, recruitment and training costs and as such should be viewed as a long-term investment in employees. There is also the issue of ensuring that all customers are treated alike, otherwise resentment may result if customers perceive that a particular customer is receiving special treatment. Bowen and Lawler caution that empowerment may not necessarily be the best choice for some companies and recommend that managers should assess the pros and cons of empowerment for their business before launching into a full-blown empowerment strategy.[37]

Developing internal marketing relationships

It is suggested that the basic philosophy of internal marketing, which has led to numerous writings on the subject, is basically sound: the idea of developing and maintaining an internal service culture and orientation; the notion of the internal customer and the part-time marketer; the role that employee satisfaction and retention plays in determining external customer satisfaction and retention. These all contribute to a greater understanding of the concepts of internal marketing and how it impacts on corporate performance. The difficulties of applying traditional marketing concepts to the internal market have already been discussed, as have some of the pitfalls of adopting a functional approach and defining the boundary of internal marketing. To date, however, there is still no consensus on the implementation of internal marketing. What is clear is that internal marketing should not be solely the domain of marketers applying traditional marketing concepts and tools. To do this would destroy the very nature of what is meant by relationship marketing and would not take into account the needs of the internal market. The model of internal marketing as proposed by Varey[35] (Figure 5.3) as a process or mechanism for integrated market-oriented management would seem to offer the most in terms of enhancing the organization's capability. Further work, however, needs to be undertaken to empirically test this model with organizations and determine what it takes in operational terms to ensure internal marketing success. In the interim, here is a range of inter-related internal marketing activities which are thought to be critical in helping to implement internal marketing.

- *Organizational design* must be conducive to internal marketing concepts and philosophies. Market-facing organizations which draw key employees together in multidisciplinary teams, which seek to marshal resources to achieve market-based objectives, provide the ideal environment for developing an internal marketing orientation.
- *Regular staff surveys* which assess the internal service climate and culture and obstacles to service quality improvements are essential in determining the focus for future actions.
- *Internal customer segmentation* by, for example, level of customer contact[1] in order to target specific service provisions and enhance both internal and external service quality.
- *Personal development and training* should be focused on core competencies for internal marketing. Performance data can be collected to determine internal and external customer requirements which will shape the nature of the training and development.
- *Empowerment and involvement* enables staff, within clearly defined parameters, to use their discretion to deliver a better quality of service to their customers. Self-directed work teams, employee involvement activities and job design all help to remove barriers between people and give them the authority to act and improve service quality.
- *Recognition and rewards* based on employees' contribution to service excellence. Rewards and recognition are critical to determining employee behaviour and should be based on careful consideration of the likely impact on behaviour and the attractiveness of the rewards and recognition for the individual. For example, Credit Card Sentinel has a range of formal rewards and recognition for employees, but their 'caught in the act' award, which is a postcard sent by the managing director to employees who have been caught by their colleagues giving exceptional service to customers, is particularly valued.
- *Internal communications* provides a mechanism for cross-functional participation in the coordination of activities in the organization. It is a valuable method for reinforcing service quality and ensuring that everyone knows what to do and their role in the activity.
- *Performance measures* should be visible and should measure each person's and department's contribution to the achievement of performance objectives for each key success factor.[35]

- *Building supportive working relationships* for employees should be a key issue when developing an internal marketing approach. Employees should be able to provide each other with consideration, trust, warmth and support, which helps to break down barriers within and between departments. This enhances internal communications and the likelihood of achieving internal and external service quality.

Effective recruitment and management of employees has become a powerful strategic weapon in an increasingly competitive environment. Companies which are able to harness their resources and implement strategies and plans to recruit, train, motivate and retain employees are much more likely to be able to compete more effectively and provide long-term service quality for both internal and external customers. The next section of this chapter will look at four case studies which all address key issues relating to the recruitment and internal market domains.

References

1 Judd, V.C. (1987). Differentiate with the 5th P: People. *Industrial Marketing Management*, **16**, 241–247.
2 Schlesinger, L.A. and Heskett, J.L. (1991). Breaking the cycle of failure in services. *Sloan Management Review*, Spring, 17–28.
3 Kuemmler, K. and Kleiner, B.H. (1996). Finding, training and keeping the best service workers. *Managing Service Quality*, **6**, No. 2, 36–40.
4 *IRS Employment Review* (1996). Policy and practice in recruitment: An IRS Survey, September, 5–13.
5 Rodger, A. (1975). Interviewing techniques. In Ungerson, B. (ed), *Recruitment Handbook* (2nd edn), Gower, Aldershot.
6 Lewis, C. (1985). *Employee Selection*, Hutchinson, London.
7 Newell, S. and Shackleton, V. (1994). The use (and abuse) of psychometric tests in British industry and commerce. *Human Resource Management Journal*, **4**, No. 1.
8 Fletcher, C. et al. (1990). Personality tests: The great debate. *Personnel Management*, September.
9 Parkington, J.J. and Schneider, B. (1979). Some correlates of experienced job stress: A boundary role study. *Academy of Management Journal*, **22**, 270–281.

10 Kelley, S.W. (1990). Customer orientation of bank employees and culture. *International Journal of Bank Marketing*, **8**, No. 6, 25–29.
11 Schneider, B. (1986). Notes on climate and culture. In Venkatesan, M. et al. (eds), *Creativity in Services Marketing*, American Marketing Association, Chicago, 63–67.
12 Wright, M. and Storey, J. (1994). Recruitment. In Beardwell, I. and Holder, L. (eds), *Human Resource Management*, Pitman, London.
13 George, W.R. and Gronroos, C. (1989). Developing customer-conscious employees at every level – internal marketing. In Congram, C.A. and Friedman, M.L. (eds), *Handbook of Services Marketing*, AMACOM, New York.
14 Gronroos, C. (1990). Relationship approach to marketing in service contexts: The marketing and organisational behaviour interface. *Journal of Business Research*, **20**, No. 1.
15 Gummesson, E. (1987). Using internal marketing to develop a new culture – the case of Ericsson quality. *Journal of Business and Industrial Marketing*, **2**, No. 3, 23–28.
16 Carlzon, J. (1987). *Moments of Truth*, Ballinger, New York.
17 Thomson, K. (1990). *The Employee Revolution*, Pitman, London.
18 Foreman, S. and Money, A. (1995). *Internal marketing – concepts, measurement and application*, Henley Management College Working Paper Series No. 9529.
19 Schneider, B., Parkington, J.J. and Buxton, V.M. (1980). Employee and customer perceptions of service in banks. *Administrative Science Quarterly*, **25**, 252–267.
20 Clark, M. (1997). Modelling the impact of customer–employee relationships on customer retention in a major UK retail bank. *Management Decision*, June, 293–301.
21 Rafiq, M. and Ahmed, P.K. (1993). The scope of internal marketing: Defining the boundary between marketing and human resource management. *Journal of Marketing Management*, **9**, No. 3, July, 219–232.
22 Petty, M.M., McGee, G.W. and Canender, O.W. (1984). A meta-analysis of the relationship between individual job satisfaction and individual performance. *Journal of the Academy of Management Review*, **9**, No. 4, 712–721.
23 Organ, D.W. (1988). *Organizational Citizenship Behavior: The good soldier syndrome*, Lexington Books, Lexington, MA.
24 Buchanan, R.W.J. (1900). *Customer retention: The key link between customer satisfaction and profitability*, unpublished paper, Bain & Company.

25 Bowen, D.E. and Schneider, B. (1988). Services marketing and management: implications for organisational behaviour. In Stow, B. and Cummings, L.L. (eds), *Research in Organisational Behaviour*, **10**, JAI Press, Greenwich, CT.

26 Gronroos, C. (1981). Internal marketing – an integral part of marketing theory. In Donnelly, J.H. and George, W.R. (eds), *Marketing of Services*, American Marketing Association, Chicago, 236–238.

27 Christopher, M.G., Payne, A.F. and Ballantyne, D. (1991). *Relationship Marketing: Bringing quality, customer service and marketing together*, Butterworth-Heinemann, Oxford.

28 Heskett, J.L. (1992). *A service sector paradigm for management: The service profit chain*, Proceedings of the Management in Services Sector Symposium, Cranfield School of Management.

29 Reynoso, J. and Moores, B. (1996). Internal relationships. In Buttle, F. (ed), *Relationship Marketing – Theory and Practice*, Paul Chapman, Liverpool.

30 Parasuraman, A., Ziethaml, V. and Berry, L. (1988). SERVQUAL, a multiple theme scale for measuring consumer perceptions of service quality. *Journal of Retailing*, **64**, 12–40.

31 Piercy, N. and Morgan, N. (1991). Internal marketing – the missing half of the marketing programme. *Long Range Planning*, **24**, No. 2, 82–93.

32 Berry, L.L. and Parasuraman, A. (1991). *Marketing Services: Competing through Quality*, Free Press, New York.

33 Bateson, J.E.G. (1991). *Managing Services Marketing* (2nd edn), The Dryden Press, Fort Worth, TX.

34 Gilmore, A. and Carson, D. (1995). Managing and marketing to internal customers. In Glynn, W.J. and Barnes, J.G. (eds), *Understanding Services Management*, Wiley, Chichester.

35 Varey, R.J. (1995). A model of internal marketing for building and sustaining a competitive service advantage. *Journal of Marketing Management*, **11**, 41–54.

36 Levitt, T. (1972). Production-line approach to service. *Harvard Business Review*, September/October, 41–52.

37 Bowen, D.E. and Lawler, E.E. (1992). The empowerment of service workers: what, why, how and when. *Sloan Management Review*, Spring, 31–39.

38 Hart, C.W.L., Heskett, J.L. and Sasser Jr, W.E. (1990). The profitable art of service recovery. *Harvard Business Review*, July/August, 147–156.

Chapter 5 case studies

Case 5.1: Club Med

This case was prepared by Professor Christopher Hart with the research assistance of Dan Maher from Harvard Business School.

Abstract

Club Med is the second case in a two-part case series. The first case addresses issues of competitive advantage in the rapidly growing American subsidiary of Club Mediterranee, an international all-inclusive vacation resort company. This case describes employee turnover problems in the company and focuses on building a service organization and improving service quality for Club Med.

A large part of Club Med's successful competitive positioning can be attributed to a large extent to the value added through interactions between employees, particularly front-line staff and customers. Employee morale and interpersonal skills are, therefore, essential ingredients in determining customer satisfaction for the company. The Club Med case charts employee turnover in the American zone and implies that the high turnover may be a cause and an effect of poor morale among American employees. The case documents how high turnover can affect team performance by dampening morale of the employees who stay with the company and lead to the hiring of less experienced and less qualified staff. The case concludes with Jacky Amzallag, director of Human Resources for Club Med's American zone, considering a series of options available to the company on how to reduce employee turnover and improve the recruitment process for Club Med.

Learning points

This case provides an excellent example of how a rapidly growing company finds it increasingly difficult to manage its relationships with its internal and recruitment markets. Specifically, maintaining an appropriate organizational climate and culture whilst rotating employees between the villages, and managing cross-cultural issues, are major causes for concern for the internal market. The company also failed to manage its recruitment activities in terms of selection, placement and training of new recruits. Broad strategic issues relating to the many learning points within the case are as follows:

Broad marketing and strategic issues:

- improving service quality
- managing rapid growth
- managing a diverse range of cross-cultural issues, particularly the dominance of French nationals in management positions in the American zone
- examines how organizational policies affect the attitudes of customer contact staff
- managing customer expectations
- organizational structure and design.

Relationship marketing issues:

- managing recruitment markets, including:
  - selection of staff
  - placing staff/job rotation
  - training staff
- managing internal markets
- how employee morale influences customer satisfaction in service businesses
- the financial and strategic costs of high employee turnover
- the impact of employee retention on customer retention
- employees as part-time marketers and primary sources of variation in service quality
- managing the organizational climate and culture and implementing change in a company with a strong corporate culture
- failing to adopt a relationship marketing approach.

This case may also provoke some discussion about the ethics of over-hiring and the merits of having a director for human resources in the American zone who doesn't have the authority over promotions and management assignments. There are also linkages with other relationship markets, including referral and customer markets.

Case 5.2: Euro Disney: The first 100 days

Research Associate Robert Anthony prepared this case under the supervision of Professors Gary Loveman and Leonard Schlesinger at Harvard Business School.

Abstract

This case focuses on the problems of adapting a service concept across cultural boundaries, particularly when the concept may conflict with local traditions and expectations of behaviour from both the employee and customer points of view.

The first part of the case describes the key ingredients behind Disney's success – specifically, the personality of Walt Disney himself, the innovative 'theme' park design, the 'rich heritage' of the cartoon characters, the unique role that visitors play in the theme park, the continually updated theme parks, and the consistently meticulous service delivery system. The focal point of the service delivery system is Disney University, the company's in-house personnel development organization, which is described in some detail in the case. It is because Disney recognizes the key role that employees play in ensuring successful visitor experiences that Disney University was set up to ensure consistency of service across the theme parks.

The next part of the case discusses how the Disney concept was adapted for the Japanese market. Tokyo Disneyland is described as a tremendous success and virtually identical to Disney's Southern Californian park. In the case it is suggested that Tokyo Disneyland's success is attributed to the 'match' between various aspects of the Disney and Japanese cultures. The case then goes on to describe the planning, design, start-up process and management problems associated with the opening of Euro Disney and particularly the adaptations that were made to fit with the European location. The case closes with a number of decisions facing Euro Disney; in particular, management must decide whether or not to invest in the second phase of development of the theme park.

Learning points

This case shows how Euro Disney has difficulty in managing its relationships with internal, recruitment and customer markets in an unfamiliar cultural setting. It documents how a well tried and tested service concept which has already been successfully transplanted to

Japan faces problems in the French market. It considers how the company initially failed to make any allowance for cultural differences in the internal, recruitment and customer markets and questions the extent to which service delivery systems are culturally specific.

Broad marketing and strategic issues:

- internationalization of service businesses
- the extent to which service delivery systems are culturally specific
- minimizing risk and failure in cross-border service opportunities
- the importance of researching international markets
- managing poor financial performance
- understanding the difficulties of managing across national and cultural boundaries.

Relationship marketing issues:

- managing internal markets in an unfamiliar cultural setting
- managing recruitment markets in an unfamiliar cultural setting
- the role that 'in-house universities' play in ensuring service quality
- managing customer markets in an unfamiliar cultural setting
- failure to develop a relationship marketing approach.

With the exception of the supplier market, this case illustrates many of the issues involved in developing a relationship marketing strategy. In addition to the internal, recruitment and customer markets, it also considers the need to develop appropriate influencer and referral marketing strategies, particularly when marketing a service concept across national and cultural boundaries.

Case 5.3: Nordstrom Inc.

This case was prepared by Richard Pascale, who is a consultant, academic and author. He is on the faculty at Stanford School of Business, Stanford University.

Abstract
Nordstrom stores are fashion speciality stores which specialize in apparel, cosmetics and jewellery. Despite being a late entrant in

retailing, Nordstrom has survived intense competition and gone from strength to strength. By the end of 1990 it had experienced an historic annual growth rate of 20 per cent per year and was the top-ranked department store in the United States.

The majority of this case is devoted to an analysis of Nordstrom's success, specifically, how Nordstrom's culture and values are at the heart of its sustainable competitive advantage. Examples of some of the other factors which contribute to Nordstrom's success include: how Nordstrom defies conventional wisdom and consistently does the opposite to its competitors; suppliers are keen to supply them; the complementary styles of the co-chairmen; differentiation through service, including responding to unreasonable requests; attracts top quality staff; recognition of the internal customer; identification of two key stakeholders, the customer and the salesperson; clear career development, with employees only promoted from within; organization reflects a network approach; and finally rewards, recognition and peer pressure.

The final part of the case reflects on the 'dark side' of the Nordstrom empire and considers the pressures which are brought to bear on employees to bring their behaviour in line with the Nordstrom culture. It is perhaps not surprising, therefore, that while the company is considered a preferred employer, employee turnover is 50-60 per cent, about typical for the industry.

Learning points

The previous two case studies illustrated how failure to manage internal and recruitment markets can have a detrimental impact on the business. The Nordstrom case study, however, is an excellent example of how a company has successfully managed its relationships, not only with the internal and recruitment markets but also with the customer market. It is a particularly interesting case and there are many learning points relating to both strategic and marketing issues as well as relationship marketing issues.

Broad strategic and marketing issues:

- service quality as a key differentiator
- managing rapid growth
- examines how organizational policies affect the behaviour of customer contact staff
- employee motivation

- organizational structure and design.

Relationship marketing issues:

- managing internal markets
- managing the organization's climate and culture
- employee turnover
- managing recruitment markets:
  - selection of staff
  - placing staff/job rotation
  - training staff
- managing customer markets
- how employee morale influences customer satisfaction.

This case highlights many of the issues involved in developing a relationship marketing strategy. In particular, it considers the importance of culture and values as being critical to organizational success. Traditionally much debate has surrounded the mystique and magic of the Nordstrom formula. However, this case goes a long way towards explaining some of the key factors which have contributed to Nordstrom's meteoric rise to success.

Case 5.4: Digital Equipment Corporation: Counting the real cost of employee turnover

This case was written by Helen Peck, who is a Research Fellow at Cranfield School of Management.

Abstract
This mini-case documents the full cost of employee turnover for the Digital Equipment Corporation. The case begins in 1987, when Digital Equipment Corporation was Britain's fastest growing computer company. However, due to the booming British economy and the skills shortage in the M4 corridor, the company was finding it increasingly difficult to recruit new employees. The case goes on to discuss how the company decided to try to minimize the need for further recruitment. It did this by undertaking an in-depth study of why employees were leaving the company and endeavoured to track the costs of replacing these employees. The company hoped

that by undertaking these research initiatives they would be able to develop employee retention initiatives. Unfortunately by 1991 the recession hit and Digital was forced to abandon its 'no layoffs' policy and the employee retention initiatives were abandoned.

Learning points

This case shows how a company which has been failing to manage its internal and recruitment markets embarks on a course of action designed to ultimately improve employee retention.

Broad marketing and strategic issues:

- managing environmental changes in a dynamic industry
- developing and implementing HR policy.

Relationship marketing issues:

- measuring the financial and strategic costs of employee turnover
- managing internal markets
- linkages between employee satisfaction and turnover and customer satisfaction and turnover
- understanding the requirements of different employee segments in the company.

This case provides an opportunity to discuss possible strategies available to the company in terms of employee retention. It also provokes discussion on the advantages that new employees bring with them when they join the company. These advantages are not quantified in this case but should nevertheless be considered.

Case 5.1 Club Med

Professor Christopher W.L. Hart prepared this case with the research assistance of Dan Maher as the basis for class discussion rather than to illustrate either effective or ineffective handling of an administrative situation.
© Copyright 1986 by the President and Fellows of Harvard College.

Jacky Amzallag, director of Human Resources for Club Med's American zone, smiled warmly at his visitors and waved them in as he spoke rapidly over the telephone to Carlos, chief of the village at Playa Blanca, Mexico. "Another GO has quit?* That's two that you've fired and two who have quit so far this season – in just four weeks! Why did this one quit?" asked Jacky with a pained look. "Too many hours? No? Maybe he missed his sweetheart back home." Jacky winked at his visitors. "Bad *attitude*?" He looked astonished. "I don't understand! He had an excellent attitude. I interviewed him myself! What happened to him?" Jacky's head jerked back as he moved the handset a foot from his ear. A booming voice from several thousand miles away, obviously agitated, filled the room. Jacky's face took on a pained expression. "OK, OK. I'll get you a replacement windsurfing instructor. How soon? As soon as possible, of course. How soon? Impossible! I know, I know ... I *have* been in your position! You were my chief of sports in Brazil – back in '82. Some friend. I teach you everything I know – all my tricks [another wink at his visitors] – and here you yell at me because your windsurfing instructor quits. Let me find out who is available."

Suddenly, Jacky leaned forward, listening intently, smiling, with a look of anticipation. Howling with laughter, he asked, "Where did you hear that one? Have you told it to anyone here? Good. I'll get good mileage out of that one! *Salut, mon copain.*" Whirling around in his chair, Jacky pressed down on the intercom. "Debbie, would you pull the active résumés from the windsurfing-instructor file?"

Jacky, smiling at his guests, leaned back in his chair and reflected. "Voltaire wrote, 'We cannot always oblige, but we can always speak obligingly.'" Then he sighed:

> I have calls like that one almost every day. Despite our best efforts, there are problems in this business that just can't be avoided. It is, after all, a service business, and my department is fairly new ... lots of bugs to iron out. It is

---

*The term "GO", pronounced *gee-oh*, stands for *gentil organisateur* in French, *congenial host* in English. Each Club Med village employed a team of about 80 GOs who handled all jobs other than housekeeping and groundskeeping. GOs organized the activities in a Club Med village and mingled freely with guests (called "GMs", pronounced *gee-ems*: *gentils membres* in French, *congenial members* in English).

vital to the Club's growth that we find these bugs and exterminate them. One of my biggest problems involves the turnover of newly recruited GOs, which is a responsibility of my department. Their turnover has been very high over the past several years – nearly 50% – and it is getting worse. What is the problem? Recruiting? Selection? Training? Maybe we have to learn more about the North American GOs and how to manage them. With the huge number of applications we get for GO positions, you would think that turnover would be much less than it is – about 25% per season, including all GOs, both old and new. This is almost twice as much as our GO turnover in Europe. Maybe 25% is not so bad, though, and we just have to learn to live with it. [See Table 5.1.1 for information on turnover.]

Jacky continued: "The Club is growing rapidly in the American zone, and my department is crucial to its success. My immediate objective is to get this GO turnover problem straightened out. I must provide my clients, the chiefs of the villages, with good service. I have some ideas about how to do

Table 5.1.1   Information on GO turnover

Number of GOs, by nationality, summer season 1986, American zone

American/Canadian	738
French/European	454
Local	134
Other	15
Total	1,341

American/Canadian GO turnover, summer season 1986, American zone (quit or fired or 'retired')*

Quit during season	74
Fired during season	66
Quit after season	55
Fired after season, due to bad report	9
Total	204

American/Canadian GO turnover by experience: historical averages, 1980–85, summer and winter seasons (%)

Turnover during first season	46
Turnover during second and third season	37
Turnover during fourth to tenth season	23
Turnover after tenth season	11

*French/European GO turnover was approximately 50% of American/Canadian GO turnover during the same time period. (French/European GO turnover in the American zone was essentially zero.)

this, but I'm not sure what my overall strategy should be. One thing is certain: I am committed to do whatever it takes … I love the Club. *J'aime trop le Club.*"

Company background and history

Club Mediterranee (often referred to as "the Club") was the ninth-largest hotel company in the world in 1986. It was founded by a group of friends in 1950 as a non-profit sports association. The group was led by Gerard Blitz, a Belgian diamond cutter and water-polo champion, who, like the others in the group, loved sports and vacations in scenic seaside locations. Members slept in sleeping bags in tent villages and took turns cooking meals and washing dishes. As the size of the association grew, running it as an informal, loosely organized group became increasingly difficult. In 1954 Blitz invited his close friend, Gilbert Trigano, an active association member whose family business had been supplying the group with US Army-surplus tents, to join the association on a full-time basis. Trigano, who saw commercial potential in the concept, became managing director and set out to turn the association into a business. By 1985 Club Mediterranee SA, a publicly owned company traded on the Paris Stock Exchange, had more than 100 resort villages throughout the world and hosted more than 800 000 vacationers.

History of the American zone

In 1972 Club Med, Inc., was formed as a US subsidiary of Club Mediterranee. The subsidiary sold the company's vacation packages and operated its resorts in North America, the Caribbean, South America, Asia, and the South Pacific. What was referred to in Club Med as the "American zone" comprised the 17 villages highlighted on the map in Figure 5.1.1.

Club Med's expansion into the Western Hemisphere was not without difficulty. The company's first village in the American zone, Buccaneer's Creek, located on the Caribbean island of Martinique, developed a "swinging-singles", sex-oriented reputation when it was first opened in the early 1970s. This image was in stark contrast to Club Mediterranee's family-oriented image in Europe. Management became so concerned that the village was temporarily closed; however, despite intensive company efforts to stamp out the "wild and crazy" reputation, the image persisted.

Gradually, however, the image was changing. By the middle of the 1980s, demographics on Club Med's clientele revealed the average age to be 37, the median income to be $39,000, and the average income to be $54,000. More than 70% were aged 25 to 44, and half were married. More

Figure 5.1.1 Location of Club Med villages in the American zone, 1986. *Note*: A new village was scheduled to open in Florida in 1987.

than 75% were college graduates, and 28% held advanced degrees.

Another source of problems in the American zone had been cultural and language differences between Europeans and North Americans. Initially, almost all GOs in the American zone were French (or French-speaking). Misunderstandings frequently arose between European GOs and American GMs – and between European and North American GOs as well. After examining the problem in depth, management decided, in the early 1980s, to dramatically increase the percentage of North American GOs, resulting in major recruiting drives in large US and Canadian cities. Based on a substantial increase in guest-satisfaction ratings among American zone villages, this policy was judged to be a success: in Jacky's view, it met the need for North American GOs while retaining the international flavor of the Club.

Some Club Med managers, however, voiced concern over continuing problems rooted in ethnic differences. Because the number of GOs recruited from Europe was much larger than that from North America, and the turnover rate for European GOs was roughly half that for American GOs, the number of European village chiefs and chiefs of service was disproportionately high.* (Table 5.1.2 shows the approximate ethnic composition of village chiefs in the American zone and chiefs of service throughout the company.)

Finally, there had been problems at times with some of the company's personnel policies. In the 1970s, decisions concerning all GOs throughout the world were centralized in the personnel department at Paris headquarters. One of Club Mediterranee's long-established policies had been to rotate GOs to different villages every six months. A company tradition was to bring the GOs to the home base in Paris at the end of each season so they could visit headquarters, reestablish old friendships, and find out where they would be assigned the next season. By 1977 the number of GOs had grown to more than 5,000 and this tradition had become a logistical nightmare. Consequently, it was stopped. To relieve the pressure on the personnel department created by the greater number of company villages and their wider geographic dispersion, a 1985 reorganization established four

Table 5.1.2 Ethnic composition of village chiefs and chiefs of service, 1986 summer season

Nationality	Chiefs of the village (American zone)	Chiefs of service (worldwide)
French	13	1006
Other European	1	224
Moroccan	1	39
North American (US and Canada)	1	52
Australian/New Zealander	0	5
Other	1	143
Total	17	1469

*"Chief of the Village" was the equivalent of a general manager, in American hotel parlance. Each village chief had seven assistants, called "chiefs of service", who managed specific functions such as sports, dining, entertainment, and so on. The chiefs of service managed the GOs in their functional areas.

GO departments in Paris, each headed by a person known as *un parrain* – a "godfather".

Under the direction of Michel Perchet, director of human resources for Club Mediterranee, and his two assistants, the four "godfathers" served as functional area heads. Each "godfather" was responsible for those GOs (and chiefs of service) throughout the world who fell into one of four functional areas: sports; entertainment; maintenance; and administration. The new organizational setup improved the process of assigning GOs to new villages and gave GOs in each of the four functional areas the feeling that they had a special connection to Paris.

Although the new structure was judged to be an improvement, rapid growth in the number of American GOs created new pressures on Paris; this ultimately led, in September 1985, to the formation of a New York-based, American zone "GO Village" at Club Med's headquarters on 57th Street in New York City. Jacky Amzallag, a French Moroccan who had joined the Club in 1966 and risen through the ranks, was chosen to head this new department, which was responsible for recruitment, hiring, orientation, placement, and performance appraisal of GOs in the American zone.

The Club wanted to increase the number of American chiefs of service, both in the American zone and in the rest of the world. Therefore, after three or four seasons, Jacky recommended promising American GOs to Michel Perchet and the functional "godfather" for promotion. These American GOs were given language goals, among others, as hurdles to jump before being promoted. Once promoted, a new American chief of service was eligible to be assigned to any Club Med village throughout the world.

Village organization

Each Club Med village was organized similarly. As shown in Figure 5.1.2, the chief of the village had overall responsibility for village operations. Reporting to the village chief were seven chiefs of service, each of whom was in charge of a different functional area. The 80 or so GOs in a village reported to their respective chiefs of service. Such back-of-the-house activities as housekeeping and groundskeeping were handled by non-GO workers who lived in the local area.

Chief of the village
The title "chief of the village" accurately described the position. Because of the remoteness of many villages and the Club's culture, the "chief of the village" position had evolved as the key leadership position in the organization. Although largely a formality, even the chairman of the board

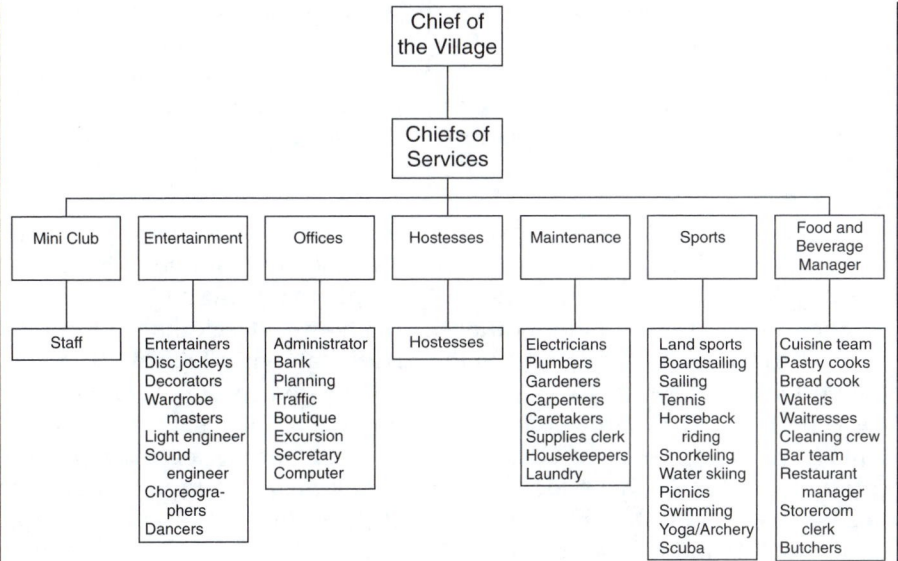

Figure 5.1.2 Organization of Club Med villages.

requested permission from village chiefs before visiting their villages. Chiefs were required to adhere to certain Club policies, but had wide latitude to set village policies that affected GMs' vacation experiences.

The chiefs were a colorful and varied cast of characters; one American GO described them as "one part Napoleon, one part Santa Claus, one part Jerry Lewis, and one part Long John Silver." Examples of the creative ways they had dealt with impending disasters and major incidents were legion, creating an almost reverential respect for the village chiefs throughout the organization. (Most of Club Med's top-level managers had been village chiefs.)

For example, Sylvio de Bortoli, vice president of sales, who had been a village chief for many years, recalled:

> When I was in Cancun one season, I received word that a charter from New York was experiencing horrible problems and was going to arrive ten hours later than scheduled – at 4:30 A.M. I knew that the planeload of GMs would be miserable, probably angry, maybe even hostile.
>
> So I rounded up my GOs and we formulated a plan. We met the GMs at the airport, whisked them through customs, and had drinks waiting in the buses. On the ride to the village, our guests told us that they had experienced a horribly rough landing – so rough, the oxygen masks descended! Then the pilot announced over the PA, "If you think *that* was bad, wait until you see how I taxi up to the gate!'

When we arrived at the village, our chefs had laid out a lavish buffet, complete with champagne. The GOs put on an abbreviated, high-spirited show. A couple of hours after their arrival – with the sun coming up – our weary GMs wandered off to their rooms with *smiles* on their faces. It was the beginning of a *great* week – for our guests *and* for the GO team.

GOs

The Club sought individuals for GO positions who were young, personable, enthusiastic and athletic. GOs were expected to put in long hours on the job, although many in the Club felt strongly that being a GO was not a job, but a way of life – and that the 80 or so hours a week that GOs interacted with GMs was not "work" in the traditional sense. Table 5.1.3 shows the posted GO schedule for a typical day. Not listed on the schedule were the 6 to 7 hours all GOs spent in their assigned jobs (e.g. restaurant hostess, tennis instructor, cook, bartender). These hours would be interspersed throughout the day, depending on the requirements of a particular GO position. Also not shown in the schedule was the casual contact time between GMs and GOs at such times as meals and sitting around the pool. The Club philosophy was for GOs to enjoy themselves and, in a sense, for them to be on vacation with GMs.

Rotation of GOs

A Club Med practice that other hotel companies found incredible was the rotation of GOs to different villages every six months. Rotation included all village chiefs and chiefs of service. The only exceptions were the chiefs of maintenance and maintenance GOs; they stayed in a village for one to two years. Many questioned the efficiency of rotation, primarily on the grounds

Table 5.1.3 Typical daily schedule: GO duties in addition to regular job tasks

Time (Sunday)	Event	Number of GOs	
7:15 A.M.	Theatrical breakfast welcome	7	
9:00 A.M.	Information meeting (for GMs)	5	
12:00 noon	T-shirt presentation (for GMs)	5	
12:15 P.M.	Sports demonstration around pool	30	(sports GOs)
1:15 P.M.	Rehearsal for evening Broadway show	40	
7:30 P.M.	Makeup for theatrical welcome at dinner	15	
9:45 P.M.	Dancing around pool with GMs	10	
10:15 P.M.	Evening Broadway show	40	
11:15 P.M.	GO–GM basketball game	19	(players)
		10	(watching)
12:00 mid.	Initiate dancing at disco	20	

- "I like to move to different villages."
- "I've never been to Europe and would love to work at Corfu [in Greece] for a season."
- "I want to experience different cultures around the world."
- "I miss my Michel, my first chief of service; he was great!"
- "I need a change. I'm bored. There's no place to go around here except the Club."
- "I wish our entire team could be rotated together to a new village. This team works together better than any of the ones at the other five villages I have been."
- "The major reason for becoming a GO is to see the world."
- "I can't wait for my next village. The atmosphere in this village isn't too great."
- "My girlfriend and I ... she is a GO ... wish we could stay together."

Figure 5.1.3   GO comments about rotation.

that it broke up groups that had become cohesive teams. Moreover, rotation often resulted in a temporary dip in guest-satisfaction ratings during the two four-week periods in the year when GOs from old teams were phased out and new GOs were phased in. Rotation of the over 1,300 GOs in the 17 villages located in the American zone also entailed substantial direct costs and logistical difficulties. Comments from company management tended to support rotation, however, although no figures pertaining to the direct costs of rotation had ever been calculated. Jacky stated that "after six months you need to break the routine. This is a motivator for the GOs. They can see different countries and experience different cultures. They learn to adapt to many types of people." (Comments from GOs are listed in Figure 5.1.3.)

Club Med American zone office, New York City

Jacky Amzallag's office, consisting of Jacky and five assistants, was charged with staffing the 17 American zone villages with all GOs except village chiefs and chiefs of service. Additionally, Jacky's office tracked GO development and made recommendations concerning which GOs should be promoted to chiefs of service. GOs who were promoted were no longer the responsibility of Jacky's office; Michel Perchet and the four "godfathers" at Paris worldwide headquarters handled all staffing decisions concerning village chiefs and chiefs of service.

According to Jacky, this had created difficulties. One major problem had been that more American chiefs of service were needed in American zone villages but were not being assigned to them. Additionally, Jacky stated:

American GOs who are promoted to chief of service should stay in the American zone for at least a season or two, to ease their transition into management. That way, they have many American GOs reporting to them – and they can focus on learning to *manage*. If they have mostly European GOs under them, their job becomes much more complex – the language, customs, habits, and so forth are so much different. For example, we had a windsurfing instructor; she was the world freestyle champion two years in a row – for both men and women! Different village chiefs watched her closely during her first few seasons as a GO to assess her potential to become a chief of service. When she was promoted, I argued with Paris that she should stay in the American zone for at least one season. But she was assigned to a village in Israel, where the GOs reporting to her were from many countries and she was in a foreign culture. This added great pressure to her job. But I understand Paris's reasons. They saw her as a great attraction for the village. The world champion from America! The same is true of other Americans who become chiefs of service.

Recruiting in the American zone

In 1985 Club Med received about 10,000 résumés in response to ads for American GOs placed in major magazines. Each ad resulted in a flood of letters, often numbering 80 a day. Jacky's office was responsible for screening the résumés and notifying those who would be given interviews (1,300 candidates per season, 2,600 per year).

"It is harder for us to assess the résumés here in America," Jacky explained, "because we cannot ask, for example, for a photo, or the person's age – because of the discrimination laws. In France, we don't have such laws, and recruitment is generally easier. Sometimes, we give interviews to people who really don't fit the ideal Club Med profile. For example," he laughed, "I interviewed one man who wanted to be an entertainer. The problem was, he had a horrible stutter. If I had hired him, it would have been a terrible experience for him and the guests. Luckily, he had ability in other areas and I was able to hire him to be a GO in another department. However, I usually cannot switch people around so easily." Jacky paused for a moment. "In that sense, we interview many more people here than we would in Europe, where we don't have discrimination laws."

Jacky continued, "On the other hand, I have seen so many résumés that I can usually tell the bad ones very quickly. It only takes five minutes to read a résumé. And my assistants know what to look for," Jacky continued, "because it's really not so difficult once you do it for a while."

Comments from others indicated that accurate assessment of résumés was problematic because North Americans' résumés tended to be "creative". Many were designed to attract attention; others overstated applicants' qualifications. One GO, a veteran of five seasons, stated:

> I wanted the job very badly, and knew the odds I was up against. So I wrote on my résumé that I was an expert in all kinds of activities, figuring that if I was hired and sent to a village, I could learn whatever I needed to learn before anyone found out what I didn't know. I was hired and made a sailing instructor. In my first week on the job, I was asked if I could take out 50 or so GMs on the village's 46-foot schooner. What was I to do? I said, "No problem." Things went great the first day, but a major storm cropped up on my next voyage. I had visions of Gilligan's Island, but didn't panic and, luckily, we had plenty of wine and food on board. I guess I lost my bearings, because we ended up sailing into the cove of another hotel – on an island 30 miles away!

After the initial screen of résumés by Jacky and his staff, those who were not invited to interview were sent rejection letters. The candidates who were granted interviews were notified by mail or by telephone with details about time and place. Interviews were conducted in major cities and on a few college campuses in the United States and Canada by Jacky and by the Club's regional sales managers. Arizona State and, to a lesser extent, UCLA and USC, had been fertile recruiting grounds for good GOs. A self-perpetuating grapevine had developed at Arizona State, which was located close to the company's national reservations center in Scottsdale. The candidates, many of whom traveled fairly long distances, arrived up to four hours early for their interviews (9 A.M. for morning interviews, 2 P.M. for afternoon interviews). Jacky or the regional sales manager talked to all applicants, as a group, about the merits of the Club before the individual interviews began. Included in the 45-minute presentation was a short, glamorous movie on Club Med, filmed in several exotic locations, showing GOs at work and at play. (GOs were allowed to use the village's sports equipment when they had time.) Jacky said:

> I do a good job when I give the presentation. I don't hear any complaints. I tell them the truth about the difficulties of being a GO, but in a funny way. I tell them they will get responsibility right away. When the new GOs land in the village they are immediately "on stage" – our term for GO interaction with the guests. We say "the curtain goes up" when the GOs leave their rooms in the morning, and "the curtain comes down" only when they go to sleep. I believe on-the-job training is the best way for new GOs to learn to interact with the guests; it is very difficult to simulate what goes on in a Club Med village. We can teach them to become "technicians" – windsurfers, for example – but we cannot teach them to become human beings.

The candidates were interviewed by Jacky and two other Club Med personnel who had been GOs (and often had been village chiefs as well). Each interviewer conducted a 30-minute interview with each candidate and asked questions that sized up a candidate's talents, attitude, and potential. Ten days later, letters were sent to everyone interviewed. Jacky estimated that it took approximately two minutes for one of his staff members to prepare a notification letter to be sent to a GO applicant. If 300 new GOs were needed, 450 to 500 were told they would be hired, to provide a reserve pool for GOs who quit or were fired during the season. The letters told those who were accepted to sign employment contracts (these were provisional upon the person actually being assigned to a village), and that they would be notified two weeks prior to departure to their assigned village. This could be a few months later for those who departed during the regularly scheduled rotation period, but could easily become four to six months for the "extra hires" used to replace GOs who quit or were fired during a season.

Jacky was concerned about the time delay between recruitment and "shipment". "My staff spends 40% or more of their time on the phone with new recruits who want to know when and to which village they are going. They spend 15 minutes on each call. The problem is that *we* don't have the answers to their questions. My staff is too busy anyway – and this makes more work for everyone, especially me."

When the time for departure arrived, a member of Jacky's staff talked to the new GOs about flight information, their villages, clothes to bring, and their new bosses, the chiefs of the villages.

"I think the Navy has copied our philosophy," laughed Jacky:

> You know how they say "It's not just a job, it's an adventure"? When I interview around the country, it seems the people don't care so much about the pay [about 400 US dollars per month] and benefits [room and board, bar allowance, medical plan], or the long hours I tell them they will have to work. They want to see the world, and they know they can with us.

"You don't earn a lot of money, but if you don't drink all your salary at the bar, at the end of the season you come back with a lot," said Patrice Prual, 34, a village manager at Cherating [Malaysia]. Patrice recalled that when he was a schoolteacher in Paris, "at the end of the month I had zero."

The stress of paying bills, furnishing a home, owning a car, commuting to work – the average GO has none of this. "I prefer to work at Club Med with palm trees, coconuts, sun, sea," said 30-year-old Tazuko Shimamuda of Japan, who has been doing just that for six years. "It's not perfect, but it's better than taking the metro to work every day."*

After each recruiting season, résumés of the newly "hired" GOs were kept on file in the New York office until Jacky and the chiefs of the village

---

*Barry Kalb, "Play Is Hard Work for Club Med Staff," *The New York Times*, September 24, 1986, p. A-33.

decided together which GOs would go to which villages. Jacky and his staff selected groups of new and old GOs for each village, based on such factors as experience, personality, age, physical characteristics, special talents, and performance-appraisal reports already on file. Although the chiefs had the ultimate authority to pick their teams, the dossiers compiled by Jacky's office for each village were usually accepted after a round of haggling that, in some respects, was more ceremonial than substantive. "Naturally the chiefs all want the best team members," commented Jacky, "and they also might want a few of their favorite GOs from previous seasons. They are always allowed to keep a few chiefs of service whom they have worked with. Some of them have worked together for years. But we obviously cannot give all the best GOs to certain chiefs. I would be killed by the other chiefs!"

Village chiefs had complete authority to dismiss GOs or chiefs of service and request immediate replacements. The Club's offices in New York and Paris were often faced with the difficult task of locating qualified individuals who could depart to a particular village immediately.

"I was on the phone with Paris the other day, telling them I need more American chiefs of service," Jacky said with a sigh. "We have sometimes the miscommunication between the chiefs of service and their GOs. Let me give you an example. Suppose I am the chief of service and see that a wall needs to be painted. If I see a French GO, I tell him 'Go paint the wall'; he finds the paint and the brush, and he paints the wall." He continued: "But, with an American GO, I must explain to him *why* I want him to paint the wall, what changes that will make, and why it is a change for the better. Then I must tell him where to find the paint and brush – he will not look for it himself. And he wants to finish what he is doing first. Finally, he wants feedback about the quality of his work." Jacky's face grew serious. "The European chiefs of service often don't understand how American GOs have to be managed, which can create serious morale problems. Additionally, the chiefs often have difficulties with the English language and accent, so they tend not to give as much training and feedback as they might. But we need the Europeans to keep the Club's international ambience. This is particularly true since the number of Europeans visiting our American zone villages is booming."

Since the American zone had 17 villages, each with about 80 GOs, the turnover problem complicated the already complex village-assignment task. Assuming that suitable replacement GOs could be located in the files, the next step would be a flurry of telephone calls to determine availability. "For example," Jacky said, "halfway through last season I needed a tennis instructor for the village in Martinique. So Debbie, in my office, called the first person in the tennis instructor file. No answer. The second one said she had been working somewhere else for two and a half months and was very happy with her job. The third one was very angry at not having been contacted earlier and told Debbie something that cannot be put in the case study."

"Finally, we found someone who wanted to go – but she wanted to give two weeks' notice to her current employer. Debbie asked her to please give shorter notice ... that we needed her in Martinique now. She agreed after a while, but then she thought she was obligated to help her apartment mates find someone to take her place. They talked some more, and finally Debbie convinced her that her apartment mates could find someone without her help. I tell you ..." Jacky shook his head, "that's a lot of work to replace a tennis instructor in Martinique."

Turnover occurred in four ways: (1) a GO quit before the end of the season; (2) a GO was fired before the end of the season; (3) a GO quit at the end of a season; or (4) a GO was given a poor evaluation by the village chief and was not asked back for the next season. Jacky's lead time to supply new GOs in the first two cases was zero.

In the third and fourth cases, the chiefs rated their GOs in mid-season in two categories: technical ability (i.e. in windsurfing), and attitude and comportment with GMs. The rating scale was 1, 2, or 3 points, in ascending order of excellence. Six was the highest possible score; GOs who were rated "3" were marginal: unless their performances improved, they would be allowed to continue the current season, but would not be invited back for another season. At the end of the season, a more complete performance appraisal was sent to Jacky. The evaluation forms had considerable space for "additional comments", but most chiefs were not inclined to elaborate on their GOs. (The space was often left blank.)

During the 1986 summer season, 66 GOs were fired by the 17 chiefs of the American zone villages. Another 74 quit. (See Figure 5.1.4 for GO comments on why they quit or were fired; Figure 5.1.5 shows comments from chiefs and GOs who did not leave.) Thus, Jacky's office had to replace 130 GOs from the reservoir of hired-but-not-assigned GOs. The GOs who quit or were fired usually were not given an exit interview.

Options

Jacky had several ideas about how to improve the system. First, some village chiefs fired two or three times as many GOs as other chiefs. "Why should that be?" he queried. "If they had to live with the aggravation it caused, perhaps they would try to work with their new GOs instead of executing them right away. Why not have chiefs more involved in the recruiting process, maybe taking time off from their villages to visit cities and conduct interviews? Then they would have a greater appreciation of how difficult it is to find good GOs." In fact, one chief was said to have recruited a number

- "They told me I could use the sports equipment, but who has time? All we do is work and sleep."
- "I met a GM from Connecticut who offered me a good job."
- "Some of the European GOs had a condescending attitude – they thought we had no idea of what the Club was all about. And we hardly got any positive feedback."
- "How many times can you tell GMs how to sign up for snorkeling before you go bonkers?"
- "No television, no radio, no magazines, no newspapers – I'm a college graduate, like most of the American GOs. My mind was going to waste down here on 'Gilligan's Island'."
- "When I arrived they showed me my room and then told me to 'go on stage'. What did that mean? No one showed me what to do. So I took advantage of the Club Med life and tried to live like a guest. Two weeks later, the chief canned me."
- "The lifestyle just wasn't for me, but I guess I had to experience it first-hand to know that."
- "I figured being an American in a French company wasn't going to do much for my career."
- "The European GOs have such a double standard – I think they're living in the eighteenth century."

Note: No exit interviews were conducted.

Figure 5.1.4   Comments from GOs who quit or were fired.

of GMs who had stayed at his village and who ended up wanting to work for Club Med.

There was also talk about opening up a GO training school at the Club's new Sandpiper village in Florida. "I spoke with Marc Tombez, the director of training in Paris. He likes the idea of setting up a school where new GOs would be trained for a week or two before they go to their first assignment," said Jacky.

Jacky wanted to take a more personal approach in tracking the best GOs for advancement within the company. "I will make it a point to meet each GO who is identified as having excellent potential during my visits to the American zone villages. I believe my taking the time to talk to each of them individually will give them a better sense of our interest and commitment to them – leading to reduced turnover among the new GOs who are good. Maybe I will miss some, but over time, this system will be an improvement. GOs who leave now because they are looking for advancement will see that the Club can be a career – and they will know that I know who they are."

Another of Jacky's ideas on how to better manage the recruiting process was to recruit and assign GOs to villages throughout the year, not on a seasonal basis. Approximately 10% of GOs would rotate (or have their con-

- "I had to fire those two GOs – they were drinking way too much."
- "That guy couldn't teach a fish to swim ... he was useless."
- "She was really good; I wish she wouldn't have quit. But better that she should leave than stay and be unhappy."
- "That GO had plenty of enthusiasm – to *complain*! If she hadn't been fired, the other GOs would have drowned her, I think."
- "Two left today. They wouldn't have made it through the season. Two weeks was enough for them. They thought they were on vacation."
- "I know a guy who just quit after a few days. He was disillusioned ... it wasn't at all what he expected."

Figure 5.1.5   Comments from village chiefs and GOs about GOs who quit or were fired.

tracts end) every month. A benefit would be less disruption in the villages at season's end. The reaction from chiefs of the village was that their teams would be disrupted constantly – that it would be hard to build their GOs into teams. A very practical concern of the chiefs was that they would have serious trouble putting on quality stage shows at night if GOs were constantly changing.

There even had been discussion in the Club about the effectiveness of rotation itself. Jacky pointed out, "We decided several years ago that the maintenance GOs should rotate every one or two years. Maybe it is time to consider rotating other GOs in a different pattern. Would it be so bad, for example, to have the chiefs stay in a village for a year at a time? Then a process of gradual GO rotation wouldn't be such a big deal."

Jacky continued, "I don't know exactly when I will have this turnover problem solved. I don't really think the way I recruit now is ideal. I know it can be done better, and I'm working on it. Maybe there are other approaches that I should consider." Then his face turned very serious. "However, there *is* one thing I'm convinced of – the North American GOs are a big part of the future success of the internationalization of the Club."

Case 5.2 Euro Disney: The first 100 days

Research Associate Robert Anthony prepared this case under the supervision of Professors Gary Loveman and Leonard Schlesinger as the basis for class discussion rather than to illustrate either effective or ineffective handling of an administrative situation. The case was prepared from published sources, and the Walt Disney Company is in no way responsible for the completeness, accuracy, or fairness of presentation of any information contained herein.

This is the most wonderful project we have ever done.
Michael Eisner, CEO, The Walt Disney Company[1]

A horror made of cardboard, plastic, and appalling colors; a construction of hardened chewing gum and idiotic folklore taken straight out of comic books written for obese Americans.
Jean Cau, French critic[2]

April 12, 1992 was a cool and hazy day in Marne-la-Vallee, France, home of the Euro Disney Resort complex. Built on a site one-fifth the size of Paris and 20 miles to its west, boasting scores of rides, attractions, hotels, restaurants, entertainment facilities, a campground, and even a championship golf course, Euro Disney opened that day on time and within its $4.4 billion budget.[3]

Roy Disney, nephew of the founder of The Walt Disney Company, addressed the opening day crowd from a platform half way up *Le Chateau De La Belle Au Bois Dormant* (The Sleeping Beauty Castle). He described the complex as an emotional homecoming for the family, which traced its roots to the French town of Isigny-sur-Mer. However, notwithstanding a $10 million ad campaign in anticipation of the opening, attendance at the event was less than some had expected. As evidence of a cool French reception to Euro Disney, commuter trains leading to the park were on strike, protesting staffing and security problems, residents of nearby villages demonstrated against the noise, and a terrorist bomb had just missed disabling nearby electrical facilities the night before.

On June 9 Disney reported that attendance for the park's first seven weeks had been over 1.5 million.[4] While the company previously had projected 11 million in attendance for the first year, it was thought likely that the majority of visitors would be attracted before the wet and colder fall and winter seasons. Also, research showed that the attendance of nearby French residents, who were projected to account for half of the park's attendance, was running well below the expected rate.[5] In New York shares of The Walt Disney Company dropped 5% following the June attendance announcement.

On July 24 Euro Disney announced that revenues for its first quarter of operations were $489 million ($451 million at April 12 exchange rates), but that it would incur a loss for the fiscal year ending September 30, 1992. The company blamed the loss on the fact that it had geared up for a higher level of operations than had actually been attained. Attendance had been 3.6 million through July 22. Shares of Euro Disney, which traded on the French Bourse, dropped 2.75% following the announcement, capping a 31% drop since the opening of the park.[6]

Disney managers remained optimistic that Euro Disney would prove to be a dramatic extension of its founder's dream to "make people happy." Chairman Michael Eisner defended the performance of the park by stating that attendance at Euro Disney exceeded that of Disney's other three theme parks at comparable points in their history.[7] Euro Disney President Robert Fitzpatrick, who had predicted that Europe would become as important to the future success of the company as America,[8] stated that it was impossible to extrapolate meaningfully from the attendance figures at such an early point in the history of the complex.[9]

Still, after five years of controversy over whether various aspects of Disney's traditional approach would fit with French culture, prompting one critic to dub the project a "cultural Chernobyl,"[10] there seemed reason to wonder whether the magic of Disney's famous Magic Kingdom would be replicated in France.

Walt Disney Attractions

Disney theme parks
The Walt Disney Company, founded by Walt Disney and his brother Roy in 1923, consisted of theme parks and resort complexes, motion picture and television production and distribution, consumer products licensing, publishing and retail, and other limited entertainment ventures. Table 5.2.1 provides aggregate financial data for the Walt Disney Company and Table 5.2.2 provides segment data. Walt Disney Attractions consisted of theme parks, hotel and conference facilities, retail complexes, and other recreational properties. In 1991 71% of Walt Disney Attractions' revenues were derived from theme parks, 21% from hotels, and 8% from other sources.[11]

Disney's largest property was Walt Disney World Resort, located on 29,000 acres in Orlando, Florida and boasting three separate theme parks. The 98-acre Magic Kingdom theme park, opened in 1971, featured 45 attractions in seven themed lands and was the site's original park. The 110-acre Disney-MGM Studios Theme Park featured 13 attractions centered around Hollywood's movie industry, as well as containing a working film and television production facility. The EPCOT Center combined the educational Future World, which featured 14 educational and entertainment-

Table 5.2.1 The Walt Disney Company financial summary ($ millions, except return on equity percentages)

	1991	1990	1989	1988
Revenues	6,182	5,843	4,594	3,438
Net income	637	824	703	522
Return on equity	17%	25%	26%	25%
Capital spending	1,425	1,352	1,414	1,043

Sources: The Walt Disney Company Annual Report (1991); The Walt Disney Company Fact Book (1991).

oriented attractions in eight pavilions, with the culturally themed attractions of the World Showcase, consisting of six attractions in 11 "country pavilions."[12]

Disneyland, which opened in 1955, was the company's first theme park. Located near Los Angeles, California, it featured over 50 attractions in seven themed lands. Tokyo Disneyland was designed by Disney but owned and operated by the Oriental Land Company. Its 114 acres was one and one-half times as large as the Disneyland in Southern California, but it was substantially similar in concept to this property.[13]

By early 1992, the company owned and operated hotel properties consisting of 17,000 rooms and 580,000 square feet of meeting space, through its development of its Florida property.[14] In 1990, over 50 million people visited Disney parks,[15] and 1991 attendance showed a slight decline due to economic recession (estimates for 1991 attendance ran as high as 57 million, including Tokyo Disneyland's 16 million).[16] In 1989, which was likely a typical year in recent experience, roughly twice as many people visited the larger Walt Disney World park than the Southern California

Table 5.2.2 The Walt Disney segment financial data ($ millions)

	1991	1990	1989	1988
Revenues:				
Theme parks, resorts	2,865	3,020	2,595	2,042
Filmed entertainment	2,593	2,250	1,588	1,149
Consumer products	724	574	411	247
Operating income:				
Theme parks, resorts	617	889	785	565
Filmed entertainment	318	313	257	186
Consumer products	230	223	187	134

Sources: The Walt Disney Company Annual Report (1991).

Disneyland. In addition, it was estimated that 90% of theme park visitors were repeat customers,[17] and 5%, or well over two million people, flocked in from Europe annually.[18] The majority of Disney visitors were adults, many of whom were in their late twenties and had young children. In 1991 a day pass at Walt Disney World cost $34.75 and at Disneyland cost $27.50.[19] It was estimated that a typical family of four spent $30 per day, per person on meals, snacks, and souvenirs while on vacation at Walt Disney World.[20]

The core of Disney's success defied easy characterization. As one observer noted, "The difference that is Disney goes (very deep) into the American consciousness, for this is a company that sells myth and fantasy."[21]

In one sense, the Disney tradition of creative imagination drew its energy from the personality of Walt Disney himself. Walt was legendary within the company for his obsessive focus on creating products and experiences for his customers which epitomized "fun," and his life was an enthusiastic quest for new technologies, plans, and possibilities which would make this a reality. At the same time, the Disney magic had seemingly been institutionalized in a creative process and meticulous service delivery system which was able to consistently perpetuate a fantastic experience for each of millions of theme park visitors every year. Twenty six years after the death of its founder Disney still had as a primary objective "preserving the basic Disney values – quality, imagination, [and] guest service."[22]

At the center of the Disney theme park experience was the "theme." Disney parks were subdivided into a number of "lands," each of which revolved around a single motif in the nature of its rides and attractions, the costumes of employees, the architectural style of its buildings, and even the food and souvenirs sold within its boundaries. More than a simple decorative device for visitors, however, once within one of the lands at a Disney park, visitors were completely enveloped within its theme. A themed land was truly a carefully planned and orchestrated imaginary world where visitors could escape the themes of the "real" world.

Within each Disney park themes were chosen to appeal to a wide variety of interests and tastes. Lands which the parks had in common included Main Street, Frontierland, Tomorrowland, Fantasyland, and Adventureland. Encompassed within these were images of the most treasured elements of America's past, the fascinations of technologies which were shaping the future, and the myths which had helped shape the American cultural heritage. The images also were brought to life in a variety of ways. Typically, each land contained adventurous, roller coaster-like rides, more subdued rides where the themes were portrayed and observed in interesting detail, and spectator films and shows. The rides and attractions had been crafted by professional "Imagineers" whose goal was to make each completely unique to the Disney theme park experience.

Another cornerstone of the Disney theme park franchise was the rich heritage of the company's cartoon characters. Developed in films which were re-released roughly every five years to acculturate a new generation of patrons,[23] the characters were active in the theme parks in a variety of ways. Costumed characters roamed the park in search of photo opportunities with young visitors, were the subject of rides and attractions within Fantasyland, and most visitors left the park having purchased some piece of memorabilia which featured the characters.

Disney characters had become staples of the American youth experience. They were colorful, fun, highly visible, and had been merchandised into the psyche of children through ubiquitous product licensing. Disney characters also represented rich experiences which helped account for the depth of their appeal. Included in the cast of characters were: "Mickey Mouse, a scrappy rodent with a 'nice guy' personality; Donald Duck, known for his flights of volcanic but ineffectual rage; Snow White and the Seven Dwarfs, each with his own particular slice of the human condition; Pinocchio, the wooden boy, so easily led astray, and his wise sidekick, Jiminy Crickett; Peter Pan, the perpetual boy; Goofy, the floppy dog full of clumsiness and wild emotion; and scores of pirouetting elephants, dancing skeletons, dandified pirates, and water-toting broomsticks." Each had been born in "ancient tales about what it means to be human, to struggle and bear scars and fight the inner wars – tales that bore the weight of myth."[24] Each, also, was alive, well, and eager to please young theme park customers.

A third element of Disney's success was the unique role that visitors played in the theme park. Once inside the park's gates, visitors were not merely spectators or ride-goers. They were considered by Disney to be participants in a play. Every need and desire was carefully planned for, and frequent interactions with staff were considered an integral part of a visitor's experience. Through the lead of the staff member, in a sense, visitors were drawn into interacting in a particular genre of history or fantasy. In addition, many Disney attractions, such as Tom Sawyer's Island at Disneyland, came to life only through use. Most rides were designed to thrust participants into the heart of the theme itself, so that it could be seen from the inside out.

Disney's theme parks were continually updated. New rides and attractions were planned every year, and major investments in facilities were made. For instance, in 1991 the company added six new rides and attractions to Walt Disney World, and it soon planned to announce a second theme park at its California site. Some additions were spectacular applications of new ride technologies. Others were based upon new character developments, such as the addition of "Beauty and the Beast" and "The

Little Mermaid" stage presentations at the Disney-MGM Studios Theme Park, and a "Muppet Vision 3D" attraction which plunged audiences into the middle of a rowdy Muppets adventure.

Service delivery

Nothing was left to chance at Disney theme parks. Standards of service, park design and operating details, and human resource policies and practices were integrated to ensure that the Disney "play" would be flawlessly performed day in and day out at each location. Known for its aggressive management of operational details, Disney's stated goal was to exceed its customers' expectations every day. As a result, one national survey conducted in 1991 measuring how consumers perceive the quality behind 190 different brand names found that Disney was the most highly regarded brand in the country, surpassing such well known names as Mercedes-Benz, Hallmark, and Rolex.[25] It was one of two service companies listed in the top 15, the other being Cable News Network (CNN).

Service delivery at Disney theme parks had been under constant refinement since the first park opened in 1955. The focal point of the service delivery system was "Disney University," the company's in-house personnel development organization with units specific to each site. Because of the nature of the Disney "play" it was the attitudes and competence of the thousands of Disney park employees which accounted for the experience of visitors. It was at Disney University that new employees were oriented to Disney's strict service standards, received on-going communication and training, and joined for frequent recognition and social events. Conceived in 1955 by Walt Disney, Disney University was opened officially in 1961.

Disney University modeled the attitudes required to re-create the desired level of service in the park. As one Disney University manager put it, "Walt felt that you couldn't have a supervisor yell at you and then walk through the front door and greet a guest as if nothing were wrong … He knew that you need to treat employees in the same way that you want them to treat guests."[26]

Disney's employee population was diverse across its various locations, administrative functions, and creative roles. The majority of Disney theme park employees were young, many of high school and college age. Park workers were paid hourly, and tasks could be routine and repetitive. Still, Disney maintained very high expectations for their performance. Consistent with Disney's entertainment concept, employees were called "cast members" even those who worked "backstage" in operations. They wore "costumes" not uniforms, and were "cast in a role" instead of given job duties. Park visitors were called "guests."

Cast members had to meet stiff dress and grooming requirements. These were communicated to potential employees at initial interview sessions, and Disney relied on self-selection as a first employment screen. Following an initial contact, Disney used a peer interview process to select cast members. Three potential hires would meet with one Disney personnel manager for a 45-minute interview session. Applicants were watched closely for how well they listened to their peers, how well they responded to questions, and whether they smiled and maintained an appropriate attitude.[27]

An extensive orientation programme was the first step for both individual cast members and the company's quality assurance efforts. The orientation consisted of indoctrination in Disney's service standards (based on the principles of Safety, Courtesy, Show and Efficiency), classroom instruction in Disney's policies, facilities, resources, and procedures, and extensive on-the-job training. Trainers were themselves cast members who had proven to be exceptional in their roles.

Certain messages were continually reinforced throughout the initial training process. First, it was stressed that happiness was measured differently by every guest and was a challenge to create. Second, trainees learned that customer perceptions were extremely fragile. Finally, it was emphasized that employees were "on stage" at every moment and should look to provide service. As one cast member described, "You don't just make good food and pass it over the counter. It's the idea of extending yourself to guests."[28] In addition, fixing customer problems was given top priority. Employees had wide latitude to "act as a company" when responding to a customer concern.

Employees were evaluated by supervisors based upon their energy, enthusiasm, commitment, and pride. The company maintained a variety of recognition programs for outstanding service delivery, including service recognition awards, milestone banquets for 10, 15, and 20 years of service, and informal recognition parties. Traditionally, the theme parks would re-open for a night during the Christmas holidays, and management would operate the park for the benefit of cast members and their families. At one such event, Chairman Michael Eisner was on hand to serve hamburgers and hot dogs.

Beyond the management of its cast to provide exceptional service, Disney was religious in managing every detail of its theme parks to exceed customer expectations. For instance, in anticipation of guests from different parts of the country asking questions about the flowers in its Florida park, Disney maintained a small instructional garden outside of its employee cafeteria at the site. Also, each park contained dozens of phones connecting to a central question-and-answer hot line, so that employees could find the answer to any question immediately. An average of 610,000 customer letters were received by Disney every year. Each one was read, and a summary report was written monthly for top management, who acted to correct any significant problems noted. As a final assurance of service quality Disney maintained an active mystery shopping program.[29]

Tokyo Disneyland

Disney products, including films and television shows, had been sold in Western Europe for over 50 years. In 1988 European sales accounted for 25% of all Disney product licensing sales,[30] and in 1991 international revenues accounted for 22% of all Disney revenues. Walt Disney Attractions had a major international presence through Tokyo Disneyland. Officially opened in 1983, attendance at Tokyo Disneyland exceeded 16 million in 1991, a record year, when it also welcomed its 100 millionth guest to the park (25 million guests had attended by August of 1985 and 40 million guests had attended by February of 1988).[31] Attendance had exceeded 10 million during each year of operation. Fiscal year 1990 revenues were $988 million at then current exchange rates.[32]

The Oriental Land Company owned and operated Tokyo Disneyland. Disney designed the park and licensed the use of its characters in return for 10% of admissions revenues and 5% of food and souvenir revenues. At the time of the arrangement, cash generated was used to help fund the Epcot Center, which was under construction. In 1991 discussions were underway between Disney and the Oriental Land Company regarding building a second theme park near the first.

Tokyo Disneyland was considered to be a tremendous success from the time of its opening. It appeared to benefit from a strong Japanese appetite for American styled popular entertainment and an increasing trend in Japan towards leisure. As one American magazine put it, "Japan has always looked to America for its popular culture: James Dean, Levi's, McDonald's. Surfer boys and Madonnas are everywhere. So nobody complained about cultural pollution when Disney's ships sailed into Tokyo Bay."[33]

In 1988 10% of the park's visitors were school children and 75% were repeat visitors, largely from metropolitan Tokyo.[34] The design of the park was virtually identical to Disney's Southern California park, and the Oriental Land Company had aggressively added new attractions each year. Virtually all signs and logos in the park were written in English, as were the name badges of cast members. While most cast members primarily spoke Japanese, most live shows and attractions were conducted in English. Of 30 restaurants in the park, only one sold Japanese food. This was because many of the park's older visitors from the Tokyo area had been slower to adapt to the American taste in food. In all other respects, the park was as American as the American parks themselves.

There was some evidence that Tokyo Disneyland was a special cultural haven in Japan, despite stylistic differences between the Disney approach and the Japanese way of life. The company noted that in a country which actively resisted many US products, there was tremendous appeal for Disney's brand of entertainment. This was evident in public transportation leading to and from the park, where normally reserved individuals were openly enthusiastic and usually carried a number of souvenirs.[35]

Also, the company's celebration of New Year's Day markedly broke from Japanese tradition. While the new year is traditionally a serious time within Japan, there was an annual festive party at Disneyland on that day and evening. Extremely popular, the event drew 139,000 visitors in 1991.[36]

Visitor experiences of Tokyo Disneyland were overwhelmingly positive. Comments often revolved around the cleanliness of the park and the efficiency and politeness of staff members. One American tourist familiar with the US theme parks said, "We had great fun. It was exactly the same as the US Disneylands. It was a little funny to see a Japanese Snow White, and the food wasn't very good, but otherwise we thought we were in Florida."[37]

Another tourist commented on locals' fascination with Mickey Mouse. "The shops were mobbed. Everybody buys souvenirs, particularly Mickey Mouse things, and there is a larger selection than in the States. Japanese culture is oriented towards giving gifts, and I think a gift from Disneyland is 'in.' And the people are every bit as good at running the park, even though it was quite crowded. It was so clean it was almost sterile."[38]

An American living in Tokyo accounted for the success of Disneyland in Japan by comparing the Disney experience with Japanese culture. "Young Japanese are very clean cut. They respond well to Disney's clean cut image, and I am sure they had no trouble filling positions. Also, young Japanese are generally comfortable wearing uniforms, obeying their bosses, and being part of a team. These are all parts of the Disney formula."[39]

She added, "Tokyo is very crowded, and Japanese here are used to crowds and waiting lines. They are very patient. And above all Japanese are always very polite to strangers. I have been welcomed into elevators. They give it and expect it, and Disney is a natural." As another observer put it, "Tokyo, Tokyo Disneyland. It's hard to tell where one leaves off and the other starts (parts of Tokyo look more like Tomorrowland than the real thing). This is a match made in Walt Disney's heaven."[40]

Euro Disney

Project overview
The idea of a European theme park and resort complex had been germinating within Disney since the early 1980s. In 1981, the company began an international bidding process for locating Euro Disney, initially involving Germany, Spain, France and others.[41] It felt that the success of the Tokyo park proved the international appeal of the concept. Spain and France were considered most seriously for the project, which would provide more than 30,000 jobs for the host country.[42] The advantage of Spain was thought to be the weather, and the advantage of France was thought to be its central location.

In 1987 Disney signed an agreement with the French government to locate the complex in the farming community of Marne-la-Vallee, just outside of Paris. The company was highly optimistic that this site would turn out to be a winner. One reason was the access to the site by the European population, which exceeded that of the United States by 150 million in roughly one-half of the land mass. Seventeen million people lived within two hours of the site by car, 109 million people lived within six hours of the site by car, and 310 million people could reach the complex by plane in less than two hours. The planned opening of the Euro Tunnel in 1994 would make Euro Disney accessible from England in four hours by car.[43]

Secondly, France, and particularly Paris, was already a highly popular vacation destination. Roughly 50 million tourists visited France annually, spending an estimated $21 billion.[44] Also, Disney hoped to benefit from European vacation practices. Europeans typically took upwards of five weeks of vacation a year, whereas most Americans took only two or three.[45] In what looked like a confirmation of Disney's decision, a poll conducted in France in 1988 revealed that 85% of the population welcomed Euro Disney.[46]

Disney downplayed concerns about the weather in central France, where winter temperatures could reach 23 degrees Fahrenheit (Table 5.2.3). Again, Disney pointed to the experience of Japan. "If Tokyo had not taught us that the parks are weatherproof," said Robert Fitzpatrick, "we might have chosen to go to Spain because of the warmer climate."[47] In Tokyo covered waiting lines and additional indoor heat had proved to be adequate buffers against inclement weather. These precautions were planned at Euro Disney, which also added an outdoor skating rink at the Hotel New York for additional winter appeal.

Contractual concessions made by the French government made the project attractive (although Spain had reportedly tabled an even more generous offer). France agreed to extend highways and the metropolitan railway to the site, build a high speed TGV train extension at their own

Table 5.2.3 Marne-la-Vallee seasonal weather averages: Average temperatures and rainfall (degrees Fahrenheit)

	High	Low	Rainy days
Winter	49	27	16
Spring	58	33	16
Summer	73	58	12
Fall	60	48	14

Sources: The HarperCollins Guide to Euro Disneyland (1992). *Note:* A rain day may be a full or partial day of rain.

expense, reduce the value-added tax on goods sold from 18.6%, to 7%, and provide over $700 million in loans (over $960 million was committed by the end of the project) at the subsidized rate of 7.85%, with no repayment for five years. In addition, France agreed to artificially value the land at $5,000 per acre, its value as agricultural land in 1971, and it guaranteed the valuation for tax purposes for 20 years.[48] This would lower the amount of taxes Disney would pay to local government for services it would provide, such as maintenance of the water supply and fire protection. A portion of the site had been expropriated from local farmers by the French government.[49]

Euro Disney was 49% owned by The Walt Disney Company (42% after adjusting for a convertible bond issue outstanding) and 51% owned by a separate company called Euro Disney S.C.A., which traded on the French Bourse. In accordance with an agreement between Disney and the French government, all shares of Euro Disney were initially offered to European investors. The Walt Disney Company had invested a reported $160 million in the equity of Euro Disney,[50] in which it had three revenue streams in addition to its equity position. These were management fees of 3% of gross revenues for the first five years and 6% thereafter, royalty fees of roughly 7.5% on gross revenues, and a hefty incentive management fee based on the cash flow of the park. One analyst estimated that Disney would capture 75% to 80% of the pre-tax income of the complex.[51]

Euro Disney financial goals for the first year of operation included attracting 11 million visitors and achieving operating income of $373 million at April 12, 1992 exchange rates[52] (Table 5.2.4). Euro Disney's financial projections were based upon a detailed study conducted by the consulting firm of Arthur D. Little ("ADL").[53] ADL developed attendance projections and tested the reasonableness of pricing and operating assumptions. Cost estimates largely were determined by comparison with the experience of the other Disney theme parks.

Admission to the park cost $41 for adults and $27 for children at April, 1992 exchange rates. Hotel accommodations ran from $130 per night to $350 per night during the peak season, and roughly 25% less during the off-season. Camp sites cost roughly $47 per night. The Walt Disney Travel Company offered discount travel packages to the complex. The company anticipated visitors would spend roughly $30 on food, merchandise, and parking, per person, per day, growing by 5% annually.[54]

The capacity of the park was 50,000 visitors, and admission gates were closed after this figure was reached.[55] As visitors left the park, then, additional visitors would be admitted for the balance of the afternoon and evening. For instance, on one occasion in the opening three months the gates of the park were closed from 11 A.M. to 3 P.M. because the park had reached capacity, and a large number of additional guests subsequently were admitted.[56] In fact, a Euro Disney spokesman reported in May of

Table 5.2.4 Euro Disney financial projections ($ millions)[a]

	1992	1993	1994	1995
Revenues:				
Magic Kingdom[b]	774.8	849.8	985.5	1,068.1
Resort development	226.6	391.2	642.3	926.5
Total	1,000.4	1,241.0	1,624.8	1,994.6
Operating expenses:				
Magic Kingdom[b]	482.3	517.6	576.8	615.0
Resort development	145.3	273.9	443.6	542.0
Total	627.6	791.5	1,020.4	1,157.0
Operating income	372.8	449.5	604.4	837.6
Other expenses:				
Royalties	55.1	60.8	70.6	77.0
Management incentive fees	10.0	31.2	87.0	175.7
Other	243.6	244.7	297.1	279.0
Pretax profit	61.1	112.8	149.7	305.9

[a] converted at April 12, 1992 exchange rate of FF 5.48 : $1; [b] includes theme park and 500-room Disneyland Hotel.

1992, "There have been between 20,000 and 60,000 visitors per day (and) many times there were more than 60,000 per day."[57]

ADL estimated that initial attendance could be as low as 11.7 million visitors and as high as 17.8 million visitors (Disney used a target of 11 million because, as Robert Fitzpatrick put it, "I prefer to under-promise and over-deliver."[58] Subsequent attendance growth was projected to average 2% annually for 20 years, compared to average growth of 3.8% at Disney's other parks.[59]

ADL's methodology involved identifying individual target-markets by distance from the site and population, estimating penetration rates for each market, and estimating the average number of annual visits per guest for each market. ADL assumed that the design and scope of Euro Disney would require visitors to either plan extended stays or return trips, and that the capacity and quality of the hotels would encourage this.

Per capita spending assumptions were consistent with Disney's other parks, and ADL considered them to be reasonable, given local market conditions. With respect to admission prices, the firm reviewed prices charged by entertainment options which were considered competitive with Euro Disney. Euro Disney prices were higher than other European theme parks, which were perceived to be of inferior quality. However, they were lower than prices charged in the Paris region for quality adult-oriented entertainment and in line with prices charged for family-oriented attractions.

Food and beverage prices also were compared with those of tourist destinations in the Paris region, as well as other theme parks, and were found to be reasonable.[60]

The company planned a Phase II of the Euro Disney project, which would include a Disney-MGM Studios Park and an additional 13,000 hotel rooms. Entrance to the second park would require a separate admission ticket, and Euro Disney projected 8 million visitors during its first year of operation.[61] Disney budgeted $3 billion to complete Phase II.[62] Originally planned to open in 1996, at one point Disney moved the opening to "1995 or even 1994."[63] However, following the opening of Euro Disney, the scheduled start date of the second park was set back to 1996.[64] In addition, Disney planned to build out the Marne-la-Vallee site for 25 years. Additional rides and attractions would be built, and the company also planned new office complexes, apartments, and perhaps other residential housing units(Table 5.2.5).

Theme park design
Euro Disney, Phase I, consisted of a theme park and extensive lodging and recreation facilities. The theme park included 29 rides and attractions, and it was somewhat smaller than Disney's Florida parks. The balance comprised six themed hotels with 5,200 rooms designed to meet a variety of budgets, a 595-site "Davy Crockett" campground which included 414 cabins, a 27-hole championship golf course, and a variety of restaurants, shops, and live entertainment options, many in the large Festival Disney entertainment center.

Table 5.2.5 Euro Disney property development plans (units listed, except as noted)

	Phase I	Long-term	Total
Theme parks	1	1	2
Hotel rooms	5,200	13,000	18,200
Campsite plots	595	1,505	2,100
Entertainment center[a]	22,000	38,000	60,000
Office space[a]	30,000	670,000	700,000
Corporate park[a]	50,000	700,000	750,000
Golf courses	1	1	2
Single family homes	570	1,930	2,500
Retail shopping space[a]		95,000	95,000
Water recreation area	1	1	
Multi-family homes		3,000	3,000
Time-share units		2,400	2,400

Source: S. G. Warburg Securities (1989).[43]
[a]in square metres.

The park was intended to continue Disney's traditional design. It shared the themed lands of the other Disney parks and featured most of the same rides and attractions. Still, the design of the complex departed in some ways from the traditional formula in an effort to accommodate the preferences of European guests and certain French cultural requirements. Market research was used to set the tone of the resort. Cultural requirements, involving such things as park design, grooming standards for employees, and eating habits, were expressed by vocal French intellectuals, French government officials, local trade unions, and local press.

Research Disney conducted on European travel to the United States showed that the three things tourists were most interested in seeing were New York, Disneyland, and the western United States. As a result, the complex was the most "Western American" of all of Disney's parks. Three of the six hotel properties, the "Cheyenne," the "Santa Fe," and the "Sequoia Lodge" had distinctly western flavors. An attraction which had been called "The Rivers of America" in other parks was called "The Rivers of the Far West" at Euro Disney, and a ride which in the US and Tokyo had been set in a New Orleans styled mansion, was in France set in a mining town of the old west.

The company also responded to concerns that the experience would be too "Americanized." France's intellectual community, particularly of the Left, voiced especially harsh criticisms. They decried what they considered to be the "cultural imperialism" of Euro Disney.[65] They felt it would encourage in France an unhealthy American brand of consumerism. For others, also, Euro Disney became symbols of America within France. On June 28 a group of French farmers blockaded Euro Disney in protest of farm policies the United States supported at the time.[66]

Subsequent to concerns raised by the French government, Disney assured that French would be the first language of the park. Still, most signs would be bilingual, as would be the park's employees. Disney also promoted the benefit of an English speaking destination in France in its American tour literature.

In other respects Disney attempted to imbue the park with a European flavor. In Fantasyland it was stressed that Disney characters had their roots in European mythology. They were portrayed as such in attractions ("European folklore with a Kansas twist," as Michael Eisner called it).[67] The Peter Pan attraction featured Edwardian-style architecture, Snow White had her home in a Bavarian Village, and Cinderella lived in a French Inn. The Alice in Wonderland attraction was surrounded by a 5,000 square foot European hedge maze. In Discoveryland Euro Disney featured tributes to European Renaissance heroes and to France's Jules Verne. Adventureland would invoke the imagination of famous European adventure tales such as Sinbad the Sailor, Arabian Nights, and the Thief of Baghdad.

A number of other concerns relating to the design of the park had been expressed in the French press. One of the most noted was a flap over Disney's decision not to serve wine at the park, consistent with its policies in the US and Tokyo. It was felt by many that this was a departure from important French tradition and lunch habits, as well as a snub to the country's reputation for excellence in wine making. Visitors who wanted alcoholic beverages congregated at Festival Disney, an entertainment complex outside of the theme park and adjacent to the Hotel New York. There they were "supervised by unsmiling security men and CRS riot police with guns."[68]

Disney addressed a concern that French visitors would not tolerate long waiting lines. The company planned films and other entertainment diversions for guests in line for a ride. Some also pointed out differences between European and American eating habits. They pointed out that Europeans were not accustomed to eating fast meals at off hours, sometimes while walking, as were Americans, and predicted that dining facilities would have problems serving peak demands. A small sample of visitors to Euro Disney confirmed that this had, indeed, become a problem, although they did not cite it specifically as a cause of dissatisfaction with the park.[69]

In anticipation of concerns about food, Euro Disney featured foods from around the world at its many themed restaurants and snack bars. This was in contrast to the strictly American flavor of Tokyo Disneyland. Disney also claimed that the food was of higher quality than at its other parks. In an effort to boldly demonstrate its claim, Disney even invited top Paris chefs to visit and taste it. Visitors to Euro Disney could not confirm a marked improvement in the food, however.[70]

While Euro Disney was controversial in the French press, not all French intellectuals criticized Disney. For instance, philosopher Michel Serres noted, "It is not America that is invading us. It is we who adore it, who adopt its fashions and above all, its words."[71] For his own part, a French critic who had been vocal in his opposition to Euro Disney did publicly lament that his young son loved Disney characters.[72]

Another American observer responded to the controversy by saying that "Euro Disney is an imaginary place, a culture without sin," and he commended American culture for producing such creativity.[73] Euro Disney's Robert Fitzpatrick took a somewhat more combative tack when he said, "We didn't come in and say 'O.K., we're going to put a beret and a baguette on Mickey Mouse ...' We are who we are."[74]

The start-up process

Disney met a monumental challenge in readying the park for its April 12 start date, which involved completion of the second largest construction project in the history of Europe, as well as preparing operationally for the launch. In addition to the task of marketing the park, Disney hired and trained 14,000 employees to fill 12,000 jobs in anticipation of the opening.[75] Another 5,000 temporary jobs were filled by the peak July season.[76]

Euro Disney was aggressively marketed by Disney as well as other firms. Disney successfully encouraged dozens of articles on the complex in magazines throughout Europe. Prior to the opening it sent a model of The Sleeping Beauty Castle around Europe to dramatically publicize the park. An extensive Europe-wide ad campaign was launched to market the opening celebration, which was broadcast live across Europe. In addition, Swiss food giant Nestlé sponsored extensive cross-promotions of Euro Disney at its own expense.[77]

Perhaps the biggest challenge was preparing operationally to provide Disney's standard of customer service. To accomplish this task, Robert Fitzpatrick announced that a leading priority was to indoctrinate all employees in the Disney service philosophy, in addition to training them in operational policies and procedures.[78]

Disney opened a special center at Euro Disney's new Disney University in September of 1991. Its goal was to select 10,000 employees within six months while maintaining selective applicant-to-hire ratios. A staff of 60 interviewers had been assembled for that purpose.[79] Stated selection criteria were applicant friendliness, warmth, and liking of people. The company attempted to hire employees of nationalities proportional to expected visitor counts. Its initial objective was to hire 45% French employees, 30% other European, and 15% from outside of Europe,[80] but by opening day the cast was 70% French.[81] Europe had recently entered a recession, in which it remained at the time of opening, making it somewhat easier to attract an applicant pool. Most cast members were paid roughly $6.50 an hour at April 12, 1992 exchange rates, which was 15% above France's minimum wage, and shifts were generally 169 hours per month.[82]

At the same time Disney aggressively cross-trained managers and supervisors to ensure service quality. Prior to opening, 270 managers were cross-trained in the Disney methods at the company's other three theme parks. Also, another 200 managers were imported from the other parks to work at Euro Disney.[83]

Disney encountered difficult resistance in the hiring process, for which it was criticized by applicants, the press, and French unions. The controversy revolved around Disney's grooming requirements. Disney strictly enforced a dress code, a ban on facial hair, a ban on colored stockings, standards for neat hair and fingernails, and a policy of "inappropriate undergarments." However, applicants and labor leaders in France felt the requirements were excessive, being much stricter than the requirements of other employers. They hoped to force the company to loosen its standards, but they were unsuccessful.[84]

Another problem Disney faced was that of staff housing. The agricultural Marne-la-Vallee did not have apartment space for the thousands of Disney's workers at the complex, and the jobs generally did not pay well enough to make decent Parisian housing affordable. At the time of the opening an estimated 4,000 staff members were affected by the housing shortage. By building its own apartments and renting rooms in local homes, Disney was adding rooms at a rate of 100 per week.[85]

Disney successfully staffed and trained cast members for the complex by the time of the opening. However, within the first nine weeks of operation roughly 1,000 employees left Euro Disney, about one-half of whom left voluntarily.[86] Under French employment law an employee could be terminated during their first two years with little difficulty, but after the two year period performance documentation, notification requirements, and severance requirements became stringent. The long hours and hectic pace of work at the park were cited as the reasons for the turnover. "A lot went because it was chaotic at first," said one English waitress.[87] Disney conceded that employees had worked under "tough conditions" at the time of opening.[88]

One example of a cast member who left was a 22-year-old medical student from a nearby town who signed up for a weekend job. After one weekend of "brainwashing," as he called it, and one weekend of training, he went to work at a Fantasyland shop. One day during his first weekend he worked 11 "frantic" hours straight, and by the next weekend the entire shop personnel had changed. He left after a dispute with his supervisor over the timing of his lunch break.[89]

Another cast member, a waiter in one of the better hotels, blamed communications problems between supervisors and workers for the difficulties. "I don't think they realized what Europeans were like," he said, "that we ask questions and don't think all the same." Still, he added that, "it's getting better; they're listening more to the staff."[90]

Visitor reactions

From a small polling sample visitor experiences of Euro Disney were mixed. Many visitors found in Euro Disney everything for which they had hoped. Others complained that the park did not meet the US standard, suffering from long lines, poor service, and operational glitches.

One family which had driven in from northern Europe was thrilled with the experience, because they were able to interact with the Disney characters they had always revered. They said they could never afford to come to the United States to do so. Another visitor from nearby Paris was simply impressed by the scale of the complex. He had already visited the complex twice in its first three months of operation.[91]

As reported in the French press, an 11-year-old visitor named Vincent exclaimed, "I loved everything. There was nothing I didn't like." Thirteen-year-old Cyndie said, "I asked my parents if we can come back. We just didn't have enough time to see everything."[92] In addition, after opening day, one London newspaper reported that a group of German visitors had all "had a great time," despite considerable frustration with waiting lines.[93]

Others were not as impressed, however. Themes echoed by visitors less enthusiastic with Euro Disney included a lack of appreciation of cast member performance, the difficulties associated with the multi-cultural nature of the park, and the high cost of a day at the park. In addition, a number of observers noted that Euro Disney represented a departure from a traditional French entertainment experience.

One British journalist wrote, "Cast members taken on to work at Euro Disney are mostly nice enough. 'Mostly,' because even on opening weekend some clearly couldn't care less ... My overwhelming impression of the ... employees was that they were out of their depth. There is much more to being a cast member than endlessly saying '*bonjour*'. Apart from having a detailed knowledge of the site, Euro Disney staff have the anxiety of not knowing in what language they are going to be addressed ... Many were struggling. One cast member, who has worked for the company in the US, candidly volunteered that service at Euro Disney falls way short of the standards at the American parks."[94]

An American visitor to the park agreed that the experience fell short of what she had come to expect from Disney. "They compete with their own high standards," she said, "but they are not winning in France. Most of the workers are simply not aiming to please, even though they are thrilled to have jobs in the rotten economy. They are playing a different game than their American counterparts. They are acting like real people instead of 'Disney' people. Unfortunately, you get the feeling that the whole thing is not yet under control."[95]

Another American visitor voiced concerns over the international flavor of Euro Disney. "The park has kind of a strange feel to it. They haven't yet figured out whether it is going to be an American park, a French park, or a European park. This is in the atmosphere of the park itself, and it is compounded by the behavior of visitors from various parts of Europe, which can be quite different. Little things, like the attitudes of different nationalities with respect to disposing of trash, are very noticeable. And difference in waiting-line behavior is striking. For instance, Scandinavians appear quite content to wait for rides, whereas some of the southern Europeans seem to have made an Olympic event out of getting to the ticket taker first." He went on to describe that there generally was considerable restlessness with extensive waiting lines, even though he did not perceive the park to be terribly crowded (on crowded weekend days visitors complained that lines averaged between one and two hours for rides which they perceived to average 10 to 15 minutes in length). However, he added that "even at its worst the service at Euro Disney was better than the best I encountered in Paris."[96]

The difficulties in accommodating the cultural diversity of the complex were also noted by the popular press. In a report on the opening of Euro Disney one newspaper asked, "Can an American theme park in Europe please all ages and nationalities? And in what country, if any, is this fantasy never-never-land which started with a Hollywood mouse? It is not, except in the most literal sense, France."[97] British advertising executive David Moutrie agreed. He observed, "I think as far as the management is concerned [Euro Disney] just happens to be in the middle of Europe handy for a big population. If somebody said to me when we get back, 'have you been to France?' I'd be tempted to say no."[98]

On this theme, some observers felt that the idea of Euro Disney was out of character for the French population. Comments included claims that many French were too individualistic and private to appreciate the standardized and crowded Disney theme park experience. Others felt that the French tended to enjoy entertainment which was more intellectual in nature than Euro Disney.[99]

And not only were there questions of whether Disney could be enjoyed by the local population, but some also felt that it was the character of the European labor force, rather than experience or training, which would account for less than perfect service at Euro Disney. Wrote one journalist, "The Disney style of service is one with which Americans have grown up. There are several styles of service (or lack of it) in Europe, unbridled enthusiasm is not a marked feature of them."[100]

The cost of the experience was thought to be an issue for some. While little dissatisfaction with admissions prices among those in attendance at the park was reported, it was reported that many French visitors knew people who had been deterred from coming by the cost.[101] There were also grumbles about the cost of Disney souvenirs. "I refuse to pay Ffr49 (roughly $9) for a little Mickey Mouse statue," said one representative Parisian visitor.[102]

For its own part, Disney acknowledged that it was still working out the details of its operations. It felt that it was unreasonable to expect a project of the size of Euro Disney to be perfect. As one senior Euro Disney manager put it, "You don't get it right the second you start."[103]

Decisions

There was precedent for believing that a rocky start was not catastrophic in the theme park business. Universal's Florida theme park had had a disastrous opening due to technical difficulties, but it quickly rebounded and was considered successful. Euro Disney was far from a disaster. It was too early to tell what the impact of poor fall and winter weather would be, but the attendance figure of over 30,000 per day was respectable. If this number were annualized, then Euro Disney's projection of 11 million visitors during the first year of operation would be met. Although the local

French population had not attended as planned (the company claimed that Parisians were "postponing their visits" until the fall), visitation from the rest of Europe was running higher than planned.

Profitability was another matter. Even if revenues could be brought in line with projections for the balance of the year, the announcement that the park would not be profitable for the five and one-half months ending September 30 was sobering news (analysts estimated that the losses could be as high as $60 million).[104] Observers could not assess with certainty whether the shortfall was due to the level at which the company geared up operations, as it claimed, or some other set of reasons. Whatever the cause, the poor profit picture would constrain Euro Disney's options for fixing any operating problems it had and initiating programs to bring in visitors.

The coming winter months were clearly critical to Euro Disney's chances for financial success. Here, also, there was cause for concern. Prior to 1987 (the last time such information was made publicly available), an average of 65% to 70% of attendance at the warm weather US parks came during the April through September period.[105] Accordingly, one knowledgeable US analyst estimated that seven million of the complex's projected 11 million visitors would attend in the five and one-half months from the opening to the September 30 fiscal year end.[106] Travel agents representing Euro Disney reported that, while demand was very strong for the balance of the summer months, advance bookings for the end of the year were much lower.[107] Analysts estimated that hotel room occupancy was running at 90% during the July peak season, and they estimated occupancy had averaged 68% for the April to July period.[108] On the other hand, one travel agent reported that less than 20% of its projected September bookings, 12% of its projected November bookings, and 10% of its projected December bookings had materialized. Other agents were in similar situations.[109]

Agents did not know whether to attribute the low level of forward bookings to lack of advance planning or more fundamental problems with the park, because they lacked experience against which to benchmark. Moreover, even if travel to Euro Disney declined, local visitation could pick up the slack. Perhaps waiting-line-cautious French simply planned to wait for crowds to thin.

Even as summer was at its peak Euro Disney management took actions to improve its attendance outlook and profit position. By the time of its opening, Euro Disney had slashed rates at its least expensive hotels by 25%. In July it confirmed that some rooms were being offered at $73 a night for the winter season at current exchange rates.[110] On July 31 it was reported that The Walt Disney Company, headquartered in Burbank, California, would slash 300 to 400 jobs from its Imagineering unit. It cited the completion of Euro Disney, Phase I as the reason that such a cut was possible.[111] Presumably, then, a portion of the cost savings would be passed on to Euro Disney.

Euro Disney management still faced serious issues. Not the least of these was prioritizing its objectives in the face of a somewhat conflicting problem set, consisting of an uncertain revenue outlook, cost problems, and mixed reviews of its service delivery system. Compounding Euro Disney's situation was the fact that it appeared to be in the spotlight of the national and international press. Given the controversies which swirled around the opening of the park, its every action could prove to be newsworthy.

One set of decisions facing Euro Disney concerned getting the service system up to the standards and cost levels of the other Disney parks. Such issues as waiting lines, consistent cast member courtesy, and employee turnover deserved immediate attention. While practice would surely help, perhaps there were other things Euro Disney could do to speed the process. A second set of decisions was the marketing of the park to achieve winter attendance targets, particularly in light of the visibility of Disney's critics in France. It needed to find a way to promote the park in such a way that there would be minimum costs in terms of public relations. Pricing, communications, special events, and tie-ins with other parts of The Walt Disney Company were all levers available to Euro Disney management.

Finally, Disney had planned major investments in Phase II of the park. The level, timing, and nature of the investments still were at issue. Perhaps there were significant lessons to be learned from the Phase I experience which Disney could apply to Phase II and improve its chances for success. Issues Disney would surely review in planning for Phase II would be whether the adaptation of the Disney entertainment concept to European, and specifically French, culture had been overdone or underdone, and whether its management policies and training methods had been appropriate to its task.

References

1 The Walt Disney Company Annual Report (1991).
2 *New York Times* (1992). Only the French elite scorn Mickey's debut, 13 April.
3 *op. cit.*
4 *Wall Street Journal* (1992). Euro Disney draws over 1.5 million in first 7 weeks, 10 June.
5 *Ibid.*
6 *New York Times* (1992). Euro Disney sees loss; Disney profit rises 33%, 24 July.
7 *Minneapolis Star Tribune* (1992). Mickey Mouse diplomacy, 19 June.

8 *New York Times* (1991). Playing Disney in the Parisian fields, 17 February.

9 *Wall Street Journal* (1992). Euro Disney draws over 1.5 million in first 7 weeks, 10 June.

10 *London Financial Times* (1992). Mickey Mouse lures the stars in Paris, 11–12 April.

11 The Walt Disney Company Annual Report (1991).

12 The Walt Disney Company Fact Book (1991), 31 August.

13 *Ibid*.

14 The Walt Disney Company Annual Report (1991).

15 Flower, J. (1991). *Prince of the Magic Kingdom: Michael Eisner and the remaking of Disney*, John Wiley and Sons, Inc., New York.

16 The Walt Disney Company Annual Report (1991).

17 *London Financial Times* (1992). Culture shock for the Mickey Mouse outfit, 23 April.

18 Flower, *op. cit.*

19 The Walt Disney Company Fact Book (1991), 31 August.

20 *op. cit.*

21 *Ibid*.

22 The Walt Disney Company Annual Report (1991).

23 Collis, D.J. (1988). *The Walt Disney Company*, Harvard Business School Case No. 2-388-147.

24 Flower, *op. cit.*

25 *Los Angeles Times* (1991). Disney tops poll of best brand names, 10 July.

26 Solomon, C.M. (1989). How does Disney do it? *Personnel Journal*, December.

27 *Ibid*.

28 *Ibid*.

29 *Ibid*.

30 The Walt Disney Company Annual Report (1988).

31 The Walt Disney Company Annual Report (1991).

32 *New York Times* (1991). Playing Disney in the Parisian fields, 17 February.

33 *Travel and Leisure* (1992). Tokyo: Mickey's first trip abroad, August.

34 The Walt Disney Company Annual Report (1988).

35 The Walt Disney Company Annual Report (1987).

36 *Travel and Leisure* (1992). Tokyo: Mickey's first trip abroad, August.

37 Case Writer interview, July 1992.

38 *Ibid*.

39 *Ibid*.

40 *Travel and Leisure* (1992). Tokyo: Mickey's first trip abroad, August.

41 *Invest in France Agency* (1992). Mickey does Marne-la-Vallee, May.

42 *Business Week* (1985). Monsieur Mickey or Señor Miqui? Disney Seeks a European Site, 15 July.

43 S.G. Warburg Securities (1989). *Euro Disneyland SCA: Offer for sale*, 5 October.

44 *Time Magazine* (1991). Monsieur Mickey, 25 March.

45 Gould, A.S. (1992). *Euro Disney SCA*, Dean Witter Equity Research, 12 June.

46 *The Economist* (1987). Mickey hops the pond, 28 March.

47 *New York Times* (1991). Playing Disney in the Parisian fields, 17 February.

48 Flower, *op. cit.*

49 *Minneapolis Star Tribune* (1991). Will French culture make room for mouse?, 19 May.

50 *New York Times* (1991). Playing Disney in the Parisian fields, 17 February.

51 Gould, *op. cit.*

52 S.G. Warburg Securities, *op. cit.*

53 *Ibid.*

54 *Ibid.*

55 Invest in France Agency (1992). Mickey does Marne-la-Vallee, May.

56 *London Financial Times* (1992). Queuing for flawed fantasy, 13 June.

57 *op. cit.*

58 Gould, *op. cit.*

59 S.G. Warburg Securities, *op. cit.*

60 *Ibid.*

61 *Ibid.*

62 *New York Times* (1991). Playing Disney in the Parisian fields, 17 February.

63 Euro Disneyland Annual Report (1990).

64 Gould, *op. cit.*

65 *New York Times* (1992). Only the French elite scorn Mickey's debut, 13 April.

66 *Investors Business Daily* (1992). French farmers blockade Euro Disneyland, 29 June.

67 *op. cit.*

68 *London Financial Times* (1992). Queuing for flawed fantasy, 13 June.

69 Case Writer interviews, July 1992.

70 *Ibid.*

71 *New York Times* (1992). Only the French elite scorn Mickey's debut, 13 April.

72 *Wall Street Journal* (1992). As Euro Disney braces for its grand opening, the French go goofy, April.

73 *Minneapolis Star Tribune* (1992). The French turn up their noses at Disney? Well, excuse Mouse!, 26 April.

74 Associated Press (1991), 19 May.

75 *London Financial Times* (1991). Disney's cast of thousands, 18 February.

76 *New York Times* (1992). Euro Disney sees loss; Disney profit rises 33%, 24 July.

77 *Wall Street Journal* (1992). Disney gets many helping hands to sell the new Euro Disneyland, 1 April.
78 Euro Disneyland Annual Report (1990).
79 *London Financial Times* (1992). Disney's cast of thousands, 18 February.
80 *Minneapolis Star Tribune* (1992). Continent will get Goofy (and Mickey and Donald) with Euro Disney opening, 29 March.
81 Invest in France Agency (1992). Mickey does Marne-la-Vallee, May.
82 *London Financial Times* (1992). Queuing for flawed fantasy, 13 June.
83 Euro Disneyland Annual Report (1990).
84 *Minneapolis Star Tribune* (1991). Costume requirements at Euro Disneyland called a Mickey Mouse idea, 10 March.
85 *London Financial Times* (1992). Queuing for flawed fantasy, 13 June.
86 *Wall Street Journal* (1992). Euro Disney's Fitzpatrick denies report that 3,000 workers quit over low pay, 27 May.
87 *op. cit.*
88 *op. cit.*
89 *London Financial Times* (1992). Queuing for flawed fantasy, 13 June.
90 *Ibid.*
91 Case Writer interview.
92 *Travel and Leisure* (1992). Euro Disney: Oui or Non?, August.
93 *London Financial Times* (1992). Queuing for flawed fantasy, 13 June.
94 Case Writer interview, July 1992.
95 *Ibid.*
96 *op. cit.*
97 *London Financial Times* (1992). Queuing for flawed fantasy, 13 June.
98 *Ibid.*
99 Case Writer interview, July 1992.
100 *London Financial Times* (1992). Mickey Mouse outfit suffers culture shock, 23 April.
101 *London Financial Times* (1992). Queuing for flawed fantasy, 13 June.
102 *Ibid.*
103 *London Financial Times* (1992). Culture shock for the Mickey Mouse outfit, 23 April.
104 *New York Times* (1992). Euro Disney sees loss; Disney profit rises 33%, 24 July.
105 The Walt Disney Company Quarterly Reports.
106 Gould, *op. cit.*
107 *London Financial Times* (1992). A question of the Mouse's attraction, 12–13 June.
108 *New York Times* (1992). Euro Disney sees loss; Disney profit rises 33%, 24 July.
109 *op. cit.*
110 *op. cit.*
111 *Boston Globe* (1992). Lay offs loom at Disney, 31 July.

Case 5.3 Nordstrom Inc.

This case was prepared by Richard T. Pascale as a basis for classroom discussion. Used with the permission of the author.

In the fall of 1990, Nordstrom opened its first store in the New York area at the Garden State Plaza in Paramus, New Jersey. Industry experts voiced scepticism. How could a parochial West Coast chain make inroads in the East Coast market where entrenched competitors like Macy's and Bloomingdale's held a solid franchise? How could Nordstrom launch an expansion programme during the trough of a national recession when all retailing was depressed and the Northeast particularly hard hit? How could Nordstrom's distinctive trademark of unfailingly polite salespeople be transplanted to a region where manners of both customers and sales personnel tended to be more perfunctory and at times abrasive?

Nordstrom had overcome similar obstacles in the past. Starting from a base of 27 shoe stores in Oregon and Washington in 1963, they had expanded into a full line of apparel, cosmetics and jewellery. (See Figure 5.3.1 for a brief history of the firm.) Soon Nordstrom department stores dominated the Northwest market. Against a tide of disbelief, they invaded California in 1978 and grabbed a 30 per cent market share in a 10-year period. In 1988, Nordstrom chose the Washington DC suburb of Tysons Corner for its first East Coast foray. Its stunning success helped knock Garfinkels, a dominant local retailer, into bankruptcy. By the close of 1991, Nordstrom had become a leading player in the capital area with three department stores ringing Washington DC. A fourth and fifth are scheduled for Baltimore and Annapolis by 1995.

The Nordstrom blitzkrieg had all the appearances of repeating itself in the Northeast. One year after opening the Paramus store, sales reached $100 million–120 million over its projections. By way of contrast, one mall competitor, Bloomingdale's, sold $55 million through a store that was 20 per cent *larger* than the Nordstrom outlet. Its main competitor, Macy's, suffered flat sales notwithstanding a major remodelling and aggressive price promotions. Macy's seven-year track record at the mall had been expected to give it a strong advantage, yet defecting customers flocked to Nordstrom's doors.

In the fall of 1991, Nordstrom opened its second store in the Menlo Park Mall, Edison, New Jersey. By all early indications, it too was a smashing success. Nordstrom's long-term plans, extending through 1995, committed to a total of seven stores in the Northeast, including possible outlets in Freehold, Short Hills or Livingston, New Jersey, and White Plains. Nordstrom's expansion plans also included stores in Chicago, Minneapolis, Denver, Indianapolis and Boston.

How could Nordstrom, a late entrant in retailing, go head-to-head with

In the late 1880s, 16-year-old John W. Nordstrom left northern Sweden for the United States, taking with him a small inheritance from his father and his first real suit of clothes. He arrived in New York over two weeks later with five dollars in his pocket, unable to speak a word of English.

The young immigrant's determination to succeed would lead him through many trials and adventures. He laboured in mines and logging camps across the United States to California and then to Washington, before heading north to the Klondike gold fields in 1897. Two years later he returned to Seattle with a $15 000 stake, ready to settle down.

Carl F. Wallin, a Seattle shoemaker he met in Alaska, offered him a partnership in a shoe store. The young Nordstrom had invested some of his money in a few parcels of property in the city, and eagerly accepted the opportunity to go into business. Together they opened their first store in 1901, with a 20-foot frontage on Fourth and Pike streets. The first day's receipts came to $12.50.

In 1929, Wallin sold his interest in the company to Nordstrom and in 1930, John W. sold the company to his sons: Elmer, Everett and Lloyd. These three men built the single shoe store into the largest independent shoe chain west of the Mississippi – a 27-unit operation grossing $12 million by 1963.

At this point, the brothers expanded into the fashion specialty business, with the purchase of Best's Apparel. By 1971, the company, now in the hands of the third generation, had seven full-line fashion specialty stores, and sales of nearly $80 million.

Today, Nordstrom is the largest fashion specialty store in the nation, with nearly 60 stores in operation, and annual sales over $2 billion. These stores include the Place Two and Rack stores. The Place Two stores are smaller versions of our Nordstrom stores. They are located in the Northwest. Their merchandise is primarily men's and women's apparel and shoes, with a small representation of accessory merchandise. The Rack stores are our version of a 'self-service' retail concept, providing the customers with marked down items from the regular stores, as well as off-priced items from our manufacturers. You will find at least one Rack store in almost every one of our regions.

Surely if John W. Nordstrom were here today to see what his modest shoe store has become, he would be amazed. But despite the modern and efficient business practices and buildings, he would still find the same philosophy he practised back in 1901: offering the customer exceptional selection, quality, value and service – the principles that have built the Nordstrom tradition.

Figure 5.3.1 Nordstrom history.

well-established retailers and succeed so handily? At face value, its competitors had formidable advantages – knowledge of local tastes, wages that are typically 15–20 per cent below Nordstrom's rates, real estate that was long since paid for and largely depreciated fixed assets. Notwithstanding, Nordstrom prevailed. States one industry expert: 'All retailers in America have awakened to the Nordstrom threat and are struggling to catch up. Nordstrom is the future of retailing ... it is the most Darwinian of retail companies today.'[1] Analyst Margaret Gillian of the First Boston

Corporation examines the sustainability of Nordstrom's prowess: 'Nordstrom is developing a critical mass of stores in the Capitol region and in the New York and New Jersey areas. It is a company possessing enormous customer appeal. Its concept cannot be readily duplicated. Such combinations are rare in retailing, where good things are predictably copied.'[2]

Bucking the trends

The decade of the eighties witnessed a major convergence of strategies among most of the US department-store chains. Hostile takeovers by the Campaneau Group and Hooker of Australia led to massive consolidations. Independent retailers such as Dillards and Dayton-Hudson embarked on aggressive acquisition efforts. Nordstrom, in the meantime, stuck to its guns and built slowly but deliberately through *de novo* expansion. The primary reason for this approach was Nordstrom's belief that its culture and values had to be integrated into each new unit from infancy and that, in the last analysis, the Nordstrom culture *was* its sustainable competitive advantage. Growth was largely funded by retained earnings, keeping Nordstrom's debt relatively low. (See Tables 5.3.1 and 5.3.2 for balance sheet and income statement.)

An essential ingredient to its expansion formula was Nordstrom's insis-

Table 5.3.1 Nordstrom Inc. and subsidiaries, consolidated statements of earnings

Year ended 31 January	1991 ($000)	% of sales	1990 ($000)	% of sales	1989 ($000)	% of sales
Net sales	2 893 904	100.0	2 671 114	100.0	2 327 946	100.0
Costs and expenses:						
Cost of sales and related buying and occupancy costs	2 000 250	69.1	1 829 383	68.5	1 563 832	67.2
Selling, general and administrative	747 770	25.8	669 159	25.1	582 973	25.0
Interest, net	52 228	1.8	49 121	1.8	39 977	1.7
Service charge income and other, net	(84 660)	(2.9)	(55 958)	(2.1)	(57 268)	(2.4)
Total costs and expenses	2 715 588	93.8	2 491 705	93.3	2 129 514	91.5
Earnings before income taxes	178 316	6.2	179 409	6.7	198 432	8.5
Income taxes	62 500	2.2	64 500	2.4	75 100	3.2
Net earnings	115 816	4.0	114 909	4.3	123 332	5.3
Net earnings per share	1.42		1.41		1.51	
Cash dividends paid per share	0.30		0.28		0.22	

Table 5.3.2   Nordstrom Inc. and subsidiaries, consolidated balance sheets ($000)

31 January	1991	1990
Assets		
Current assets:		
Cash and cash equivalents	24 662	33 051
Accounts receivable, net	575 508	536 274
Merchandise inventories	448 344	419 976
Prepaid expenses	41 865	21 847
Total current assets	1 090 379	1 011 148
Property, buildings and equipment, net	806 191	691 937
Other assets	6019	4335
Total assets	1 902 589	1 707 420
Liabilities and shareholders' equity		
Current liabilities:		
Notes payable	149 506	102 573
Accounts payable	204 266	195 338
Accrued salaries, wages and taxes	128 697	122 607
Accrued expenses	34 668	29 080
Accrued income taxes	24 268	12 491
Current portion of long-term liabilities	10 430	27 799
Total current liabilities	551 835	489 888
Long-term debt	457 718	418 533
Obligations under capitalized leases	21 024	22 080
Deferred income taxes	45 602	43 669
Contingent liabilities (Note 11)		
Shareholders' equity	826 410	733 250
Total liabilities and shareholders' equity	1 902 589	1 707 420

tence that every new store have a critical mass of Nordstrom managers, buyers and salespeople. When staffing the Paramus, New Jersey store, 5000 applicants competed for 800 openings. Over 200 of these slots were filled by former Nordstrom employees transferred in from across the United States. The company wanted a sufficient number of role models to guarantee replication of the 'Nordstrom Way'.

The trend toward consolidation by Nordstrom's competitors had been driven by a variety of perceived economies of scale. In particular, the promise of sophisticated management information systems appeared to offer these far-flung retailing conglomerates a means of making the most efficient use of inventory, enhancing stock turnover, fine-tuning buying at centralized command centres, and maximizing purchasing clout. Also appealing was the promise of lowered overhead by spreading corporate staff functions over a large number of outlets.

With hindsight, none of these 'economies' seemed to have lived up to their promise. The Federated empire, including Abraham & Straus, Lazarus, Rich's, and Bloomingdale's, entered Chapter 11. The same fate befell Allied stores, which owns Jordan Marsh, Bon Marche and Sterns. Macy's, which also owns Bullocks and I. Magnin, teetered near insolvency under the debt load of its 1986 leverage buyout. Carter Hawley Hale, which owns The Broadway, The Emporium and Weinstocks, also found itself in a weakened financial condition. Smaller players, including Bonwit Teller, Bergdorf Goodman, Altmans and Garfinkels, filed for bankruptcy and largely disappeared. Brueners, Saks Fifth Avenue and Marshall Field, sold at a premium as a takeover defence, had slipped in execution and merchandising identity. Even those that remained independent, such as Dillards and Dayton-Hudson, found themselves somewhat distracted in digesting their acquisitions. (See Table 5.3.3.)

In the meantime, Nordstrom continued to flourish, defying conventional wisdom at every turn. While others strove for scale economies through centralized buying, Nordstrom continued to decentralize buying decisions to the lowest level possible. While others acquired and diversified, Nordstrom did not. Competitors invested hundreds of millions in computerized information systems; Nordstrom remained relatively unautomated and added new systems only when the people closest to the customer believed it enhanced the customer's shopping experience (see Figure 5.3.2). While competitors aggressively sought to cut costs, Nordstrom *added* costs through imaginative customer services and higher staffing levels (nearly twice Macy's). As noted earlier, Nordstrom's base pay is typically 10–20 per cent higher than prevailing retail wages of its mall competitors. At the Paramus store, Nordstrom paid $9.50 per hour; Macy's and Bloomingdale's averaged $6.00–$7.00 per hour.

Nordstrom also has a profit-sharing programme, which contributes an additional 8 per cent per year to the income of each employee, phased in on a graduated basis and fully vested after seven years.

Nordstrom can afford its higher hourly wage rates because of its aggressive use of commissions. High sales quotas require employees to earn their wages by moving a high volume of merchandise. Employees who fall below quota for three consecutive pay periods are dismissed. States one former salesman: 'The structure of their commission system makes the $9.50/hour a draw. If you survive at Nordstrom, you actually have to earn *more* than $9.50/hour just to meet your quota.'[3]

The average Nordstrom salesperson earned $23 000–$26 000 per year while the industry average for retail clerks stood at $12 000 per year. In 1990, Nordstrom sales per square foot (the most widely used yardstick for

Table 5.3.3 How the stores stack up – department stores

Retailer headquarters	Operations	Retail sales (millions)	Pre-tax profits (millions)	Interest expenses (millions)	5-year growth rate (%)	Comments
Allied Stores Cincinnati	Bon Marché, Maas Brothers, Jordan Marsh, Stern's	$2974	($194)	$251	n/a	The stores are caught in the Campeau cash crunch.
BAT Industries London	Breuners, Ivey's, Marshall Field's, Saks Fifth Avenue	$2643	$177	n/a	n/a	Stores are for sale as part of takeover defence figures.
Carter Hawley Hale Los Angeles	Broadway, Emporium, Thalhimers, Weinstocks	$2787	$12	$160	n/a	Another LBO overhaul. Still struggling to increase sales.
Dayton Hudson Minneapolis	Dayton Hudson, Mervyn's, Target	$12 204	$472	$231	1.0	Well positioned in two niches: discount and department stores.
Dillard Dep't Stores Little Rock, Ark.	Dillard's	$2558	$173	$81	24.0	This profitable traditionalist could pick up new stores cheaply.
Fed. Dep't Stores Cincinnati	Abraham & Straus, Bloomingdale's, Burdines, Lazarus, Rich's	$4542	($209)	$651	n/a	Campeau's cash cow is weak from the milking. An underperformer.
Hess Dep't Stores Allentown, Pa.	Hess's	$614	$13	$17	n/a	Developer-owner didn't bleed the stores. Looking to make a big purchase soon.
R.H. Macy & Co. New York City	Bullock's, Bullock's-Wilshire, I. Magnin, Macy's	$6974	($47)	$707	n/a	The winners of the battle for Federated, which sold it Bullock's and I. Magnin.
May Dep't Stores St. Louis	Filene's, Foley's, Hecht's, Kaufmann's Lord & Taylor, May Co., Robinson's	$11 525	$781	$254	18.0	A middle-brow moderate that is dumping its discount store business.
Nordstrom Seattle	Nordstrom, Nordstrom Rack	$2328	$198	$45	22.0	For service it is without peer in the industry.
J.C. Penney Dallas	J.C. Penney, Thrift Drug	$14 833	$1192	$351	8.7	Woke up and smelled the coffee. Boldly moving into apparel.

Source: Fortune, 18 December 1989.

retailing) was $380. This is roughly twice the industry average of $194.[4] Bloomingdale's, Saks Fifth Avenue and Macy's rank near the retail norm.

Given Nordstrom's size and prominence, clothing and apparel manufacturers compete to position their products in Nordstrom stores. Says one: 'Their ethic of putting the customer first is infectious. They expect us to treat them as they treat the customer. It leads to remarkable feats on our part – like hand-manufacturing a pair of shoes in a size we don't produce for a customer with very small feet. It cost us $600 to do this once last year. We sold them to Nordstrom at our *normal* wholesale price. I think many vendors find themselves being more lenient in accepting returns from Nordstrom because of the store's lenient return policy with its customers. In numerous ways, Nordstrom gets better service from its suppliers than most other retailers in the industry.'[5]

By the end of 1990, Nordstrom was the top-ranked department store in

---

Maybe more information isn't always better.

Two new economic studies provide fresh evidence that, contrary to expectations, computers don't increase productivity – at least partly because the added data delivered to workers don't lead to better decisions.

Catherine Morrison and Ernst Berndt, economists at Tufts University and the Massachusetts Institute of Technology, respectively, studied the performance of 20 US industries from 1968 to 1986. They found computerization tended to reduce productivity growth and, as an investment, was less efficient than other capital spending.

Gary Loveman, an economics professor at Harvard Business School, saw a similar pattern when examining the operations, from 1978 to 1984, of 70 large manufacturers based in Western Europe and the US.

The studies lend credence to fears that organizations often incur a heavy burden when adjusting to new computer equipment. They come at a time when many corporate customers are complaining that computer makers don't pay enough attention to the issue of worker productivity.

Mr Loveman faults computer buyers for having too much faith in the value of information. Employees compound the problem, he says, by larding pedestrian memos with facts gleaned from costly computer searches and spending hours playing with spreadsheet graphics or preening word-processing documents. "We don't have a good way to make decisions about how much [information] is enough," he says, "so people devote too much time to further analysis."

Computers also have a limited impact on the productivity of service companies, Mr Berndt says, because their benefits in service fields often are qualitative rather than quantitative.

Moreover, computers seldom confer a competitive advantage because rivals "end up doing the same thing," Mr Loveman says. "So computers simply raise the ante for playing in a market."

Figure 5.3.2 Computer data overload limits productivity gains

the United States, with 64 stores, annual sales of $2.89 billion and an historic annual growth rate of 20 per cent per year. More remarkable yet, in a decade of increasingly intense competition, Nordstrom had doubled its space but tripled its sales.[6] Former sceptics were now believers: Nordstrom was not only the most Darwinian of retailers but its formula seemed close to invincible.

The Nordstrom formula

At face value it seems easy: value your customers, differentiate through superior service and motivate employees with commissions. As the Nordstrom juggernaut moved south through California and then east, many competitors scrambled to replicate their model. Intense in-house charm schools on service quality were offered by rival department stores. Recast vision statements were unveiled, trumpeting the importance of customers and the value of service. Commissions and incentives were introduced. Without exception, these failed to close the gap. Illustrative of this pattern was The Broadway, which introduced sales commissions and offered personalized shopping services. Executives later conceded that it was virtually impossible to reach Nordstrom standards. 'Service is an attitude of caring on the part of everybody in the store. It takes a lot of coaching and leadership,' said Phillip M. Hawley, Chairman of The Broadway's parent company, Los Angeles-based Carter Hawley Hale Stores.[7] The programmes were discontinued.

Boston analyst Margaret A. Gillian observes: 'A few years ago, a lot of retailers thought that if they put their people on commission – as Nordstrom does – good service would automatically follow. It did not and I am convinced that Nordstrom provides service *despite* the fact that people are on commission, not because of it. Most commission sales people give preferential treatment to those customers making major purchases while neglecting others. That does not happen at Nordstrom. Nordstrom people are recognized with citations and cash prizes when they perform superior customer service and they have the opportunity to move up the ladder.'[8]

Nordstrom executives observe a pattern among the dozens of companies, reporters and analysts who visit them each year. 'They attribute our success to coaching, caring, leadership, commissions, and service,' states co-president, John Whitacre, 'Somehow it all largely misses the point.'[9]

Nordstrom's underlying culture (or organizational paradigm) is at the heart of things. It is more important than its product/market strategy, organizational structure or compensation system. States one observer: 'Nordstrom's secret seems to be largely inaccessible. According to managers interviewed, visitors strive to isolate one or two key ingredients – silver bullets to make their competitive problems go away while preserving a traditional management culture. It's the same pattern that we see among manufacturing companies where efforts to emulate Japanese rivals have led to the proliferation of Just-in-Time inventory systems, Total Quality programs and corporate vision statements.'[10]

The three secretive Nordstroms, all co-chairmen and usually known inside the company as Mr Bruce, Mr John and Mr Jim, usually refuse to explain to outsiders how their company works. When pressed, they proffer self-effacing comments such as 'We live for the customer' or 'There is nothing special or difficult about what we do; it's just a lot of people doing a lot of things on their own.'[11] 'Don't be taken in by this,' says Francine Schwedel of the *Wall Street Journal*. 'The Nordstroms are shrewd businessmen who are also masters at motivating their troops.'[12]

Complementary styles at the top have benefited Nordstrom. Mr Bruce serves as 'spokesman and ambassador'; Mr John is the 'feisty merchant' who can play the heavy; Mr Jim is the 'focused operator' who has a knack for persevering toward long-term goals. 'They go into a room and grapple with a big issue,' observes Nordstrom co-president, Daryl Hume. But no matter how much fighting occurs behind closed doors, they always come out with one game plan – never several different game plans. 'It saps the strength of a company when the leaders aren't really aligned. The troops can smell it a mile away.'[13]

Nordstrom's model rests on: (1) attractive surroundings, (2) competitive prices, (3) a wide selection of merchandise, and (4) superior customer service. In the first two areas, Nordstrom strives for parity with the best of competitors. With respect to the third element, it stocks 50 per cent more inventory than competitors both in terms of the depth of selection and across the full range of sizes. The fourth arena provides a particularly formidable competitive advantage. 'All this, of course, comes at a price,' states Bruce Nordstrom. 'High levels of inventory and customer service add expense; we compensate by generating volume ... we depend on good sales people to help us do it.'[14]

Nordstrom is not just a high-end store. Its popular priced departments, such as Town Square and Point of View, are very big in many locations. The firm tends to layer in higher priced departments once a store is established. It is store policy not to jack prices up and down with one-day sales events. But when a competitor runs one-day specials, Nordstrom always matches their prices not just for that day, but forever. The company strives to offer the best quality available in a category by using its own labels at prices lower than comparable branded items.[15]

Intricacies

Look closely behind the self-effacing rhetoric and glib explanations. At the heart of Nordstrom is a finely tuned operating system that is both intricate and difficult to copy.

Nordstrom keeps its eye squarely on the two stakeholders that drive the business: the customer and the floor salesperson. Unlike most stores, where the customer is often secretly regarded as a nuisance, Nordstrom makes the customer the centrepiece. Shoppers experience Nordstrom as a better place to shop. Says one national footwear manufacturer familiar with Nordstrom and its competition: 'Look at the situation from the customer's point of view. First, Nordstrom carries a wide selection, so you are exposed to more options there than elsewhere. Second, they carry your size – and if they don't, the salesperson tracks it down and delivers it to you personally, if necessary. Third, they'll meet anyone's price – just tell them or show them an ad ... if anybody sells it for less they'll match it on the spot. No hassles. Finally, they have more sales people and they are helpful. Put it all together and it's not surprising that Nordstrom wins hands down.'[16]

Harvard Business School professors Len Schlesinger and Jim Hefkett have investigated these assertions empirically. They cite a recent Department of Commerce survey that shows a close link between resolving a customer's problem on the spot and the customer's intent to repurchase. 'When customers experience minor problems,' they write, '95% say they will repurchase if the complaint is resolved speedily. If the resolution process takes even a little time, however, the number drops to 70%. A spread of 25 percentage points can easily mean the difference between spectacular and mediocre operating performance. (Indeed, studies on the effects of customer loyalty have shown that even a 5% increase in customer retention can raise profitability by 25% to 85%.)'[17]

All of this tends to be especially relevant because demographic shifts in the United States have produced a population of older customers. In particular, the 44–50-year-old segment is growing fast and has more disposable income. This population group also expects and appreciates service and is willing to pay for it.

The following is but a partial list of Nordstrom's techniques for differentiating itself through service.

● Sales personnel dress like customers and wear no name tags. 'This prevents the customer from doing half the work,' says one saleswoman. 'It is our job to find the customer and make contact. We are expected to greet *everyone* who enters our department.'[18] Bruce Nordstrom adds: 'we coach them to avoid a canned greeting like – "may I help you?" We want to recognize the customer in an individual way by saying something like "Isn't that a great color?" or "What a storm! Have you seen the great rain gear over there?"' Signs posted prominently in

Nordstrom's stock rooms confirm this point: 'Greet every customer entering your department within 60 seconds.'

- Any merchandise that Nordstrom handles will be reimbursed for cash or charge, whether used or unused, *without* a Nordstrom sales slip. If a customer indicates that an identical item is priced more cheaply elsewhere, the price is lowered on the spot. Nordstrom Executive Vice President, Bob Nunn, explains: 'Nordstrom literally grinds up truckloads of shoes each year. These are returned shoes that have been worn and cannot be sent back to vendors. At face value, it seems nuts to have a policy like this. But if we run one full page promotional ad in the Seattle newspaper, it costs us the equivalent of 500 pairs of shoes. And we don't know if the ad works. But give a customer a new pair of shoes with no hassle and it's a story that gets told and retold at parties and at the bridge table. Word-of-mouth endorsement really works. We know it.'[19]
- There are no evident security measures or security guards at Nordstrom, nor are there restrictions on the number of items taken into a dressing room. Very few items have security devices attached or are chained to racks.
- Nordstrom accepts all major credit cards.
- Live piano music plays in most stores. In colder climates, greeters may offer to check the customer's coat at the door. Diaper changing tables are installed in spotlessly maintained restrooms.
- Personal shoppers (the first Nordstrom salesperson you encounter) offer to escort you throughout the store or to assemble a complete outfit while you wait by retrieving possible items that mix and match from several different departments. The whole ensemble is handled in one check-out procedure.
- Sales personnel may offer to deliver merchandise to a customer's home personally. Some voluntarily spend their off-duty time tracking down merchandise and calling customers with results.
- Gift wrapping is performed at the department where purchases occur at no charge. Items purchased at other stores will also be wrapped at no charge and even shipped if a customer so requests.
- When an item or an ensemble that a customer desires is not carried or is out of stock, sales personnel have been known to travel to a competing store and purchase it.

Responding to unreasonable requests

Overriding these examples, Nordstrom treats all customer requests, however unreasonable, as *opportunities* to expand upon its legendary reputation and keep itself on its toes. Exceptional demands, termed 'heroics' by insiders, include such things as paying a customer's parking tickets when a shopping trip at Nordstrom runs longer than expected, warming up customers' cars on exceptionally cold days, changing a customer's flat tyre, meeting customers with lost luggage at airports with selected attire, writing thank you notes to customers with purchases over $500, and on and on.

Such demanding service standards would represent a serious burden to many. Indeed, working at Nordstrom isn't for everyone. Through careful selection, Nordstrom seeks out the dedicated go-getters who really *love* to sell. Among the thousands of applicants there is a percentage who fit this profile. If you happen to be one, Nordstrom is a retailer's nirvana. As a fundamental tenet of policy, the company *never* puts the salesperson at cross-purposes with the customer. (The ultimate sin at Nordstrom is to say 'no' to a customer.) Co-chairman Bruce Nordstrom states: 'we want the salesperson to say "yes" and put *management* in the position of occasionally saying "no". And we do say no – for example, when a person appears to be shoplifting from another store and repeatedly attempting to exchange the merchandise at our stores for cash.' As noted earlier, management rewards its employees for going the extra mile. Unusual heroics reap prizes, citations and other forms of peer recognition. 'Nordstrom's makes its salespeople the kings and queens,' states one experienced saleswoman. 'Our department managers and store manager act as facilitators and provide support. Around here, the floor salespeople really have status.'[20]

Contrast the above characterization with the plight of many sales clerks in dead-end retailing jobs. Not surprisingly, Nordstrom is considered a preferred employer. Prior to opening the Menlo Park store in New Jersey, the company screened 5100 applicants and actually interviewed 3000 people to fill 700 positions. Applicants were told that Nordstrom expects a far higher standard of sales productivity and service than they have probably experienced anywhere; they are also told that failure to achieve the demanding sales quota for three consecutive pay periods means dismissal. Many applicants remained undaunted. Some talked about how tired they were of being jerked around by absentee financial investors who know little about merchandising and care little about customers. Nordstrom stood out as an attractive alternative.

Nordstrom has a particular genius at defining a unique relationship with its 'internal customer' – the salesperson. It adheres to the philosophy that the people on the sales floor are the most important contributors and it actually displays its schematic diagram of the organizational chart upside down. (See Figure 5.3.3.) The customer is included at the top of the chart; salespeople occupy the highest internal rung. Next are buyers, merchandisers, store and department managers. At the bottom are the four co-presidents and five co-chairmen who occupy small walk-up offices on the sixth floor of Nordstrom's oldest department store in Seattle. They share secretaries, answer their own phones and generally act as high level 'go-fers' and firefighters in support of those 'above them' in the organization.

Everyone, repeat *everyone*, wishing to rise up into a buying, merchandising or managerial position at Nordstrom starts out selling. (Many experi-

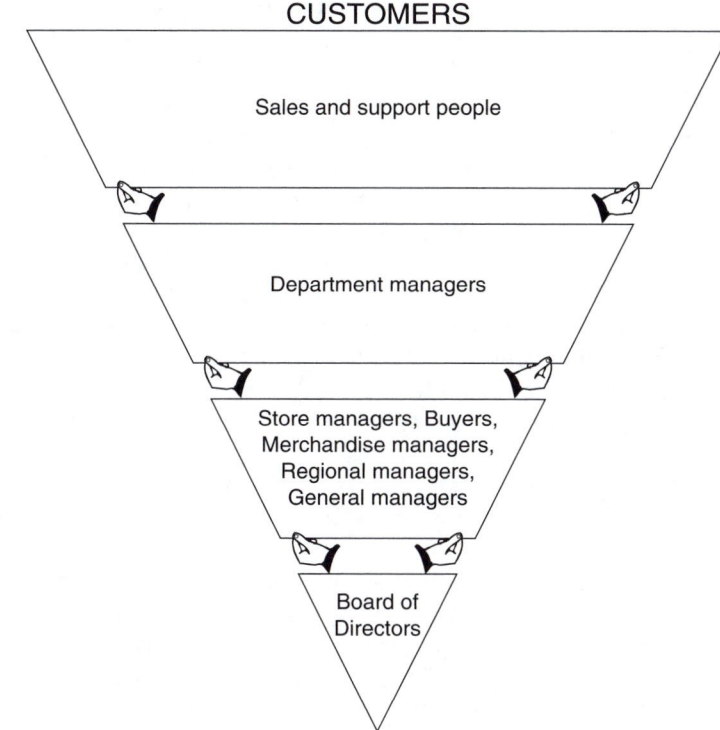

Figure 5.3.3 Company structure.

enced buyers and managers from rival stores who have sought employment at Nordstrom have been shocked when told that they must start on the floor regardless of their current position elsewhere.) Nordstrom believes that everyone, as a rite of passage, must demonstrate an ability to respond to customers and sell effectively. An aspiring employee is not regarded as knowing the culture until proficiency is proven. The company only promotes from within. (The four new co-presidents are role models, each having risen up from sales clerks through the ranks.) States one executive VP: 'It's almost impossible to overstate how important this common career track contributes to our ability to communicate within the company. It's like we all speak a common language from top to bottom. It helps us avoid the *we* versus *them* mentality.'[21]

Too confusing

Nordstrom's operating system is replete with contradiction and paradox. For years, management has refused to publish an organization chart because it would be too confusing. Each of the four co-presidents oversees a piece of the company defined by categories of merchandise – shoes, men's apparel, women's apparel and so forth. Within each of these domains there is an executive vice president, merchandisers, and below them, buyers. Buyers spend at least one day a week visiting *and selling* in the department they service. They, too, regard sales personnel as the keys to success. Beyond these generalizations there is no company pattern except that buying decisions are pushed to the lowest practicable level. Shoes, for example, are purchased by the department manager. Since there are typically three shoe departments in each store, this translates into 250 separate department heads who act as shoe buyers in the Nordstrom system. Each establishes sales targets and makes pricing decisions. At the other extreme, menswear, cosmetics and women's sportswear are purchased on a regional basis. There are typically three to seven stores within a region.

Any characterization of how Nordstrom is organized tends to obscure more than it illuminates. The main problem is that any description that conveys a hierarchy and lines of authority is woefully misleading. For years academics have talked about the obsolescence of the *command and control* model of organization and advocated a *network* model in its place. In this respect, Nordstrom may be one of the most advanced of Western companies.

Nordstrom's top executives share power. With five co-chairmen and four co-presidents, Nordstrom underscores its strong conviction that no single leader can ever be broad enough to lead an organization over time.* States co-president, John Whitacre: 'The single hero-leader doesn't have the benefit of seeing an issue from all sides. They get isolated and are all too easily buffered from reality.'[22] Whitacre continues: 'We (the four of us) have all been developed to be co-leaders since we were store managers. Because of the high degree of decentralization *and* overlapping responsibilities here, you have to manage in a participative fashion. Yes, the overlap generates heat. But we really believe a little contention spurs creativity. We're like an Italian family. We argue, get things off our chest, everybody expresses their point of view. But we remain a tight, cohesive family. A lot of organizations never let the differences out. Because retailing is a nitty-gritty environment, we get it out on the table and deal with it.'[23]

A useful building block in conceptualizing Nordstrom's operating system is to think of each individual department (e.g. Collections, Town

---

*Prior to January 1991, there were no co-presidents. The company was managed by five co-chairmen, three of which were members of the Nordstrom family. The current arrangement with four co-presidents is an interim phase in anticipation of the retirement of the five men at the co-chairman level.

> We're glad to have you with our Company.
>
> Our number one goal is to provide
> **outstanding customer service**.
>
> Set both your personal and professional goals high. We have great
> confidence in your ability to achieve them.
>
> Nordstrom Rules:
> Rule #1: **Use your good judgment in all situations.**
> There will be no additional rules.
>
> Please feel free to ask your department manager, store manager
> or division general manager any question at any time.

Figure 5.3.4 Welcome to Nordstrom.

Square) as a separate small business. Within it, the salesperson is expected to set and achieve sales goals. Nordstrom's 'policy manual' is a 3×5 card, which says: 'Rule #1: Use your own good judgment in all situations. There will be no additional rules.' (See Figure 5.3.4.) As noted earlier, sales personnel are very highly valued in the stores and broadly supported. But it would be misleading to omit the tensions that are kept in play which drive peak performance.

Sales personnel, as well as buyers, merchandisers and managers at all levels, are heavily incentivized. At the centre is Nordstrom's sacrosanct measure: sales per hour per employee, or 'SPH' as it is known. It is the primary yardstick of sales goals and sales quotas. Reports of SPH are shared publicly within each store and posted on bulletin boards, where every salesperson can see their ranking in the pecking order. Most compare themselves to others within their store as well as with like departments of other stores within their region. This same information, undistilled, is given to the co-presidents and co-chairmen [Yes, SPH by *employee*, as well as a few other key indicators, is shared from top to bottom of the corporation in a very public fashion.] (See Tables 5.3.4 and 5.3.5.) A salesperson from the San Francisco store states: 'I used to work at Macy's. They didn't share any of the numbers, so when a buyer or store manager seemed stressed we didn't know why. Here we are all working off the same playbook. I know where I stand, where my department stands and how the store stacks up in the region. It makes me feel part of the team.'[24]

While many Nordstrom employees view selling as their ultimate calling (top performers earn up to $100 000 per year), a few aspire to move up the career ladder. To do so, one must work up from the sales floor. The first

Table 5.3.4 Payroll group sales per hour

Employee number	Employee name	Store No.	Dept No.	Selling hours worked	Gross sales	Returns	Net sales	Sales per hour
75051	Arima, Mark A	6	26	86.50	2.95	40.17–	37.22–	.43–
208520	White, Gregory L	10	36	86.50	392.49	434.23–	41.74–	.48–
453456	Nichols, Nancy	335	10	86.50	.00	44.00–	44.00–	.50–
511857	Cefalu, Nicole	435	92	86.50	.00	45.56–	45.56–	.52–
807925	Kobashigawa, Kim	375	10	86.50	191.09	236.80–	45.71–	.52–
227785	Bauer, Georgia S	191	58	86.50	.00	46.40–	46.40–	.53–
92981	Yamashiroya-Welch, Celeste	435	57	86.50	.00	46.40–	46.40–	.53–
279901	Johnson, Julie J	455	9	86.50	52.00	99.00–	47.00–	.54–
865428	Costley, Alan	23	36	86.50	144.36	192.75–	40.39–	.55–
371344	Gough, Kim	191	81	86.50	208.00	256.52–	48.52–	.56–
421768	Messnick, Nancy Pooler	335	6	86.50	.00	49.00–	49.00–	.56–
335232	Dorrell, Patcharine	360	20	86.50	27.95	79.90–	51.95–	.60–
224735	Nelson, Steven	341	36	86.50	.00	52.95–	52.95–	.61–
240648	Diggs, Elizabeth A	355	81	86.50	435.90	495.44–	59.54–	.68–
338061	Starley, Laura	94	15	86.50	.00	66.00–	66.00–	.76–
331140	Zaun, Cynthia	335	92	86.50	21.60	88.00–	66.40–	.76–
403709	Loofburrow, Diane	455	14	86.50	1,041.54	1,114.00–	72.46–	.83–
173773	Garlock, Carol	455	9	86.50	495.72	575.00–	79.28–	.91–
487256	Knowles, Cheryl	355	910	86.50	.00	90.70–	90.70–	1.04–
524199	Chomeau, Patricia	335	48	86.50	31.00	121.93–	90.93–	1.05–
346106	Cooney, Nancy	355	6	86.50	169.00	268.00–	99.00–	1.14–
476630	Keski, Kristine A	191	28	86.50	450.60	552.60–	102.00–	1.17–
231118	Ekman, Anders	335	77	86.50	.00	105.95–	105.95–	1.22–
403535	Nordlund, Colleen	191	57	86.50	134.25	251.10–	116.85–	1.35–
165977	Bailey, Jerriann	96	10	86.50	29.50	155.80–	126.30–	1.46–
772160	Bailey, Ron	375	77	86.50	.00	129.90–	129.90–	1.50–

Table 5.3.5   Daily sales by department store

Dept no.	Department description	Daily net sales This year	Last year	Difference Amount	% of last year	Month-to-date cumulative net sales This year	Last year	Difference Amount	% of last year
Store 0001 Downtown Seattle									
0045	Men's Polo Shop								
0075	Men's Brass Rail								
0076	Men's Clothing								
0077	Men's Sportswear								
0078	Men's Furnishings								
0080	Luggage/Work Clothes								
Men's Wear	Total								
0024	Men's Casual Shoes								
0025	Men's Dress Shoes								
Men's Shoes	Total								
Men's Wear and Shoes	Total								
Store 0001	Total								

(difficult) rung is assistant department manager. This is a training slot where an aspiring salesperson often *requests*, on top of normal duties, that they take over certain paperwork and begin to learn the ropes of being a department manager. There is no economic reward for this extra work. Sales personnel in this position are paid a small stipend but the wage increase involved does not offset the loss of commissions from selling.

Given these *disincentives*, why would a salesperson want to be a manager? In fact, Nordstrom wishes to avoid the trap of many corporations that force their most productive employees into management in order to earn more or to be seen as successful. At Nordstrom, status clearly does not correlate with hierarchical position and pay grade. Paywise, a store manager at Nordstrom earns about as much as the store's most successful sales personnel. Only those employees who really itch to manage clear the assistant manager hurdle. It is a kind of trial by fire that truly tests commitment. For those that persevere (and who, while doing so, are adept at pleasing customers and getting their departmental sales people to play as a winning team) Nordstrom becomes, in the words of one, 'a high speed vehicle to the top'. It is not unusual at Nordstrom to be a store manager in one's late 20s and a regional manager by mid-30s. Nordstrom's four co-presidents range in age from 36 to 50 years old; the average is 43. For the individuals who self-select onto this track, few think twice about the long hours, seven-day work weeks, and the geographic dislocations that are required.

As a department manager, one truly feels in charge of one's own business, with responsibility for co-hiring sales personnel (the store manager also conducts interviews), setting budgets, sales targets, and making and influencing buying decisions that range from complete to partial control, as noted earlier. In effect, Nordstrom says to its department managers, 'We're loaning you this business. You can rise as far and as fast as you can grow it.' 'Because of the high degree of delegation,' one manager stated, 'It really feels like it's yours. It inspires and motivates. How often do you have people in companies say: "If this were *my* business I could really make a lot of money." Well, at Nordstrom they get out of your way and let that happen.'[25] Department managers receive compensation based on three components – base salary, commissions on own sales, and a bonus that is 1 per cent of sales increases of their department over the previous year.

Enter the buyer: buyers spend at least one day a week on the floor of the departments for which they do the buying. The result of their close working contact with the sales personnel permits Nordstrom to react much faster to consumer trends and to spot when shoppers are flush or trading down. While department managers officially report to the store manager, they are incented to work very closely with the buyer. The fates of the two are linked since they are given bonuses based on sales increases over the previous year. One former department manager states: 'At Nordstrom, the typical interaction is for a salesperson or department manager to pick up the phone and say to the buyer, "Hey – this is moving. Get me some more of this." Buyers often initiate similar calls on the merchandiser. If the system breaks down, our safety valve is the customer who, if frustrated, is *encouraged* by the salesperson to write or call one of the co-presidents. The top guys act like ombudsmen for the customer and the sales force. Things don't stay broken for long.'[26]

Some buyers are also given an incentive based on net margin (to penalize excessive markdowns). Department managers earn $20 000–30 000 per year, buyers and store managers earn between $40 000 and $100 000 per year. A high percentage of the compensation for all managers and buyers is tied to performance.

In a typical facility, store managers oversee $30 million in inventory and 600 employees. Owing to the strong ties of salespeople to customers and department heads to buyers, one might think the store manager's job is mostly personnel, public relations and facility maintenance. In fact, it's a great deal more; the fabric of shared power woven of solid and dotted line relationships makes it difficult to visualize in traditional authority / responsibility terms. Top management views the store manager as blending the drive for individual performance with store-wide perspective. They serve as facilitator, orchestrator and, at times, hard-nosed enforcer of relationships. Nordstrom managers and employees continually stress the fact that there is no hiding behind organizational structure: rather, the individual is given the responsibility to work things out. In this system, relationships matter for a great deal.

The regional managers have a great deal of authority in hiring personnel for new stores, selecting buyers and working closely with buyers, merchandisers and department heads to get the right mix for the store in their region. The store manager decides who to promote to department manager, is a co-interviewer of all new personnel hired, and plays a central role in hammering out sales targets, supporting sales personnel and sponsoring rallies, meetings and a variety of recognition programmes.

Built-in tensions

The core tension in the Nordstrom system stems from the drive to encourage the greatest possible degree of individual initiative, delegation and decentralization against the backdrop of team goals and corporate values. Nordstrom executives were quick to emphasize that this cannot be fine-tuned adequately with financial incentives alone. Over the years an intricate set of values and social norms have evolved which give the Nordstrom organizational paradigm its real power.

Nordstrom feeds on youth, energy and vitality. In part this is reflected in

At Nordstrom, we have a strong belief in an individualized approach to fashion. Each of our stores has been carefully tailored to reflect the lifestyles of the customers in the surrounding area. Our wide selection of shoes, apparel and accessories are showcased in a variety of distinctive 'shops' that are rich in color, texture and design. Your buyers work closely with top quality manufacturers from here and abroad to obtain the best values, the most unique items and the widest selections for our customers.

We encourage you to keep notes on your customers' sizes and preferences, and to let them know when something that may be of interest to them arrives in the store. This type of customer service is our company's greatest strength. Friendliness, courtesy and a sincere desire to help is the rule rather than the exception. A perfect example of this belief is our return policy. If one of your customers isn't completely satisfied with their purchase – for whatever reason – take it back, no questions asked. And, always remember to do it with a smile!

We also encourage you to present your own ideas. Your buyers have a great deal of autonomy, and are encouraged to seek out and promote new fashion directions at all times. This distribution of responsibility fosters enthusiasm and loyalty in all of our employees – toward both our company and our customers. Creativity and motivation are often quickly rewarded, as promotions are made almost exclusively from within.

Nordstrom has a strong open-door policy and we encourage you to share your concerns, suggestions and ideas. This results in an environment of open communication which allows us to resolve any situation internally. The information provided within this folder will help you to be successful with this company. However, nothing in this folder, nor in our guidelines, constitutes an employment contract. The information we have provided is subject to change at any time.

The Nordstrom philosophy of quality, selection, value and service began in 1901, when John W. Nordstrom opened his first shoe store in Seattle. It is the foundation of our company's growth and success. We currently have stores in Washington, Oregon, California, Alaska, Utah and on the East Coast. And with a continuing commitment to these beliefs and employees such as yourself, we can face our future challenges with confidence.

Figure 5.3.5 The Nordstrom philosophy.

the company philosophy, as presented in Figure 5.3.5, and in part it derives from a dedicated family intent on redefining the future of retailing. There is no other establishment that has sustained the passion of its vision through so many generations. Two generations is unusual, three is unheard of.[27] Nordstrom is in its fourth generation, with seven Nordstrom family members working up through key positions in the ranks.

Co-president John Whitacre states, 'This is a people-driven business. I spent most of Sunday reviewing the 20 entries for a "Make Nordstrom Special" competition in our Seattle Region. I'll go out there tomorrow, acknowledge each entry, pick the top seven and explain why, then give awards to the best and the runner-up. It takes a lot of time and energy to provide meaningful recognition but recognition fuels this place. We [co-presidents] really do "manage by walking around". I am hesitant to use an expression like that because it too easily becomes a cliché. The sincerity gets lost – and sincerity is very important around here. We discuss people and performance a lot at my level. Recognition of a hot performer spreads rapidly through our grapevine. Long before you meet a rising star, you hear about him or her. We travel together to regional meetings and spend most of the flight talking about people.'[28]

Bob Nunn observes: 'There is always a danger of a strong culture like this turning inward, getting smug and self-congratulatory. I don't rule that out. Encouraging the customer to be demanding keeps us honest. So does our own competitiveness internally. Everyone at every level is trying to excel, beat another department or another store. Our biggest competitor isn't Macy's or Bloomingdale's, it's one another. It's part of the fun of coming to work. I always try to stir things up. I regularly bet a shoe buyer a dollar or a free lunch that I can get a greater increase in shoe sales nation-wide today than he or she can get in their department. The department manager upstairs in the Brass Plum owes me $4.00.'[29]

What sets Nordstrom apart is the sophistication with which they have blended financial incentives with recognition and peer pressure. This inter-weaving achieves a fine-tuning of the motivational machinery that is seldom encountered in Western companies. Other examples include:

- Salespeople are encouraged to keep scrapbooks of letters from customers. These are reviewed when promotions are considered. Exceptional acts of heroics earn citations, peer recognition, deep discounts on Nordstrom merchandise and monetary awards. At the monthly regional meeting of buyers, store managers and department managers, several sales personnel are invited to acknowledge heroics which triggered a particularly positive customer response. Their feats are described and awards made. One of Nordstrom's co-chairmen or co-presidents always attends these monthly meetings. The Outstanding Store is selected monthly, based on customer praise and sales results. This monetary award is given to the store manager, who hosts dinners for major contributors or distributes spot bonuses to deserving individuals or departments.
- Sales personnel whose yearly performance ranks them in the top 9 per cent of Nordstrom system-wide are recognized as members of The Pacesetter Club. Recipients receive a lapel pin indicating the number of years as a Pacesetter and receive 30 per cent discounts on Nordstrom's merchandise. The award is made at a black-tie dinner, with members of the Nordstrom family attending. On a monthly basis, All-Stars are selected as the high performer in their store. These individuals are acknowledged at the monthly regional meeting and given sub-stantial discounts on merchandise. Most prestigious of all is the 'J.W.

Nordstrom' award given annually to one member of management below the vice president level who exemplifies the values and work ethic of the founder.
- Smaller awards are given out daily. For example, sales personnel are encouraged to be thoroughly acquainted with their competitors in surrounding stores. If they identify a competitor's price that is below Nordstrom for like merchandise, they are given $5.00. Nordstrom's prices are immediately lowered. At many stores, the store manager is on the loudspeaker before doors open, reviewing how store sales did the previous day and acknowledging the one or two departments that contributed most dramatically to the previous day's results.
- Peer pressure is particularly keen at the management level. At the climax of the yearly budget cycle, annual sales targets are set in an intense and spirited public forum. All buyers, department managers and store managers within a region attend the annual event. One by one, each department head stands up and announces sales goals for the following year. Amid suspense and the theatrical fanfare one might associate with a TV game show, the number is written upon a large card on stage. Immediately thereupon the individual's boss unveils the previously hidden goal for that manager. If the luckless participant has projected sales below the boss's target, boos and catcalls follow. If projections exceed, cheering and whistles greet the news. One observer states: 'It's a very intense but generally upbeat experience. Like a high stakes auction, peers are continuously revising their sales targets in the minutes before their name is called, gearing their projections off imputed growth rates in the numbers that have been revealed. As with all regional meetings, one of the co-chairmen or co-presidents is present – often wearing a sweater emblazoned with the letter "N".'[30]

The dark side?

In the face of so many factors driving performance, one might anticipate negative side effects. These might include grandstanding before customers in inappropriate ways to garner service awards or hanging out near the register to spot (and cash in on) the customer who has already preselected an item. (At Nordstrom this inappropriate behaviour is labelled 'sharking'.) Nordstrom's pervasive but informal management system provides a powerful deterrent to such abuse. If peers or a department head spots inappropriate sharking behaviour, the violator is soon to know. In addition, Nordstrom fields an auditing team known as 'phantom shoppers', who randomly appear in departments and act as customers. Later they meet with the department head and sales personnel to report on (1) whether they were approached within 60 seconds of entering the department, (2) friendliness, and (3) whether the salesperson seemed well informed.

Bob Nunn, an Executive Vice President of Nordstrom, states:

'Nordstrom selects for style and fit, not just performance. You don't make it here if you badmouth people or succeed by climbing over the backs of others. It's a youthful, success-oriented, do-what-it-takes-for-the-customer culture. You need to be a team player, friendly and upbeat. It sounds contrived and plastic, but when you're in it, it really works.'[31]

Nordstrom's social system 'is so thick you can cut it with a knife,' says one former employee.[32] Weekly pep rallies at stores find some personnel dressed up as raisins leading cheers: 'Vol-ume – vol-ume', 'We're No. 1', 'You Want To Do It'. Stores run smile contests and hang candid shots of winning smiles in the lunchroom. Monthly, there are 'Who Looks The Most Nordstrom' contests, awarding those wearing store merchandise. Signs on stockroom walls feature such slogans as 'Make Your Goal', 'Be A Top Dog Pacesetter', 'Go For The Golden Milkbone', and 'Don't Let Us Down'. (See Figure 5.3.6.) Free dinners are given to sales personnel with the most 'multiples' (i.e. cross-departmental sales). Elective motivational seminars teach a technique called affirmations, in which participants are encouraged to silently repeat a mantra such as 'I only sell multiples', 'I enjoy being a store manager', or 'I feel proud being a pacesetter'.[33]

Is it a cult? Some say so. In a 1989 dispute with its only union, The United Food and Commercial Workers (which represents 2000 of its 30 000 employees), some complained that Nordstrom was a coercive employer that placed unreasonable demands on its sales force. The dispute was triggered when management, in response to employee requests, proposed to the union that dues be made optional. Nordstrom also *increased* employee

> YOU HAVE 169,000 CUSTOMERS WORKING DOWNTOWN TODAY WHY NOT CALL AND INVITE ONE IN

> REMEMBER: THE CUSTOMER IS BOSS

> WHAT DID YOU DO FOR NORDSTROM TODAY

> THE DIFFERENCE BETWEEN THE ORDINARY AND EXTRAORDINARY IS THAT LITTLE BIT EXTRA

Figure 5.3.6 Signs in Nordstrom stockrooms.

vacation packages unilaterally without giving the appearance of having negotiated the improvement with the union. Rather than call a strike, a highly effective publicity campaign was launched in which the union orchestrated unflattering stories about Nordstrom in the Seattle press and the *Wall Street Journal*. The union president stated that 'blue collar employees governed by hourly wage laws were themselves having to make time-management decisions usually reserved for white collar "professionals".'[34]

Several former employees testified to Nordstrom's 'dark side'. The following excerpts were published in the *Wall Street Journal*:[35]

> The first time Nordstrom sales clerk Lori Lucas came to one of the many *mandatory* Saturday morning department meetings and saw the sign – 'Do Not Punch the Clock' – she assumed the managers were telling the truth when they said the clock was temporarily out of order. But as weeks went by, she discovered that on subsequent Saturdays the clock was always 'broken' or the time cards were not accessible. When she and several colleagues hand-wrote the hours on their time cards, they discovered that their manager deleted the hours and accused them of not being 'team players'. Commenting on the variety of tasks that implicitly had to be performed after hours, Ms Lucas said, 'You couldn't complain, because then your manager would schedule you for the bad hours, your sales per hour would fall, and next thing you know, you're out the door.'

Patty Bemis, who joined Nordstrom as a sales clerk in 1981 and quit eight years later, told a similar story:

> Nordstrom recruiters came to me. I was working at The Broadway as the Estee Lauder's counter manager and they said they had heard I had wonderful sales figures. We'd all heard Nordstrom was the place to work. They told me how I would double my wages. They painted a great picture and I fell right into it ...
>
> The managers were these little tin gods, always grilling you about your sales ... You felt like your job was constantly in jeopardy. They'd write you up for anything, being sick, the way you dressed ... the girls around me were dropping like flies. Everyone was always in tears.

Where does the truth lie? There is no denying that Nordstrom churns through thousands of employees each year in its drive to hone its competitive edge. Its highly motivated cadre of sales personnel and managers must prove themselves at every level or they are terminated. But for many, the counterpoint is captured by an articulate and highly successful women's apparel salesperson in one of the San Francisco region stores. An expatriate of Haiti, she states: 'Nordies as the Moonies of retailing? I find it amusing. It seems that some people don't want to accept that we are in an increasingly competitive world. There is no free lunch. Period. If you want to survive, you've got to hustle. I suppose that first class organizations like Nordstrom do "use you up" in a way. That's true of Japanese companies that we read about like Toyota and Honda. It's true of Intel, it's true of Federal Express, it's true of Goldman Sacks. But most American companies are staffed by people who punch the clock, put in their hours and don't really try that hard. That's been the story in manufacturing where US companies keep losing ground. Many thought we were immune from global competition in retailing. But all it takes is one Nordstrom that offers a huge improvement in service quality and the game is changed. The competitors fall all over themselves in a scramble to merge and all they did was dig a deeper hole, and haven't been able to make the mergers work and pay off the debt. They bet on efficiency and size that the customer never sees nor gives a damn about. Banking is like this today. Big mergers, lousy service. The customer is conditioned to a low standard until someone else figures out how to really do it right. Then the customers, especially the affluent ones, jump ship. But it takes relentless effort and day-to-day discipline to pull this off. Most American companies lack this. Many took pleasure when Nordstrom got the bad press because it reaffirmed "their way is better".'[36]

Another employee states: 'Winning customer loyalty spurs relentless improvement. Some burn out. Others may find it exploitative. But remaining a winner today demands that price – not just of companies but of the individuals who work for them.'[37]

Nordstrom's turnover is 50–60 per cent per year – about typical for the industry. These figures include seasonal hirings. Turnover among permanent staff is 15–20 per cent per year, slightly better than the industry average. Half of those exits are employee initiated, and are mostly due to family related matters. The balance of approximately 7–10 per cent of annual turnover is company initiated.[38]

When confronted by reporter Morley Safer during a *60 Minutes* interview, Bruce Nordstrom summarized Nordstrom management's beliefs:[39]

> The system is to have self-empowered people who have an entrepreneurial spirit, who feel that they're in this to better themselves and to feel good about themselves and to make more money and to be successful. That's the system. [We have] expectations of our people. When people apply for a job any place, they want to work hard and they want to do a good job. That's their intention. And our intention is to allow them the freedom to work as hard as they want to work.

References

1   Stevenson, R.W. (1989). Watch out Macy's, here comes Nordstrom. *New York Times*, 27 August, 34.
2   Gillian, M. (1991). *First Boston Equity Research Report*, 28 October, 5.
3   Interviews at Reebok.
4   Parpia, M. (1991). *Nordstrom*, Harvard Business School case study, Harvard Business School Case Services, Cambridge, MA, February, 15.
5   Interviews at Reebok.
6   Parpia, *op. cit.*, 23.
7   Hamilton, J.O'C. and Dunkin, A. (1987). Why rivals are quaking as Nordstrom heads east. *Business Week*, 15 June, 99–100.
8   Gillian, *op. cit.*, 4.
9   Interviews at Nordstrom, Seattle, November 1991.
10 *Ibid.*
11 Schwadel, F. (1989). Courting shoppers: Nordstrom's push east will test its renown for the best in service. *Wall Street Journal*, 1 August.
12 *Ibid.*
13 Interviews at Nordstrom.
14 Parpia, *op. cit.*, 23.
15 Gillian, *op. cit.*, 2, 4.
16 Interviews at Reebok.
17 Schlesinger, L. and Hafkett, J. (1991). The service-driven service company. *Harvard Business Review*, September/October, 75. Also see Reicheld, F. and Sasser, W.E. (1990), Zero defections: Quality comes to service. *Harvard Business Review*, September/October, 10.
18 Interviews at Nordstrom, San Francisco region, Spring 1988.
19 Interviews at Nordstrom, November 1990.
20 Interviews at Nordstrom, Spring 1988.
21 Interviews at Nordstrom, November 1991.
22 *Ibid.*
23 *Ibid.*
24 Interviews at Nordstrom, Spring 1988.
25 *Ibid.*
26 *Ibid.*
27 Gillian, *op. cit.*, 3.
28 Interviews at Nordstrom, November 1991.
29 *Ibid.*
30 *Ibid.*
31 Interviews at Nordstrom, November 1991.
32 *Ibid.*
33 Schwadel, *op. cit.*
34 Faludi, S. (1989). At Nordstrom Stores, service comes first – but at a big price. *Wall Street Journal*, 1 August.

35 Weston, H. (1991). *Nordstrom: Dissension in the ranks*. Harvard Business School case study, Harvard Business School Case Services, Cambridge, MA, January, 1.
36 Interviews at Nordstrom, Spring 1988.
37 *Ibid.*
38 Parpia, *op. cit.*, 8.
39 *60 Minutes* (CBS News documentary on national television), 6 May 1990.

Case 5.4   Digital Equipment Corporation: Counting the real cost of employee turnover

This case was prepared by Helen Peck, Cranfield School of Management, as a basis for discussion, rather than to illustrate effective or ineffective handling of an administrative situation. The author wishes to thank Digital Equipment Corporation for allowing the use of this data.
© *Copyright Cranfield School of Management, May 1995.*

There is a wealth of accumulated research to suggest that satisfied and well-motivated employees perform better than indifferent or dissatisfied ones. Furthermore, anecdotal evidence supports the notion that if employee satisfaction can be raised, then employee turnover is likely to fall, and corresponding improvements in customer satisfaction and retention can be realized. However, it appears that few companies have ever tried to quantify the full cost of employee turnover, or assess its impact on revenues. This is a rare example of one that did: the Digital Equipment Corporation.

In 1987, the British economy was booming, and in the business systems end of the computer industry, things had never been better. At the time, the UK subsidiary of Digital Equipment Corporation was Britain's fastest growing computer company. Digital's Human Resource (HR) department had, however, identified a problem that threatened to limit further growth – an acute skills shortage. The skills shortage was causing concern throughout the South East of England, with the 'M4 corridor' – epicentre of Digital's UK activities – being the area most badly affected. Digital had a reputation as a caring and enlightened employer. It readily invested in employee training and development, and offered its staff security of employment with its strict 'no layoffs' policy. Nevertheless, the company was finding it increasingly difficult to recruit new employees of a sufficiently high calibre.

The HR team decided, therefore, to do all that it could to minimize the need for further recruitment. If suitable new staff were so difficult to find, then it made sense to make greater efforts to minimize the voluntary turnover of existing employees. With this objective in mind, the HR team began an in-depth study to discover why people were leaving Digital, and (to strengthen support for future employee retention initiatives) to track the cost of their replacement.

The study took four easily defined categories of employees as its sample, two customer facing – salespeople and customer training staff – and two categories of support staff – secretarial and professional (accountants, marketing personnel, etc.). Together these groups accounted for roughly half of the company's work force.

Trainers and professionals were the most expensive employees to replace in terms of direct costs, because the company offered relocation payments of £10 000 when recruiting employees for these two categories (see Table 5.4.1). Relocation payments were not needed for salespeople because, although they sometimes switched employers, they rarely moved to a different area. Relocation payments to secretarial staff were considered unnecessary, as it was assumed that they could be replaced locally.

Table 5.4.1 Employee turnover – direct costs (£)

	Salesperson	Trainer	Professional	Secretary
Sourcing	4000	3500	3500	600
Interviewing				
Mgmt time	2000	2000	2000	2000
Admin time	300	300	300	300
Expenses	100	100	100	100
Relocation	–	10 000	10 000	–
Training	6000	2000	1000	1000
Total per emp.	12 400	18 000	17 000	4000

As the data collection progressed, it became apparent that the direct costs of replacement were only part of the equation. When the opportunity costs of lost employees were factored in, the picture changed dramatically (see Table 5.4.2). The costs of losing a professional rose from £17 000 to £27 000, while the full costs of a lost secretary was judged to be around £6600. The opportunity costs for both groups were based on estimates of the reduction in productivity that might reasonably be expected when a new recruit takes over from an experienced member of a team. The opportunity costs of losing customer-facing employees were much higher. Moreover, both groups were likely to defect to rival businesses. Trainers were considered to be revenue-generating employees, who added directly to customer value. In practice their work was rarely billable, as most of the time their services were offered free of charge to customers, by salespeople eager to clinch a deal. Trainers were in short supply, and very difficult to replace; their loss reduced the company's capabilities and its ability to 'delight the customer', which did have a small impact on sales – hence the opportunity costs of around £45 000 per trainer. That left the sales force as the only overtly revenue-generating group in the study.

A salesperson took roughly nine months to complete his or her training, and to become fully familiar with the technology and the customers. Approximately 30 per cent of the opportunity costs for this group was due

Table 5.4.2   Employee turnover – full costs (£)

	Salesperson	Trainer	Professional	Secretary
Direct cost	12 400	18 000	17 000	4000
Opportunity costs	500 000	45 000	10 000	2600
Total per emp.	512 400	63 000	27 000	6600
No. emps lost	31	11	140	248
Total cost per category	15 900 000	700 000	3 800 000	1 600 000

to reduced efficiency during these nine months. The other 70 per cent was due to customer defections. When a salesperson left and went to a competitor, a small amount of business left immediately with them, but the long-term effects were much more serious. Usually business continued to migrate for some time – often several years – afterwards. Hence the massive opportunity costs of £500 000 per lost salesperson. During the financial year 1988/89, Digital lost a total of 430 employees, at an estimated cost of £22 million. The mean average cost was £51 000 per person. Digital lost a total of 800 people that year from its UK work force. Assuming that these were all voluntary departures (and basing calculations on mean average costs), total employee turnover for 1988/89 would eventually cost the company £40.8 million.

Digital's employee turnover rate of 11.83 per cent compared badly with that of its main competitor (IBM), which enjoyed a voluntary turnover rate of only 5 per cent. The HR group decided that it would attempt to meet the IBM figure half-way, aiming to lower Digital's own rate to 8 per cent; representing a saving of 259 employees and an estimated £13.2 million.

The long-term nature of the costs associated with the loss of a salesperson meant that the study took longer to complete than anticipated, so it was not until 1991 that the findings were released within the company. The figures produced by the study were based on a number of arguable assumptions, and it has never been externally audited, but the company's own accountants did examine them and were satisfied that they were a fair representation of the costs.

The report was taken very seriously by Digital's UK board of directors. On discovering that the company was losing around £1.6 million per year just replacing lost secretaries, the board acted at once to correct the problem, by awarding them a 25 per cent pay rise. The secretarial problem was solved at a stroke, and the company was still saving money. Other measures were planned to arrest defections by other categories of workers, but these initiatives were overtaken by events.

By 1991, the developed world had slid into recession, and for the computer industry the bubble had burst. In July 1990, the Digital Equipment Corporation employed 123 500 worldwide; the following year it announced that it was hoping to reduce the work force by voluntary redundancy and natural wastage. By October 1991, it was forced to abandon its cherished 'no layoffs' policy for the first time in its history, as downsizing became the order of the day. Digital's employee retention initiatives were abandoned.

The computer industry has since gone through a traumatic restructuring, with the largest players being amongst the most badly hit. By April 1995, Digital had been forced to reduce its work force worldwide to 63 100 employees.

Chapter 6

Creating and implementing relationship marketing strategies

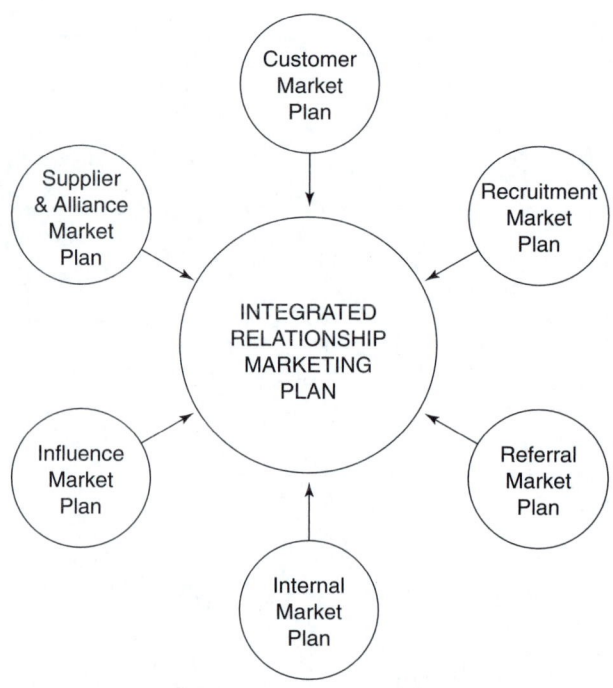

Integrating the six market domains

Introduction

Strategy is the art of creating value. It provides the intellectual frameworks, conceptual models and governing ideas that allow a company's managers to identify opportunities for bringing value to customers and for delivering that value at a profit. In this respect, strategy is the way a company defines its business and links together the only two resources that really matter in today's economy: knowledge and relationships or an organization's competencies and customers.

Normann, R. and Ramirez, R.,
From value chain to value constellation: Designing interactive strategy.
Harvard Business Review, July–August, 1993, pp. 65–77

The transition that is now underway from the classic transactional approach to buyer/seller relationships to the collaborative, interactive mode that we now call Relationship Marketing implies much more than a change in philosophy. Whereas the traditional approach to marketing was based upon the management of the so-called marketing mix, relationship marketing instead is built upon the management of a number of key stakeholder domains – the 'six markets' as we have dubbed them in this book and elsewhere.

Classic '4 Ps' marketing was based upon a mechanistic, almost manipulative, approach to consumers. It implied that if the right marketing mix could be deployed, then a sale would result. It was as if a formula could be determined to programme the marketing mix for optimal results. Indeed, in the 1960s and 1970s there were many scholarly attempts to present the management of the marketing mix as a mathematical exercise.[1–3]

It is possible that these more formalized approaches to marketing planning and strategy determination were appropriate for the environment of the time – in other words, high growth markets with consumers who for the first time had higher levels of discretionary income to dispose of than at any time in the past. Today, however, with many markets now better described as mature and in many cases closer to commodity markets and with ever higher levels of consumer awareness and sophistication, a different approach is required.

The Relationship Marketing philosophy that is now emerging is grounded on the tenet that as long as customers and consumers continue to perceive that they are receiving superior value from a specific source of supply, then they will remain loyal. Whilst it should be emphasized that the conventional marketing mix elements – product, price, promotion and place – are still critical, they need to be augmented by elements which enable customer and consumer value to be enhanced. The suggestion here is that there are three particular elements that can powerfully impact on customer and consumer value: people, processes and proactive (and personalized) service.

People

One of the most critical changes in marketing thought that has occurred over the last two decades has been the recognition that 'people are the brand'. Hence the significant attention that is now paid to attracting, training and motivating employees who can share the values of the business and project those values at every point of customer and consumer interaction. Much has been written about the connection between employee satisfaction and customer satisfaction[4, 5] and the importance of the 'internal market' is now generally recognized.

Processes

The means by which value is created for customers and consumers is through what we now call 'processes'. Processes are, in a sense, 'the way we do things'. In other words, they are linked sets of activities that enable market demand to be satisfied. The core processes of any business will include:

- The new product development process
- The order fulfilment process
- The supplier management process
- The customer management process.

Whilst individual businesses will need to redefine these processes to fit their own particular circumstances, it is self-evident that the creation of customer value is achieved through these fundamental tasks. In a sense these core processes are the means by which we attain the goal of 'the right product in the right place at the right time'. Thus, in a consumer packaged goods business the specific value-generating processes could be defined as in the example shown in Figure 6.1.

- **Consumer equity**

  Leveraging deep understanding of the consumer to define and communicate specific value proposition that meets both the needs of individual consumer segments and the corporation's strategic/financial goals.

- **Product development**

  Designing and continuously redesigning product concepts to ensure that they deliver the 'right' value proposition to targeted consumers.

- **Supply chain management**

  Sourcing, producing and delivering the right combination of benefits – both products and services – to retail customers at the lowest possible total cost.

- **Customer management**

  Leveraging deep knowledge of consumers and category dynamics to develop selling and supply chain strategies that benefit both trade customers and consumer products manufacturers.

  Source: Armstrong (1996).[6]

Figure 6.1   Core processes in consumer packaged goods.

The critical features of processes are that they are market facing, they are integrative – cutting across conventional functions and hence must be managed on a cross-functional basis – and, most importantly, they are the means by which customer and consumer value is generated and determined.

Proactive/personalized service

One of the keys to success in the demanding markets in today's developed economies is the way that customers are supported and serviced. Customer service is defined here as the totality of all the encounters between a supplier and a customer that together combine to enhance the value of the 'offer' to that customer. In other words, customer service is the result of the combined impact of all the points of contact between the firm and its customers.

Customer service is increasingly seen as a powerful differentiator in markets where the core product is essentially the same as competitors' offerings and where there are readily available alternative sources of supply. Because customer service innovations can often be adopted by competitors – note, for example, how most international airlines rapidly adopt successful service innovations created by their competitors – there is a significant advantage to be gained from seeking out the precise requirements of individual customers and then, wherever possible, personalizing and customizing the service package. It should also be recognized that the concept of proactive and customized service is as relevant to business-to-business markets as it is to consumer markets; thus companies as diverse as Procter & Gamble and Dow Chemicals seek to use different channels and service packages to meet the needs of customers with widely differing needs.

These three additional elements of the marketing mix should be seen to be just as critical to sustained marketplace success as the original '4Ps'. The challenge to marketing management is to develop a framework in which they can be integrated and focused around a relationship building strategy. Figure 6.2 shows the seven elements of the expanded marketing mix.

What sort of relationships with what sort of customers?

It should be emphasized that the philosophy of relationship marketing does not necessarily imply that the firm should seek high-intensity, interactive relationships with all its customers. Equally, it should not be assumed that all customers want or require that sort of relationship.

There are two key determinants for the desirability of high-intensity, one-to-one type relationships in a business context. The first is customer profitability and the second is the opportunities that exist

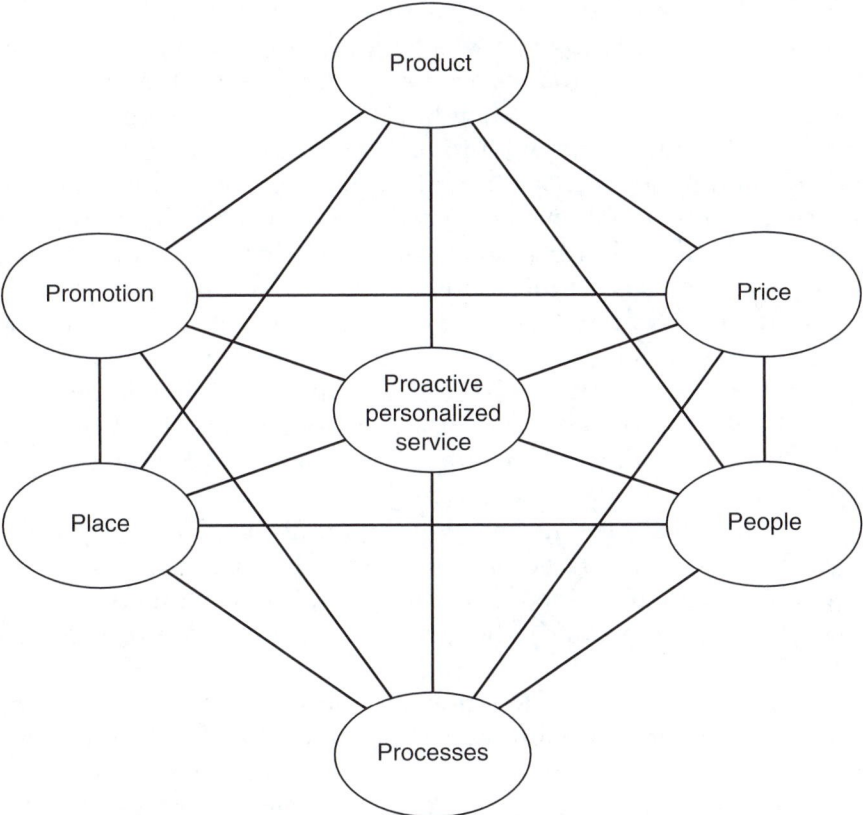

Figure 6.2 The expanded marketing mix.

for creating additional value for those customers through customizing the offer.

Customer profitability

It is perhaps surprising that most companies focus more upon identifying the profitability of *products* rather than *customers*. The paradox is that it is customers that generate profits, not products. Products create costs but customers create profits. This is more than mere semantics. What we find is that the difference between profit and loss is typically determined *after* the product is manufactured. The costs of storing, moving and supporting products are significant. Customers differ widely in their requirements for delivery service, in their ordering patterns and, indeed, in the products that they purchase. Each product has its own unique profile of margin, value/density, volume and handling requirements. Similarly, customers will order different product mixes, will have their own unique requirements as to the number of delivery points and of course the number of times they order and the complexity of their orders will differ. Putting all these factors together can produce widely differing cost implications for the supplier.

The so-called 'Pareto Law', or 80/20 rule, tells us not only that 80 per cent of the total sales volume of a business is generated by just 20 per cent of the customers, but the likelihood is that 80 per cent of the total costs of servicing all the customers will be caused by only 20 per cent of the customers (but probably not the same 20 per cent).

Figure 6.3 illustrates the shape of the profit distribution resulting from the uneven spread of revenues across the customer base. From this example, it will be seen that there is a 'tail' of customers who are actually unprofitable and who therefore reduce total profit contribution!

One of the key tenets of Relationship Marketing is that ideally the firm should only seek high-intensity relationships with those customers who are, or who have the ability to become, high performing clients in terms of profitability. A fundamental basis of a successful relationship marketing strategy therefore is the ability to create individual customer profit and loss accounts. The problem is that traditional accounting systems make it difficult, if not impossible, to identify the true costs of serving individual customers. So often it is assumed that there is an 'average' cost of serving a customer and so the ability to target those customers with the real potential for transforming the suppliers' bottom line is foregone.

Opportunities for value-adding through customization

High-intensity customer relationships are based upon long-term collaboration and interaction. The presumption is that customers will recognize the benefits that accrue from allowing the supplier to become more involved in their own value chain or, if consumers, in their 'activity chain'. These types of relationship are really partnerships or alliances and they work best when customers actively seek

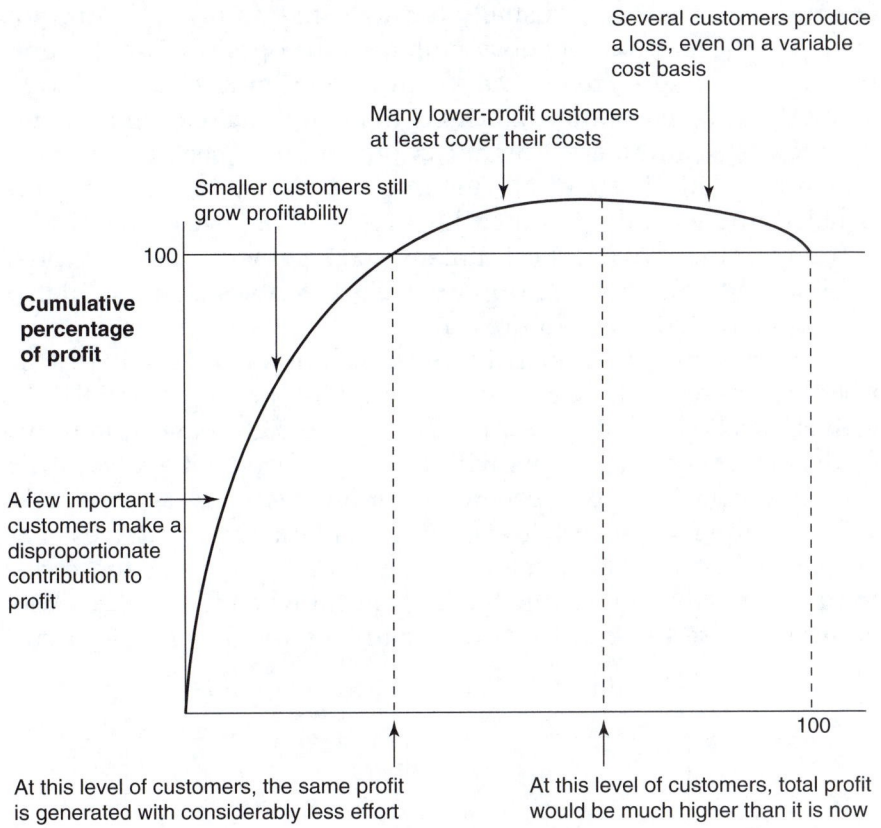

Figure 6.3 Some customers erode profitability.

tailored solutions for their own specific buying 'problems'. In other words, the customer and the supplier recognize the advantage to be gained through closer relationships.

In business-to-business marketing, this collaboration will often extend to strategic alliances and joint ventures. There will be a recognition of the mutual benefits of linking each other's value chains, in sharing resources, knowledge bases and capabilities to achieve agreed strategic goals. By definition, a high level of customization will be involved in this type of relationship.

In consumer marketing, the opportunities for customization may seem more constrained. However, in this era of what is sometimes called 'mass-customization',[7] where standard products are subsequently 'personalized' – usually through last-minute configuration from generic modules or subassemblies – the opportunities for one-to-one marketing increase. Marketing communications can also be tailored, particularly through direct mail and data-base marketing. An increasing number of organizations are now seeking to create dialogues with individual customers based upon a far more detailed understanding of their lifestyles and life-cycle stage.

Bringing these two critical dimensions together – customer profitability and value-adding opportunities – enables a simple matrix to be constructed, as in Figure 6.4.

It suggests that high-intensity relationships should really be confined to those customers with the highest profit potential who actively seek customized solutions (the top right-hand quadrant). Conversely, those customers with lower profit potential and where the opportunity or the need for customization is less should be served through low-cost 'transactional' marketing strategies. The bottom right-hand box is interesting in that it represents those customers who are amongst the top 20 per cent when it comes to profit and yet where there are few opportunities for developing a cus-

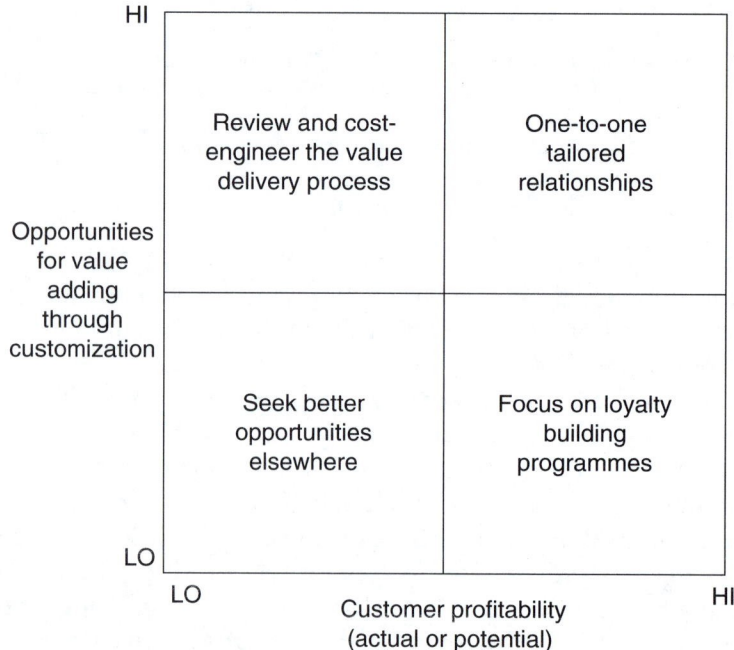

Figure 6.4 What sort of relationships are appropriate?

tomized and tailored offer. These customers present an opportunity for classic retention strategies through loyalty schemes, reward programmes, retrospective discounts and by seeking to increase switching costs. Finally, the customers in the top left-hand box currently do not justify a high level of tailoring of the product/service offer even though they might respond positively to it. The challenge is to find ways to enhance their profitability, either by cost-engineering the value delivery system or by encouraging them to move towards single-sourcing in situations where they split their purchases between multiple suppliers.

Models and matrices like that shown in Figure 6.4 should not be interpreted too rigidly. However, this one serves to make the point that it is appropriate to have different types of relationship strategy for different types of customer. Further situational variables that will affect the choice of relationship strategy are such issues as the uncertainty and volatility of the marketing environment,[8] the degree of commoditization and hence price sensitivity of the market and the magnitude of transaction costs. In particular, the choice of relationship should be specifically influenced by the over-arching business strategy adopted by the firm. In this 'contingency' approach to relationship marketing strategy, it is helpful to utilize the three 'generic' business strategies identified by Treacy and Wiersema:[9]

- Operational excellence
- Product leadership
- Customer intimacy.

Treacy and Wiersema called these three routes to success 'value disciplines'. Based upon their research, they suggested that marketplace success was usually based upon 'what kind of value proposition the companies pursued – best total cost, best product, or best total solution'. They went on to say, 'by operational excellence, we mean providing customers with reliable products or services at competitive prices, delivered with minimal difficulty or inconvenience. By product leadership, we mean providing products that continually redefine the state of the art. And by customer intimacy, we mean selling the customer a total solution, not just a product or service.'[9]

Whilst these three 'disciplines' or 'generic' strategies should not be assumed to be mutually exclusive, it will more often be the case that companies have different strengths – or weaknesses – in each of the three. The argument we make here is that each of these three generic strategies needs to be supported by an appropriate relational strategy for each of the six markets.

Developing a six market strategy

In Chapter 1 the idea of multiple market domains was examined in depth and what we have called the Six Markets model was developed. Whilst a strong position in each of these markets should be the aim of any business, it is suggested that the precise emphasis on each of the six should reflect the chosen underlying generic strategy.

Diagrammatically, the chosen Six Markets strategy is conveniently displayed by use of a 'spidergram'.[10] The spidergram summarizes, using arbitrary scales, the emphasis placed (or desired to be placed) on each of the six markets; Figure 6.5 gives an example.

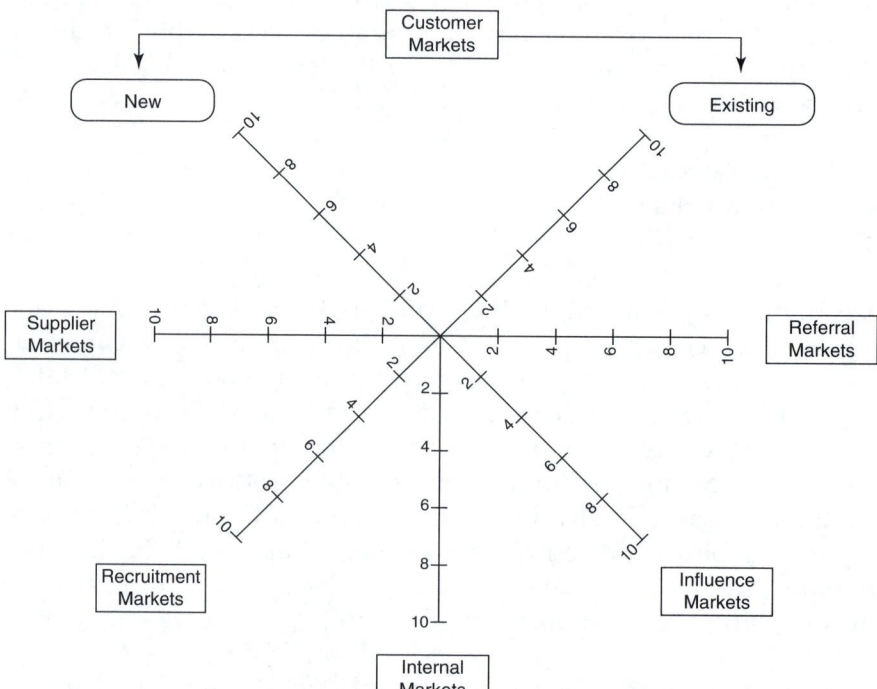

Figure 6.5   Present and desired emphasis on relationship markets.

We have found it helpful to divide the customer market into two: existing customers and new customers.

It is possible to make a connection between the three 'value disciplines' of Treacy and Wiersema and the Six Markets model. In order to achieve a leadership position through one or other of these generic strategies it will be necessary to vary the emphasis placed on each of the six markets in the overall marketing strategy of the business.

Thus, we would suggest, organizations seeking to follow the discipline of *operations excellence* will need to have an internal culture that is based on 'lean thinking'.[11] In other words, the focus on the internal marketing effort should be on continuing improvement, multi-skilling and all those activities such as quality circles that lead to greater internal efficiency. Equally, significant emphasis must be placed upon the supplier market domain since, for many organizations, the cost of materials and supplies is a major proportion of total cost. By working more closely with suppliers, many opportunities for cost reduction and quality improvements can usually be identified. In the same way, it can be argued, the interface with downstream intermediaries such as distributors and retailers will need to be managed closely. For example, through the use of Electronic Data Interchange (EDI) and other forms of electronic commerce, it will often be possible to significantly enhance the responsiveness and cost-effectiveness of the supply chain.

On the other hand, companies that seek to place the emphasis in their strategic focus upon *product leadership* will need to construct a six market strategy that will place different weights upon each domain. It might be appropriate, for example, for such companies to invest in creating an internal culture that encourages innovation, risk taking and entrepreneurship. Hence these companies will be concerned to recruit people who will be able to contribute to the innovation process – perhaps with a skills profile and experience that indicates their creativity or their in-depth knowledge of technologies or markets. It is interesting to reflect that Microsoft, an acknowledged world leader in its field, has declared that its sole criterion in recruitment is 'intelligence'.

Another key focal point for businesses that seek product leaderships is their relationship with suppliers. In many industries today a significant proportion of innovation is supplier driven. Bringing suppliers into the product development process can often lead to breakthroughs in design and functionality. Most of the innovative features that we now take for granted in today's motor car, for example, are supplier originated.

Closely related to the focus on suppliers for those companies seeking product leadership is the leverage that can be gained through developing alliances with other organizations' specific skills, knowledge bases and market understanding that they can impact.

The third 'discipline' of *customer intimacy* requires a continuing focus on the means whereby the relationship with customers can be made more intense and customized – even personalized. As such it is inevitable that the 'internal' market becomes of critical importance. Many studies now confirm the impact of employee motivation and commitment on customer satisfaction.[4, 5] Customer intimacy as a strategy also depends for its success on the identification of those customers or segments who are more likely to seek this type of relationship. Certain types of products and services and certain buying occasions are more likely to provide a successful platform for such a strategy than others. Hence customer-intimate companies tend to focus on building relationships with existing customers with the greatest potential for growth and profitability.

The relationship marketing plan

To focus and integrate strategies for each of the six markets into a cohesive whole is a challenging task. Marketing planning has long sought to bring a systematic discipline to the structuring of the marketing mix around clearly articulated strategic goals. In the new era of relationship marketing, an extra dimension to the planning process is required. Essentially organizations need to have a planning template supported by an organization structure and climate that will enable them to align and coordinate what were previously separate responsibilities. In other words, the planning process must reflect the intended contribution to the achievement of the overall marketing objectives of each of the six market domains (see Figure 6.6).

The ultimate objective of the Relationship Marketing Plan is to ensure that there is a cohesive and 'pan-company' approach to the determination and implementation of marketing strategy. Clearly there are organizational implications for this wider view of

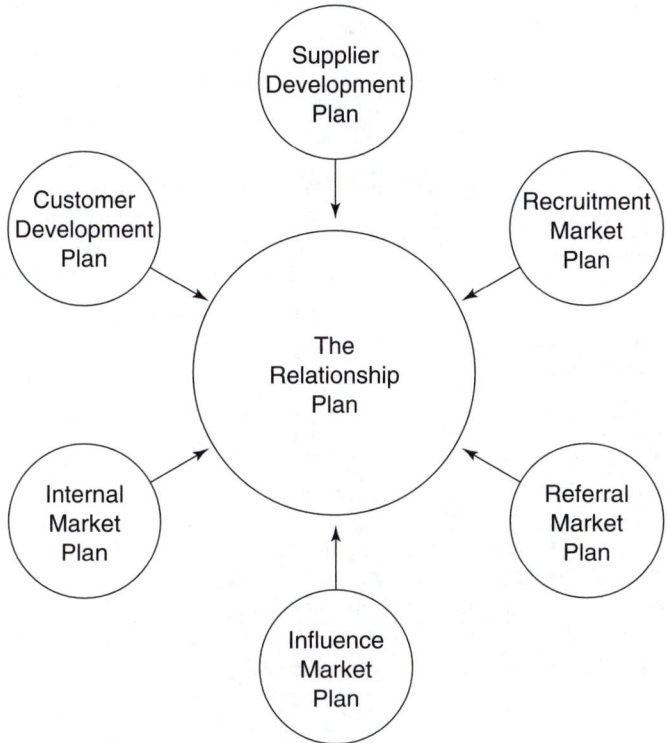

Figure 6.6 The relationship marketing plan.

marketing within the business – issues which will be considered shortly.

The relationship management chain*

Transactional marketing is, in reality, often more product focused than customer focused. The tendency has been to see the customer as someone to whom an offer is made – that offer having been predicated by the organization's own capabilities and competencies. Clearly companies must focus the offer around their own skills and

*This section draws upon material presented in the authors' book, *Relationship Marketing for Competitive Advantage*, Butterworth-Heinemann, Oxford, 1995.

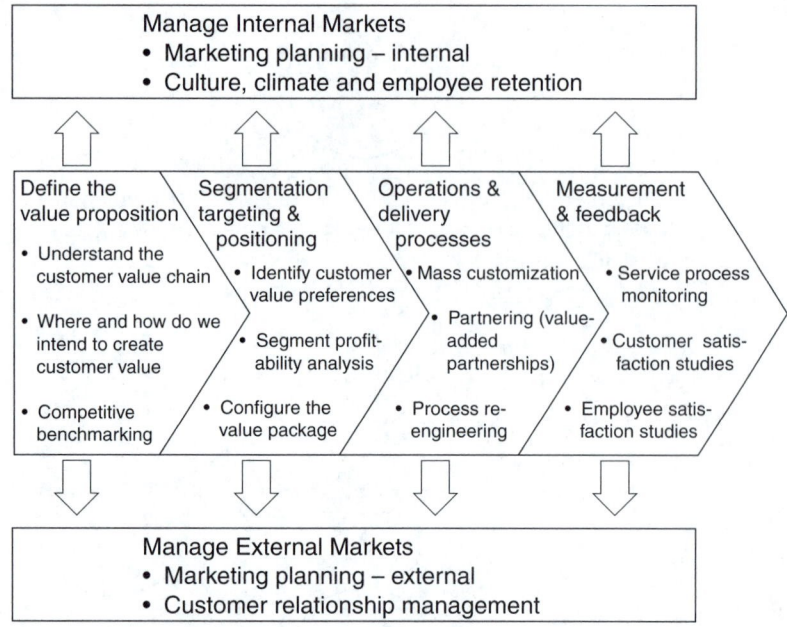

Figure 6.7   The relationship management chain.

strengths; however, true relationship strategies begin not with the concept of a bundle of features or even a brand, but rather with a clear understanding of what constitutes *value* in the eyes of the customer.

The goal of relationship marketing is to create and deliver superior customer value on a continuing basis. To help in the development of an integrated approach to the achievement of this goal we have proposed the use of a planning template known as the 'Relationship Management Chain'.[12]

The chain seeks to operationalize the Six Markets model by achieving a common focus around the creation of customer value. It is suggested that there are four distinct – but linked – elements in the chain (see Figure 6.7).

1 Defining the value proposition
2 Identifying appropriate customer value segments
3 Designing value delivery systems
4 Managing and maintaining delivered satisfaction.

Defining the value proposition

Value, in the sense most widely used in marketing, is customer specific and essentially subjective to the customer. At its most fundamental, it represents the perceived benefits that customers believe they receive from ownership or consumption of a product or service relative to the 'total costs of ownership'. Customer value can best be defined in terms of the impact that the supplier's offer has on the customer's own value chain. If the offer delivers enhanced performance, increased perceived benefits or lower customer costs, then there is a clear added-value from the customer's perspective.

The starting point, therefore, of any relationship marketing programme should be the clear definition and specification of the precise nature of the value to be delivered, market segment by market segment – or even customer by customer. This we term the value proposition or, simply, 'how to intend to create value for customers'.

Identifying appropriate customer value segments

Because customers' value requirements and perceptions will differ between customers, this provides the marketer with a powerful means of market segmentation. In-depth customer research will help reveal the salient dimensions of value, and through the use of techniques such as 'trade-off analysis' we can identify groups of customers who share common value preferences. In other words, markets can be segmented on the basis of groupings of customers sharing common value preferences. It is quite possible that the resulting segments might cut across the more traditional bases for segmentation such as demographic or socio-economic variables, but the likelihood is that the construction of marketing strategies based upon customer value preferences is more likely to succeed.

Designing value delivery systems

The means by which value is 'delivered' to customers is in itself a key element in the relationship. When we talk of delivery systems, we refer not only to the physical delivery of products or the presentation of services, but also to the marketing channels employed, the flexibility of response, the linking of buyer/supplier logistics and information systems and so on. In other words, we view the design of the value delivery system as a critical means of engineering stronger linkages between the customer's value chain and our own. Because of the increasing fragmentation of the marketplace in many industries, which has led to a demand by customers for a greater variety in the format of products or services, there is a consequent need to engineer more flexibility into our delivery systems.

Flexibility in this context refers to the organization's ability to tailors products and services to the precise needs of individual customers or segments. It will frequently involve a radical review of conventional wisdom on manufacturing and distribution – for example, focusing upon reducing batch quantities in production and in distribution, moving to a just-in-time delivery environment and delaying the final configuration of the finished product.

Managing and maintaining delivered satisfaction

Because the quality and strength of customer relationships is so critical to the survival and profitability of any business, it is essential that the processes that deliver satisfaction, as well as the customers' perceptions of performance, are regularly monitored. In the same way that it is now widely accepted that the quality of physical products is dependent upon the control of the process that manufactures them, so too the quality of customer service is determined by the extent to which the delivery process is under control. Service process monitoring should be continuous and, in particular, all potential 'fail-points' should be identified and, if they cannot be made fail-safe, should be carefully controlled. Managing the 'moments of truth' makes the difference between customer satisfaction and customer disappointment in any service process.

Employee satisfaction studies as well as customer satisfaction studies should form an integral part of the management of the service delivery process. Many companies already conduct such studies but only on an ad hoc basis. Paradoxically, those same companies might monitor brand awareness or attitudes and usage every month, but fail to pay equal attention to such vital performance indicators as employee and customer satisfaction.

Organizational issues

We earlier described the way in which organizations are increasingly becoming 'process oriented' rather than 'functionally focused'. The implications of this transition for marketing are significant. Firstly, as processes are by definition 'market facing' then, in effect, the whole business has to be concerned with the achievement of marketing goals. Furthermore, the Relationship Marketing philosophy implies that the organization as a whole seeks to establish mutually profitable relationships with customers. Thus it follows that marketing can no longer be the sole concern of the marketing 'department' but rather becomes the concern of all. This idea has always been true and few would dispute the underlying philosophy. However, the reality has been that marketing has become very much a functional concern in most organizations over the years.

If anything, the new role of marketing in the firm is more challenging than was ever the case before. Now, marketing must take responsibility for translating the overall strategy of the business into specific plans for each process. Once the organization as a whole has decided upon how it wishes to compete and the value proposition that it wants to deliver, marketing must identify and link individual process strategies so that together the corporate goals are achieved.

Figure 6.8 provides an example of how one fmcg company has redefined its business around market-facing processes and how marketing planning becomes the key integrative activity.

Whilst such a model seems enticingly attractive, it has to be recognized that its achievement requires significant organizational and cultural change. For instance, working across functions using multidisciplinary teams becomes the norm in these types of organization:[13]

> We now work much more within teams, which include marketing expertise, but also include people representing the different disciplines of the business – production, distribution, finance, R&D. We have come out of functional silos and into a flatter team-oriented structure.

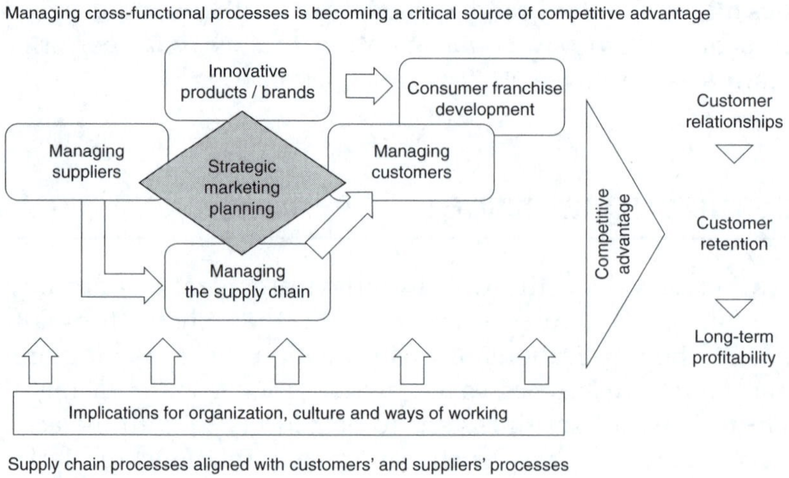

Managing cross-functional processes is becoming a critical source of competitive advantage

Supply chain processes aligned with customers' and suppliers' processes

Figure 6.8 Process orientation.

> The management of marketing has become a much more general function for the total business, led by the expertise of marketing but on a team-wide basis.

The types of skills and breadth of knowledge required to make a success of this philosophy are quite different from those possessed by the more narrow functional manager of the past. There will be a continuing need for management development in such areas as process understanding and team working/leadership skills, for example. The McKinsey 'Seven-S' model, as shown in Table 6.1 provides a helpful framework to show the many dimensions of organizational change that will be involved in moving to a process-oriented relationship marketing strategy.

It will be apparent that the changes that are implicit in the transition to Relationship Marketing management are profound. There are many obstacles to this transition, not the least being entrenched interests in preserving the status quo.

Making it happen

No marketing strategy, be it based upon the traditional model or the new relationship-oriented philosophy, can succeed without the appropriate internal culture and organizational structure. These

Table 6.1 Changing the organization

	From	*To*
Strategy	Market to major customer segments	Add value to individual customer relationships through tailored interactions
Shared values	Serve customers well	Serve customers differently; serve best customers really well
Structure	Product orientation with focus on current period economics	Customer-segment orientation with focus on lifetime customer value
Skills	Analytical orientation towards major segments and their collective behaviour over long periods	Ability to gather, analyse and interpret data and design systems to exploit a large, constantly evolving customer information base; ability to react at individual customer (or at least micro-segment) level
Staff	Marketing analysis managed statistically; information technology acts as support, but not as an active partner	Integration of marketing creativity with systems competencies to create capability that is both ideas-driven and analytically intense
Systems	Detailed, segmented, but relatively static decision support tools	Extensive, dynamic and flexible marketing support tools, programme management and execution systems, and operating links to support front-line actions
Style	Marketing plan orientation with emphasis on programmes for major segments delivered within standard period; mass media focus	Analytical approach and experimental attitude with emphasis on continuous learning (do, test, measure, fix) and value of data
	Old leading measures of success: ● Market share ● Current period profits	New leading measures of success: ● Share of most attractive customers (based on lifetime profit potential) ● Continuous learning/tailored marketing ● Large impact on a small set of customers

Source: The McKinsey Quarterly (1995) **3**.

twin pillars – culture and structure – are critical to the achievement of relationship marketing goals. The first of these, culture, has been addressed in Chapter 5, but it is important that brief mention is made of the need to have an organizational structure that is capable of supporting the company-wide concept of Relationship Marketing.

The fundamental idea that underpins the six markets framework is that marketing can no longer be seen as the responsibility of a single 'vertical' function. Instead there has to be a recognition that to be truly 'customer facing' the organization has to move from the vertical to the 'horizontal'. In other words, the focus now is upon the management of the cross-functional processes that we referred to earlier in this chapter. Managing in the horizontal organization brings with it some specific challenges:[14]

> The horizontal corporation seems to be characterised by seven main trends: organised around process, not task; a flat hierarchy; team management; measuring performance by customer satisfaction; rewards based on team performance; maximisation of contacts with suppliers and customers; informing, training and retraining of employees at all levels.

There is no doubt that one of the biggest hurdles to becoming a customer-driven business is the entrenched functional hierarchies that dominate much of industry. It is only through processes that customer value is created and hence it makes sense that the process rather than the functional task should provide the foundation for the organizational structure. Equally, the multiple levels that characterize most organizational charts encourage a 'top down' mentality and a focus more upon reporting and accounting procedures than upon responsiveness and flexibility in serving the customer. Functional 'silos' also create narrow, functional managers. However, in today's rapidly changing business environment the call is for more broadly based skills profiles whereby managers can work effectively in cross-functional teams.

Underpinning any successful relationship marketing strategy has to be a corporate culture that recognizes that the delivery of customer and consumer value is the primary purpose of the business. Obvious though this sounds, it is not always recognized by those who guide, direct and manage the business. Relationship marketing strategies can only succeed where there is a close alignment between the mission, the rhetoric and the implementation.

Not only do senior managers need to 'walk the talk' but they must constantly seek to motivate and empower those they work with to seek constantly to improve the quality of relationships with all the key market domains that we have identified in this book.

Internal communication, training and development play a key role here. Creating a sense of common purpose and establishing 'shared values' is an essential foundation for relationship-based marketing strategies.

A further and critical prerequisite for success in bringing about a relationship culture in the business is an appropriate performance measurement system. As the organization makes the transition from a vertical, functionally oriented business to a horizontal, process-focused business, then so too must the performance measures change.

People generally behave according to how they are measured and so, if the performance indicators are functional and financially directed, then managers will focus upon functional efficiency. If, on the other hand, the measures are outwardly focused on process goals and customer satisfaction, then the likely focus of behaviour will be upon process effectiveness.

We are now in the era of the 'Balanced Scorecard'[15] in which the importance of using non-financial performance indicators alongside the more traditional financial measures is now recognized. Key elements in this new multi-dimensional concept of performance measurement must be relationship-oriented metrics. For example, such measures as customer retention, customer satisfaction, perfect order achievement, complaints, customer referrals and 'share of wallet' must stand alongside the more traditional measures of performance, such as achievement against budget. Where possible, these measures should be process-based, such as 'time-to-market', 'time-to-serve' and 'cost-to-serve'. Equally, they should be widely communicated and, ideally, incorporated into financial incentive schemes such as quarterly bonuses, employee awards and the like.

Performance measurement, culture change and behaviour are all closely intertwined. The success or failure of a relationship marketing strategy will be largely determined by how well these critical issues are managed. It will be clear from the breadth of these ideas, and the business-wide implications that they have, that the vision of Relationship Marketing we have presented in this book and elsewhere is profoundly different from the classic view of marketing as a function. David Packard, one of the co-founders of Hewlett Packard, was reported to have said that 'Marketing is too important to be left to the marketing department'. In a sense that philosophy is the basic premise that underpins the new paradigm of Relationship Marketing.

References

1 Montgomery, D.B. and Urban, G.L. (1969). *Management Science in Marketing*, Prentice Hall, Englewood Cliffs.
2 Montgomery, D.B. and Urban, G.L. (1969). *Applications of Management Science in Marketing*, Prentice Hall, Englewood Cliffs.
3 Palda, K.S. (1969). *Economic Analysis for Marketing Decisions*, Prentice Hall, Englewood Cliffs.
4 Schlesinger, L.A. and Heskett, J.L. (1991). Breaking the cycle of failure in services. *Sloan Management Review*, Spring, 17–28.
5 Schlesinger, L.A. and Heskett, J.L. (1991). The service-driven service company. *Harvard Business Review*, **69**, No. 5, September/October, 71–81.
6 Armstrong, G. et al. (1996). What's wrong with the consumer goods organization? *McKinsey Quarterly*, **1**.
7 Pine, B.J. (1993). *Mass Customization: The new frontier in business competition*, Harvard Business School Press, Boston.
8 Cravens, D.W., Piercy, N.F. and Shipp, S.H. (1996). New organisational forms for competing in highly dynamic environments: The network paradigm. *British Journal of Management*, **7**, no. 3, 203–218.
9 Treacy, M. and Wiersema, F. (1995). *The Discipline of Market Leaders*, HarperCollins, London.
10 Payne, A. (1995). Relationship marketing: A broadened view of marketing. In Payne, A. (ed), *Advances in Relationship Marketing*, Kogan Page, London.
11 Womack, J.P. and Jones, D.T. (1996). *Lean Thinking*, Simon & Schuster, New York.
12 Payne, A. et. al. (1995). *Relationship Marketing for Competitive Advantage*, Butterworth-Heinemann, Oxford.
13 Bull, G. (1995). *Marketing Business*, October.

14 Castells, M. (1996). *The Rise of the Network Society*, Blackwell, Oxford.

15 Kaplan, R.S. and Norton, D.P. (1993). Putting the balanced scorecard to work. *Harvard Business Review*, September/October, 134–147.

Chapter 6 case studies

This section examines the critical issues raised in implementing relationship marketing strategies. There are many fundamental changes that organizations will need to undertake if they are to become truly relationship-oriented businesses. The two cases presented in this final section give an indication of the 'pan-company' approach that must be adopted in order to implement successful relationship marketing strategies.

Case 6.1: Relationship marketing: Lessons from Laura Ashley

This case was written by Helen Peck, Research Fellow at Cranfield School of Management.

Abstract

This case describes the attempts of the Chief Executive of Laura Ashley, Dr Jim Maxmin, to restore the financial health of an ailing business through a company-wide change programme. Whilst he was unsuccessful in his efforts to achieve a turnaround, the case provides an insight into the challenges of managing the 'Six Markets' and highlights the need to follow through from mission statement to implementation with a clearly defined relationship strategy.

Learning points

Some of the key learning points to emerge from this case are: the importance of managing all the six markets, not just the brand, in an integrated and focused way; the critical role of the 'influencer' market and the importance of managing the processes that support the delivery of value to the ultimate consumer.

Broad marketing and strategic issues:

- even strong brands need to be supported by integrated business strategy
- the critical role of the Chief Executive in translating the mission into action
- the vital role of the supply chain in retailing.

Relationship marketing issues:

- the importance of managing all 'six markets' in a coordinated and cohesive way
- the role of the business leader in creating a 'relationship culture'
- the need to recognize that any business is ultimately a 'value delivery system' and that it needs to be managed as such.

Following the departure of Jim Maxmin and his replacement by Hugh Blakeway Webb, the company continued to languish and in 1995 another Chief Executive, Anne Iverson, was appointed. Ms Iverson, an experienced retailer, felt that the brand had become tired and that significant design changes were called for. When these changes began to filter through into the stores, it was clear that this was not what the loyal Laura Ashley customer wanted - the results were disastrous. Ms Iverson left the company in 1997, the third Chief Executive in three years.

Case 6.2: Creating success through relationship marketing at British Airways

This case was written by Professor Adrian Payne, Moira Clark, Andy Coaton and Marcus Hickman whilst working for Cranfield School of Management Marketing & Logistics Group.

Abstract
This case study examines the strategy adopted by British Airways in the period following its privatization and the creation of a customer-focused approach to marketing. The British Airways story is one of large-scale corporate transformation which has resulted in a significant improvement in its marketplace standing.

Learning points
One of the key learning points in this case study is that the whole organization must be engaged if slogans like 'Putting People First'

are to become a reality. The case provides an excellent example of how relationships with each of the six markets must be planned and managed on an integrated basis.

Broad marketing and strategic issues:

- how global competitive and market trends can put a slow-moving organization at risk
- understanding that service quality is a key differentiator in a service industry
- the importance of marketing communication in supporting a brand strategy.

Relationship marketing issues:

- the critical linkage between employee satisfaction and customer satisfaction
- the importance of customer retention, particularly the most profitable customers
- to maintain a strategy of differentiation based upon service quality requires a constant focus on improving relationships with all the six markets.

British Airways has continued to build upon its success in what is still a very competitive and often unprofitable marketplace. Their progress has not been without problems and there is still a challenge to balance the need for cost reduction with service enhancement. A concern that still exists is the need to improve the level of relationships with both the internal market and the influencer market.

Case 6.1   Relationship marketing: Lessons from Laura Ashley

This case was prepared by Helen Peck, Cranfield School of Management, as a basis for class discussion rather than to illustrate effective or ineffective handling of an administrative situation.
© 1995, Cranfield School of Management.

A new Chief Executive

On 17 July 1991, Laura Ashley Plc, the British-based fashion and furnishing retail group, announced that it was about to appoint Dr Jim Maxmin to the post of Chief Executive. The announcement came almost a year after the resignation of the previous incumbent. Laura Ashley had been one of the retail stars of the 1980s. But, like so many other speciality retail success stories of that decade, the company appeared to have crumpled under the strain of high interest rates and deepening recession. Difficult trading conditions across several key markets, crippling interest payments, and order processing problems were cited in its 1990 annual report as the cause of the difficulties. Press commentary was more circumspect, remarking on how Laura Ashley seemed to have lost its way since the untimely death of its eponymous founder, some four and a half years earlier. Hopes were high that Maxmin would now provide the leadership and direction needed to revive the business.

Finding a new Chief Executive had not been easy. A firm of headhunters charged with the task had failed to find a suitable candidate. Thereafter the post was advertised in *The Economist* and the *Financial Times*. Jim Maxmin, a former Director of World Retailing at Thorn EMI, was eventually selected from over 500 applicants as the man most likely to succeed. To some onlookers though, Maxmin – an energetic American with a doctorate in philosophy and a fondness for transatlantic marketing jargon – seemed an unlikely choice for this most demure and English of retailers.

Company history

In the summer of 1953, Bernard and Laura Ashley chanced upon some attractive hand-printed head scarves during a holiday in Italy. Back home in the kitchen of their London flat, the young couple were convinced that they could produce something similar themselves. With an initial investment of £10, a batch of 20 scarves was produced and sold to a London department store. Encouraged by the early success of the head scarves, the whim became an obsession. Laura designed a range of tea towels and other kitchen accessories which found a ready market through department stores, wholesalers, and craft shops. In time Bernard Ashley left his job in the City and became the driving force behind the development of the business.

By 1960 the Ashleys had moved their now thriving cottage industry to a disused railway station in the village of Carno, Powys, in Laura's native Wales. Principally a textile design and printing operation, the business acquired a retail sideline in 1965 when it opened a small sales outlet in Mid-Wales. Three years later a Laura Ashley shop opened in London, in a quiet part of Kensington. Entry into the world of fashion followed soon afterwards, with the introduction of 'Basic Dress 1'. The simple above-the-knee housedress was to be the forerunner of the famous Laura Ashley floral frocks, long white nightdresses and Victorian-style pin-tuck blouses. Laura's countrified and quintessentially English designs conjured up rose-tinted and rose-printed folk memories of bygone days. The clothes were not made for real country people, but for city dwellers who craved a romanticized version of a rural idyll. The Laura Ashley look soon acquired a loyal following in London and the Home Counties, and attracted interest from abroad.

As the 1980s dawned, Laura Ashley shops were to be found scattered throughout the UK, continental Europe, and the US. The business went from strength to strength, fuelled by the beginning of a retail boom in the British market. The Laura Ashley look was consistent throughout the world, but brand positioning differed between the home and overseas markets. In the UK and continental Europe, Laura Ashley was a moderately priced purveyor of rustic charm to the urban middle classes. In the US it was seen as an exclusive designer label. The difference was reflected in the margins, which were as high as 18-20 per cent in North America, four times the level of continental Europe and almost double those in the UK.

By the end of 1984 the Ashleys themselves had become tax exiles. Turnover had reached £96.4 million per annum, and the company was opening one new store per week somewhere in the world, yet it still retained the culture of a close family business. In February 1985 another important milestone was reached with the signing of a joint-venture agreement with the Aeon Group, giving Laura Ashley its first toe-hold in Japan.

By now the company offered its customers a wide range of fashion and furnishings, 85 per cent (by value) of which were produced in-house. Design teams developed two fashion collections (Spring/Summer and Autumn/Winter) and one home furnishing collection each year. Work on the fashion collections started around 18 months ahead of launch, and would eventually incorporate around 125 items per season, often made up in a choice of colours or fabrics. Designs would be selected – usually by Mrs Ashley herself – from hundreds of made-up samples. The Home Furnishings collection was revised annually and had gradually expanded to incorporate wallpapers, upholstered furniture and a complementary array of paint, tiles and bedlinen. A substantial proportion of the range was new each year, but best-selling lines would be retained for a number of seasons. In addition, a separate and exclusive 'Decorator Collection' of furnishings was produced for interior designers.

Fabric, soft furnishings and wallpaper came from the company's own factories, one in Carno and one in Helmond, Holland. Meanwhile 10 factories in Wales and one in Ireland provided the shops with garments. Additional garment factories were planned for the UK and in the US. Wallpaper and fabric production capacity was also about to increase with the development of a major wallpaper and fabric plant alongside a new distribution centre at Newtown, close to the Carno facilities.

In the summer of 1985, an official announcement confirmed rumours of a public flotation. The offer was expected to raise around £2 million to fund the next round of business development. A majority holding – approximately 72 per cent of the stock – would remain under the direct control of the Ashley family. Of the shares on offer, around £1.2 million worth were to be made available to Laura Ashley staff at no cost to the employees themselves. Employees would also be given preferential rights to buy additional shares.

The fairy-tale story came to an abrupt end on the night of 8 September 1985, when Laura fell down some stairs and was fatally injured. The family had been celebrating her 60th birthday. She died of her injuries a week later.

The company's bankers suggested that the flotation should be delayed, but were persuaded by other members of the management team that Laura's death was inconsequential to the day-to-day running of the business (see Table 6.1.1 for profiles of directors, 1985/86). The Ashleys had reduced their direct involvement in the day-to-day running of the business to some extent the previous year, but Bernard Ashley scaled down his involvement further following the death of his wife. With the company valued at £270 million, the flotation went ahead. The share issue was oversubscribed 34 times.

Post flotation

From 1985 to 1988 the expansion of the retail empire continued apace, almost doubling the number of shops in the UK and continental Europe. In North America and the Pacific, the company quickly realized its ambitions to have a Laura Ashley store in every major city in Australia, and in every city with a population of over 1 million in the United States (refer to Tables 6.1.2–6.1.4 for growth of retail operations). By 1986, the North American division (helped by favourable exchange rates) was contributing about half of the company's overall pre-tax profits.

Meanwhile, behind the scenes, radical changes were occurring in the way that the products were designed and selected. Gone were the selection

Table 6.1.1   Laura Ashley – Board of Directors, 1985/86

Bernard Ashley, Chairman: Co-founder of the business and husband of the late Laura Ashley. Worked in the City of London from 1947 to 1953, then became the driving force behind the business. Joint Managing Director with John James from 1976 to 1988.

John James, Group Managing Director: Chartered Accountant who joined the group in 1974 as Financial Controller and Company Secretary. Previously worked with Unilever Plc and Fitch Lovell Plc, where he was a divisional finance director. Appointed Joint Managing Director with Bernard Ashley in 1976. (*Became Vice-Chairman and Chief Executive in January 1988; resigned 25 August 1990.*)

Adam Lofthouse, Managing Director – Production Division: Joined the group in 1984 with previous experience as a divisional director with Courtaulds Plc and with Coats Patons Plc, where he was Managing Director of a number of operating subsidiaries. (*Resigned 30 September 1989.*)

Peter Phillips, Group Finance Director; Managing Director – Group Services: Chartered Accountant who joined the group in 1982, with previous experience with The Marley Tile Company, RCA Corporation, Celanese Corporation and The Plessey Company Plc, where he was a divisional finance director. (*Resigned to join Hepworth Plc, November 1988.*)

Peter Revers, President – Laura Ashley Inc.: Joined Laura Ashley in 1970 as Marketing Manager. Responsible for the development of retail activities in the US, since opening its first shop there in 1974. (*Resigned 9 November 1990.*)

Alphons Schouten, Managing Director – Continental Europe Retail Division: Joined the group in 1981 from the Bluebell Corporation, where he was a marketing director. (*Board notified of his imminent resignation 5 July 1994; resigned 1 February 1995.*)

Mike Smith, Managing Director – UK Retail Division: Joined the company in 1981, appointed director of the UK Retail Division in March 1986. (*Acquired responsibility for manufacturing following the resignation of Adam Lofthouse in September 1989. Appointed Acting Chief Executive following the resignation of John James in August 1990. Held the post until the appointment of Jim Maxmin on 16 September 1991. Resigned 28 September 1991.*)

John Winter, Deputy Group Managing Director; Managing Director – Design Services: Joined Laura Ashley in 1984. Prior to this he was a partner in the management consulting division of Deloitte Haskins & Sells. He had been advising the group since 1980. (*Resigned to join Coats Viyella 30 November 1987.*)

Lord Hooson, Non-executive Director: Lord Hooson is a barrister, called to the Bar in 1949, becoming Queen's Counsel in 1960. He was Liberal MP for Montgomery for 17 years before being created a life peer in 1979. He was first appointed to the board as a Non-executive Director in 1985.

Table 6.1.2   Laura Ashley growth of retail operations

Shops	UK*	Europe	North America	Far East	Australia	Total
1968	1	–	–	–	–	1
1970	3	–	–	–	–	3
1977	6	5	1	–	–	12
1980	25	34	11	–	1	71
1981**	29	35	15	–	1	80
1982	39	39	28	–	1	107
1983	55	42	43	–	1	141
1984	72	43	55	–	1	171
1985	87	47	84	1	12	231
1986	109	53	113	2	15	292
1987	140	61	137	5	19	362
1988	157	66	167	12	25	427
1989	171	75	185	24	25	480
1990***	182	95	213	41	24	555

Source: Laura Ashley, Offer Document 1985 and Annual Reports 1986–91.
*Includes Republic of Ireland, **includes Homebase outlets, ***includes franchises opened 1986–90.

sessions where new collections would be chosen from hundreds of samples. To eliminate waste and cut costs, the number of designs made up was halved. The intricate Laura Ashley signature blouses and dresses, with their labour-intensive pin-tuck details, had hitherto been produced by the relatively slow 'make-through' process by which one person made the whole garment. The system was more costly to operate than a sectioned approach, but worked well for complex designs. Quality was good and the system could rapidly accommodate changes in demand. From the mid-1980s the company started experimenting with new manufacturing systems, establishing high productivity units in two of its plants. The new units enabled simple garments to be produced in large numbers. Soon all the garment factories were switched over to sectionalized working. The core products themselves were changing as accountants sought to cut costs by reducing the number of tucks and pleats and volume of fabric in each dress. In the shops, racks of long, flowing skirts were replaced with short, straight ones, which lent themselves more easily to volume manufacturing techniques. Laura Ashley designs increasingly echoed those of every other high street retailer. The Spring 1987 fashion collection, with its Caribbean prints, striped shorts and polka dot dresses, demonstrated this new and more catholic design philosophy. The management felt that the new design strategy and a further extension of product lines demonstrated that the company itself was maturing and adopting a suitably sophisticated approach to a fiercely competitive retail environment.

Table 6.1.3 January 1991 – 555 retail outlets worldwide

Division	No. of outlets
European:	
Austria	2
Belgium	4
Denmark	1
Finland	1
France	26
Germany	16
Greece	3
Holland	8
Iceland	1
Ireland	2
Italy	8
Middle East	2
Norway	2
Portugal	1
Spain	12
Sweden	3
Switzerland	5
UK	180
North American:	
Canada	11
Mexico	1
USA	201
Pacific Basin:	
Australia*	24
Hong Kong	1
Japan	37
Korea	1
Malaysia	1
Singapore	1
Total	555

Source: Laura Ashley Annual Report, 1991.
*Fabric is also distributed in New Zealand and sold through independent relailers.

The following year, in keeping with the mood of the times, Laura Ashley became a 'lifestyle concept' with the launch of two new retail ventures, 'Mother and Child' and 'Laura Ashley Home'. Mother and Child shops were opened as either stand-alone shops or in-store boutiques. Inside they offered an upmarket selection of children's clothing and matching mother and baby wear. A coordinated range of nursery furniture and furnishings completed the package. The second venture, Laura Ashley Home, was

Table 6.1.4   Turnover by region, 1985–91 (£m)

Year	UK & Eire	North America	Europe	Pacific & others	Total
1984/85	39.7	13.4	40.1	3.2	96.4
1985/86	56.4	56.2	16.1	2.8	131.5
1986/87	78.4	66.5	21.6	4.4	170.9
1987/88	95.0	74.8	24.8	6.9	201.5
1988/89	125.9	90.2	28.5	7.8	252.4
1989/90	142.6	117.2	28.6	8.2	296.6
1990/91	143.5	136.2	40.7	7.1	327.5

Source: Laura Ashley Annual Reports.

introduced first in the US in November 1988, and then in the UK and Europe in early 1988. Laura Ashley Home stores (opened in slightly less prominent city centre locations) contained a vastly widened product range which included wooden cabinet furniture and a selection of glass, lighting, and tableware. By now, though, the retail boom of the 1980s had run its course, but the programme of shop openings continued in the hope that the slowdown was a temporary pause. Profits for the group had peaked in 1987, but an underlying decline in sales per square foot was masked by an overall increase in retail space (see Table 6.1.5).

In January 1988, amid rising concern for the financial well-being of the business, Managing Director John James was promoted to Vice Chairman and Chief Executive, relieving Bernard Ashley of some of his executive duties. James's workload increased further following the departure of Finance Director, Peter Phillips, the following November. Measures to arrest the decline in sales were introduced, but as 1989 slipped by, turnover and group profitability continued to run in opposite directions. In the UK

Table 6.1.5   Retail sales per square foot, 1984–91

	Sales (£m)	PBIT (£m)	sq ft (000)	Sales/sq ft (£)
1984	96.4	14.1	259.9	370.9
1985/6	131.5	18.0	357.0	368.3
1986/7	170.9	22.5	492.1	347.2
1987/8	201.5	23.1	629.8	319.9
1988/9	252.4	20.3	737.5	342.2
1989/90	296.6	(4.7)	816.4	363.3
1990/91	327.5	(6.7)	n/a	n/a

Source: Laura Ashley Annual Reports.

and Europe, sales crept on upwards, but margins were looking increasingly sickly. Rising interest rates had frozen the UK housing market, tipping furnishing sales into a dramatic decline. To add to the woes, deliveries to the shops were becoming more and more erratic.

Financial crisis and aftermath

Gradually and inexorably, the financial situation at Laura Ashley worsened until, in February 1990, the company announced end of year losses of £4.7 million, on a turnover of £296 million. The announcement came as a complete shock to the financial institutions of the City of London. The City, though growing accustomed to increasingly lacklustre forecasts from Laura Ashley, had nevertheless expected results which were broadly in line with profit forecasts issued only a month earlier. The share price plummeted. Days later, the banks were called in following the breach of a loan covenant. The value of the company had halved in less than 12 months. While Laura Ashley (still without a Finance Director) teetered on the verge of collapse, two rival syndicates representing 11 of its banks argued among themselves over which of them should be responsible for supplying the extra money needed to keep the company afloat. Eventually the Bank of England intervened, and refinancing was agreed, once a way had been found to reconcile the competing interests of the 25 banks involved. It was against this backdrop that Andrew Higginson left Guinness Plc to take up the reins as Finance Director at Laura Ashley. The refinancing was subject to stringent conditions. The banks demanded a reduction in the company's £89 million debt, and improvement in operations. In particular, something had to be done about its appalling logistics performance. At year end Laura Ashley had £105 million tied up in stock, yet it still could not deliver to the shops on time. The North American division had performed particularly badly, with a third of the 1989 Autumn/Winter clothes collection arriving in the shops approximately three months late, resulting in immediate mark-downs.

In August 1990, Laura Ashley's immediate debt problems were alleviated, when a personal request from Bernard Ashley secured a £45 million cash injection from the Aeon Group, in exchange for an increased stake in the business. Chief Executive John James stood down a few days later. The resignation of Peter Revers, President and Chief Executive of the North American Division – the man who had overseen the US retailing operation from the opening of its first shop in 1974 – followed soon afterwards. A new Chief Executive from outside the group would be appointed in due course. Meanwhile with Mike Smith, head of the manufacturing operations and the UK retail division, as Acting Chief Executive, attention turned to matters of sourcing and supply.

Dual or multiple sourcing of all materials had become company policy in 1986. This was felt to be beneficial because it would avoid any significant disruption in production if a single supplier were unable to meet an order. The proportion of goods sourced externally had risen significantly in recent years; nevertheless, about 70 per cent of home furnishings were still produced in-house. The management, determined to retain control of its fabric design and printing, agreed that the figure should remain roughly the same. As for garments and other product ranges, they decided that the proportion of bought-in products would increase further from the existing 45 to 85 per cent, with almost immediate effect. The outsourced items would come from a host of new suppliers in the Far East (mostly Hong Kong) and Eastern Europe. Laura Ashley sold more than 3 million garments a year through its stores, and the move offered a possible saving of up to £2 for each garment sourced externally. Consequently, an announcement on 27 September 1990 informed the company's employees that a total of six Welsh sewing factories and the factories in Ireland, Holland and America were to be sold, closed or transferred to the Carno and Newtown sites. The move reduced the 8000 strong work force worldwide by 1500. A rationalization of the design studios followed soon afterwards.

A review of the business by management consultants, Coopers and Lybrand, made depressing reading, but one glimmer of hope shone out among the gloom: the Laura Ashley brand itself was still a force to be reckoned with. In recognition of this, Glenne Gibson, principal consultant of Coopers and Lybrand's retail group, joined Laura Ashley, in March 1991, as the group's first ever market planning director. In an interview with *Marketing* magazine soon afterwards, Gibson spoke confidently of 'a changing philosophy' at Laura Ashley, explaining that 'The management has a real commitment to ensure Laura Ashley is customer-led and not production driven'.[1] Ominously perhaps, the same article observed that 'Talk in the industry is not so positive ... Bernard Ashley has a reputation as a difficult boss who acts on his own whim and yet is unwilling to change.'

Under new management

Jim Maxmin formally moved into his new post on 16 September 1991. On Maxmin's appointment Bernard Ashley relinquished his executive duties, becoming Non-executive Chairman. Ashley had rarely attended board meetings over the previous year; nevertheless, as holder of 59 per cent of the company's shares he continued to cast a long shadow over its proceedings. Putting the business in order would be a formidable challenge, but the optimistic Maxmin was already planning its marketing-led renaissance. 'You have to find out what your customers want, give it to them, and just love 'em to death', enthused Maxmin to a journalist from the

Financial Times.[2] In more measured tones, he made it plain that his decisions would be guided by research and financial discipline.

In its formative years the company had successfully relied on Mrs Ashley's instinctive grasp of her customers' preferences and requirements to guide the business. Even after her death, it had never felt the need for, or indulged in, regular market research. The names and addresses of 165 000 customers were held on a database of Laura Ashley store card holders, but beyond that very little was known about the customer base. In an effort to improve the situation, questionnaires were sent to 45 000 Laura Ashley customers in the UK, France, Germany, the Netherlands and North America during 1991. Laura Ashley customers were, it transpired, predominantly female and aged between 18 and 50. They were affluent, well educated and many were working mothers. They looked for quality, service and value, and for things that would last. On average they visited a Laura Ashley store 11 times per year, but purchased on only two of these occasions. In the UK, their interests were likely to include theatre going, fine art, gourmet cooking and the National Trust.* Further research showed that the profile was remarkably similar in all 28 countries in which the company operated. When questioned about the brand, customers responded very positively. They identified with the Laura Ashley image, and its perceived brand values, though many of them felt that the clothes were 'not relevant to their lifestyles'.[3]

Maxmin's belief in the future of the Laura Ashley brand was unshakable. In accepting the post he had declared his intention to reposition the brand worldwide. In his opinion Laura Ashley was not, and would never be, a competitively priced mass-market brand. 'This company is only going to appeal to between 3 and 5 per cent of the world's population and we have got to serve them.'[4] He went on to state: 'I want Laura Ashley to be one of the UK's top international brands; a place where people want to work. And I want it to be successful because of high quality products and good culture, not financial engineering.'[5] Maxmin was convinced that the way forward for Laura Ashley was to build much stronger relationships with its customers through what he described as 'a word-of-mouth repeat-purchase kind of marketing'.[6] These sentiments were embodied in the company's new mission statement,[7] which read:

> Our Mission is to establish an enduring relationship with those who share a love of the special lifestyle that is Laura Ashley. We will act so as to protect the integrity of that relationship and to ensure its long term prosperity.

Former acting CEO, Mike Smith, resigned a week after Maxmin formally

---

*The National Trust is a charitable organization concerned with the preservation of Britain's heritage.

took up his appointment, and a few days before the group announced its half year results. Sales were down by 11 per cent for the same period last year. The time had come, Maxmin decided, for the company – the whole company – to get closer to its customers. He announced that from now on all senior executives (including full-time board members) would be spending one day every two months working in a Laura Ashley store. While there, they would be expected to serve customers, operate tills, unload deliveries and (when necessary) sweep the floors. Maxmin had already visited around 100 Laura Ashley shops himself in the preceding weeks, observing that he 'had never seen a business where so many people are able to tell you what is wrong but are not empowered to do anything about it'.[8] To prove the point, a 'Profit Improvement Programme' was introduced in the worst performing stores. Responsibility for the organization of fashion shows, local advertising, merchandising, stock replenishment, in-store entertainment and refurbishment were all handed over to the shop staff, who were allowed to keep a percentage of any increase in profits. A shop by London's Marble Arch subsequently turned in a 62 per cent increase in profit, while a Liverpool store achieved an astonishing 139 per cent improvement. The initiative was followed up by incentive schemes for all shop staff.

Behind the scenes

An interim management team had already begun to tackle some of the company's underlying problems in the year leading up to Maxmin's appointment, though much remained to be done. First on the agenda was distribution. For some reason Laura Ashley seemed incapable of getting its products to the right place at the right time. A closer examination of the company's logistics operations revealed a hideously inefficient, disjointed and tangled web. There was a sizeable in-house distribution department operating five major warehouses around the globe, using a total of 10 largely unconnected management information systems. The group used no less than eight principal linehaul carriers, and a host of other transport suppliers, to serve 550 stores in 28 countries.

The warehouses were holding over 55 000 lines of inventory (though only around 15 000 were current stock) ranging from 35-metre rolls of fabric, through to hand-made wedding gowns, bottles of perfume, wooden wardrobes, and tins of emulsion paint – only 5 per cent of lines were common to all stores. Separate stockholdings existed for each of the geographically defined retail divisions, sometimes within the same facility. The result was that when a store in Dusseldorf experienced a stock-out on an expensive bedspread, it was told that the item was unavailable, even though over 500 of them belonging to the UK division were sitting in the same warehouse.

There were problems with the order processing system too. A specially designed electronic point of sale system had been installed in the UK shops in 1984. Similar systems were installed in the US and continental Europe the following year. The EPOS systems notified the group's head office in Carno that an item had been sold; this triggered an automatic replenishment order. Optimum stock levels for each store were calculated each season back at HQ, but if shortages of any item occurred, then the system automatically gave priority to full replenishment at the larger stores. Large London shops would be replenished on a daily basis, whether the stock was selling quickly or not, but a small regional store which sold its entire allocation of an item within a day might be left for weeks with a total stock-out. A further complication was that the systems did not distinguish between real sales and goods redirected to other stores.

The recent move towards 'cheap' multiple sourcing had created further difficulties. Supply routes had become contorted as goods manufactured in Hong Kong or China were transported to Wales and then on to the US or back to the Pacific. Securing additional supplies to meet unforeseen demand became more difficult as orders were placed piecemeal with a multitude of overseas suppliers. The orders were not sufficiently high volume, high value or frequent for Laura Ashley to command flexibility or priority customer status from its suppliers. The sourcing of seasonal gift items and secondary home furnishings was particularly complicated. Lampstands, for example, were sourced with several small orders to each of 79 separate suppliers; another 80 supplied seasonal gifts.

Overall, stock availability was around 80 per cent, and in the UK alone the company was spending a small fortune handing out £25 vouchers to placate frustrated customers whose orders stubbornly refused to arrive. In North America, distribution problems were even more acute. Shipments of garments and furnishing fabrics were supposed to arrive weekly by air, but late processing at the Welsh factories meant that shipments often missed the weekly flight. Rather than putting the goods on the next available flight, freight forwarders held the consignment over to the following week so that loads could be consolidated. There were import quota problems too. While in-bound shipments originating in the UK were cleared very quickly through US customs, similar shipments from the Far East were likely to experience long delays. Service to the West Coast 'Ocean Stores' was particularly poor. Replenishment cycles which used to take around 10–15 days gradually lengthened to 39 days from factory to shop. As their performance deteriorated, the Ocean Stores sank lower and lower in priority of delivery, eventually receiving shipments of garments by sea. Exasperated sales assistants did what they could to remedy the situation, attempting to fill customers' orders by chasing goods from other parts of the country.

Simplify, focus, act

In Maxmin's eyes, the business was inefficient and over-complicated. Drastic action was needed if Laura Ashley was going to be coaxed back into profitability, and its core strengths were to be developed for the future. On 30 September 1991, Maxmin unveiled his plans to reorganize and simplify the business. Under the proposed programme, entitled 'Simplify, Focus, Act' (SFA), operating structures were to be simplified, with the removal of over 80 head office management posts, along with two layers of field management in the UK and one layer in the US. Meanwhile the US head office at Mawah, New Jersey was to be relocated to a smaller site in the same region. To improve the international coordination and integration of the business, systems, distribution, finance, buying, design and product development would be managed and run as single global activities. Managers responsible for each of these activities would come together to form a Global Operations Executive which would be responsible for overall management at Laura Ashley. To accomplish this, all duplicate and obsolete computer systems were to be written off, with a view to introducing common systems across the whole business. Financial reporting would consequently be improved, and then coordinated through Andrew Higginson. A new 'Futures Function', headed by former Bain & Co consultant David Oliver, would be established, with responsibility for basic research, business analysis, and strategic planning. The Group Marketing function would cease to exist. Instead a Group Collection Development (GCD) team would take over responsibility for the development of the Laura Ashley product offering. In particular, GCD would ensure that lead times were reduced; the Laura Ashley brand was consistently positioned throughout the world; and that duplication of activities was eliminated. Communications and public relations would be handled separately by a Central Marketing Services group. Finally, a new Director of Human Resources was to be recruited, with responsibility for the development of a global human resource strategy. As for dealing with the distribution difficulties, Maxmin had other plans.

Getting by with a little help from some friends

The recovery package won the support of the financial institutions of the City of London, so much so that an existing loan agreement was replaced with a much more favourable facility, drawn up with the company's 'seven core relationship banks'.[9] A delighted Andrew Higginson claimed that 'this represents our banks' endorsement of the longer term plans for the group. The fact that we have been able to arrange five year funding in the current financial markets, and given the state of the retail sector, is a strong vote of confidence.'[10]

Another upbeat press release was issued a few weeks later when, on 19 March 1992, Laura Ashley and Federal Express Business Logistics jointly announced that they had formed a 'Global Alliance'. Together they would restructure and manage the flow of goods and information within the Laura Ashley supply chain. The alliance, agreed in principle during a telephone call from Maxmin to a long-standing business contact, Tom Oliver (Senior Vice President of Federal Express), had taken five months to finalize. Under the terms of the agreement, Laura Ashley would turn over its entire logistics operation to its new partner. The spirit of the alliance meant that this would be an open-book agreement, with both parties sharing financial and strategic planning information. A representative of Federal Express would also be present at the meetings of Laura Ashley's Global Operations Executive. The partnership would be for an indefinite period, but for a minimum of 10 years' duration, and would be worth an estimated £150 million to Federal Express. In return, they would supply the logistics and stock management systems that Laura Ashley so desperately needed, upgrading its capabilities almost immediately and reducing its operating costs in the longer term. Products could be delivered quickly and efficiently via Federal Express's own global air network, thus enabling Laura Ashley to reduce its cripplingly high stock levels, while significantly improving the quality of its customer service.

The same spirit of partnership and long-term cooperation was evident later that year when Laura Ashley signed a multi-million pound deal with ICL, the Anglo-Japanese computer manufacturing and services company. ICL would develop the software and provide the hardware to progressively upgrade Laura Ashley's management information systems. In addition, they would assign 10 members of staff to coordinate the project from Laura Ashley's offices in Maidenhead. Laura Ashley had approached both of its existing suppliers – IBM and Siemens Nixdorf – but as Higginson explained, 'We had extensive talks with IBM and it was quite keen to do a deal. But IBM shied away from giving us a turnkey solution. We were looking for more than a "we will sell you this and walk away" attitude from suppliers.'[11] Siemens Nixdorf were rejected because of questions over their long-term commitment to the retail software market.

So far so good

In April 1992, Laura Ashley announced its trading results for the first six months of Maxmin's tenure. The results, though not dazzling, were nevertheless encouraging. Profits before tax and exceptional items were up by £9.4 million on the previous year. Restructuring costs, asset write-offs and the strategic alliance had, however, resulted in exceptional costs of £11.8 million, leaving a net loss for the year to January 1992 of £9.1 million. This was roughly in line with the City's expectations, and analysts nodded in approval at the company's much strengthened balance sheet.

In June 1992, expectations of Laura Ashley's imminent return to profitability were fuelled when Higginson addressed a conference for institutional investors in New York. Laura Ashley, he informed them, was likely to turn in net profits of £7–8 million for the year on flat revenues. Sales were not expected to rise significantly in 1992/93, partly due to the prolonged recession in all of the company's major markets, but also because Maxmin was determined to manage the business for profitability rather than sales. This policy had produced a 4 per cent growth in margins in the UK, where efforts to improve the price positioning of the brand were already well underway. As part of a cohesive programme to implement the repositioning, customers' perceptions of quality had to be raised. The shops themselves had been starved of investment for several years, so older stores were refurbished and updated. Meanwhile, separate 'sale' outlets were opened well away from main shopping centres so that excess, old or defective stock could be sold off without lowering the tone of the high street stores.

In the light of the earlier market research, thought was given as to how Laura Ashley might remain true to its brand values, updating its collections to appeal to the working women who still frequented its stores. In December 1992, top fashion designer Eric Bremner was recruited to preside over a revival of the company's original design principles. Shortly afterwards, a tie-up with students from the Royal College of Art was arranged to breathe new life into the clothes collections.

The company still had plans to extend its core product range further, but now looked towards licensing agreements as a way to widen its offer. A range of colour coordinated carpets was subsequently produced for Laura Ashley by Firth Carpets, a reputable carpet manufacturer. Leaflets in Laura Ashley stores informed customers of their availability and provided a list of approved stockists. Cross-referral schemes were set up to encourage the carpet stockists to refer their customers on to Laura Ashley shops. The system worked particularly well in North America where customers, unfamiliar with the British one-stop shopping approach to decorating, preferred to purchase home furnishings from a number of specialist retailers.

People power

As Futures Director David Oliver and his team got to grips with the improved information systems, some clear but surprising purchase patterns began to emerge from aggregated customer purchase data. It transpired that some items were actually 'trigger' purchases. An example of this was curtains, particularly made-to-measure curtains. Analysis of the purchase data showed that if a customer bought a pair of made-to-measure curtains, then as likely as not they would return to buy other home furnishings to match. A £300 purchase could easily become the first of a series of transactions with a combined value of several thousands of pounds. A study of customers' curtain-buying decisions was quickly commissioned, the idea being to trigger more of these valuable sequences. To the surprise of everyone, the study revealed that many sales opportunities were lost, the reason being that customers often came in to buy their curtains on a busy Saturday. If they were attended to by a mature and apparently knowledgeable salesperson, then they would usually place an order. If, however, they were attended to by a young 'Saturday girl', they were likely to leave without making a purchase. The problem seemed to be one of confidence. Ordering made-to-measure curtains involved some fairly complicated calculations, meaning that there was a wide margin for error.

Customers would only place an order with someone who they believed was competent enough to check their own calculations and order the curtains correctly. The curtains research demonstrated how vital it was for the company to have well-trained employees ready to serve its customers, and highlighted the dangers of over-reliance on casual staff. Under the supervision of newly appointed Global Human Resource Director, Denise Lincoln, substantial changes were made to the way Laura Ashley recruited, managed and trained its front-line staff. Firstly, it deliberately set out to recruit older store managers, with care taken to match the age and experience profile of the assistants to appropriate product categories. All were to receive extensive, on-going, product and sales training, with reward systems designed to recognize experience and expertise.

The new human resource policies improved morale dramatically, and the comprehensive training ensured that employees understood how their efforts eventually impacted on the long-term success of the business. Maxmin firmly believed that 'In today's world no company can realize its potential unless its people do. Education and personal development will become one of our competitive weapons for the future.'[12] In recognition of this, Lincoln took up a seat on Laura Ashley's main board, the first woman to do so since the death of the company's founder. Employee turnover subsequently fell, and back at the company's headquarters, corresponding improvements in repeat purchase rates were identified. Meanwhile, a loyalty programme was introduced through the Laura Ashley store card as a direct incentive to customers. The scheme offered loyal shoppers a range of benefits, including discount vouchers for customers who spent £250 or more per year on full-price Laura Ashley products, or on its newly introduced interior design services.

Operational mishaps

On 15 January 1993, just two weeks before the end of its trading year, Laura Ashley issued a profits warning to an unsuspecting City. Something had gone badly wrong with the US operations. Analysts, acting on information from Kleinwort Benson (Laura Ashley's stockbroker), slashed profit forecasts for the year from £6 million to £2 million, and waited nervously for the company to release its final figures. In the meantime, Maxmin continued to invest in the brand through a £2 million image advertising campaign. The campaign, aimed at reviving Laura Ashley's long neglected relations with the fashion press, celebrated the company's 40th birthday.

When the year end results were finally announced on 15 April, they were very much as expected. Laura Ashley was back in profit for the first time in four years – but only just – with group profits of £1.8 million before tax. Massive operating losses in the US had almost wiped out the profits generated from improved trading in the rest of the world. For the third year in succession the company paid its shareholders only a nominal dividend of 1p per share. Details of the US catastrophe had started to bubble to the surface a few days before the announcement.

The US distribution system had apparently collapsed following the relocation of its headquarters from Mawah to Boston, New Jersey. Staffing levels were reduced from 350 to 80, though only a handful made the move. The remaining staff did not know how to operate the stock allocation systems; consequently, all US stores were supplied with identical product allocations. A tiny store on the outskirts of Houston, Texas, received the same number of woollen sweaters and ballgowns as the flagship store on New York's Madison Avenue. Back in Wales, bewildered managers, anxious to reduce the company's stock levels, watched US sales plummet while stock levels soared (refer to Table 6.1.6 for details of total stocks as a percentage of sales). In desperation, they cut off all stock supplies to the US for 60 days. Meanwhile, equally desperate store managers drove around America swapping car loads of stock with other Laura Ashley stores. In June 1992, when the problems were most acute, there were over 110 000 stock swaps between the US stores, an average of 550 per shop.

The self-inflicted wound damaged Maxmin's credibility in the City and with the financial press, but following assurances that the US situation was now under control, they deferred judgement for the time being. A month later, Bernard Ashley, who had not attended a board meeting for over a year, resigned as company Chairman. Taking the title of Honorary Life President, Ashley retained a place on the board as a Non-executive Director, but reduced his share in the company to 36 per cent. Deputy Chairman and Non-executive Director, Hugh Blakeway Webb, moved up to become Non-executive Chairman. Blakeway Webb, a barrister and former partner of accountants, Deloitte Haskins & Sells, joined the

Table 6.1.6 Stock levels as % of sales, 1980–94

Year ending	Sales (£000)	Stock (£000)	Sales (%)
Dec 1980	25 393	7225	28.5
Dec 1981	34 153	8880	26.0
Dec 1982	44 556	10 652	23.9
Jan 1984	66 701	17 053	25.6
Jan 1985	96 448	22 732	23.6
Jan 1986	131 513	35 603	27.0
Jan 1987	170 892	45 521	26.6
Jan 1988	201 477	66 824	33.2
Jan 1989	252 431	75 790	30.0
Jan 1990	296 608	104 804	35.3
Jan 1991	327 533	64 642	19.7
Jan 1992	262 757	57 239	21.8
Jan 1993	247 793	76 436	30.8
Jan 1994	300 387	70 802	23.6

Source: Laura Ashley Annual Reports.

company as its Deputy Chairman in 1990, but as tax adviser to the Ashley family had been associated with the business for many years. It was Blakeway Webb, therefore, who announced Laura Ashley's half year results in September 1993. Sales were up by 23 per cent, but pre-tax profits to 31 July were a mere £1.3 million, 30 per cent down on the previous year. The US operations remained a bleeding sore, turning in half year operating losses of £3.3 million. Maxmin was nevertheless determined to continue with the planned investment, meaning that profits for the full year were unlikely to be significantly higher than the year before. He confessed that 'If I have been guilty of anything, then I have been guilty of being too optimistic about the rate of turnaround in North America.'[13] The news sent the share price tumbling, finishing the day 22p lower than its opening price of 87p. Only one institutional investor sold, but patience was wearing thin (refer to Figure 6.1.1 for stock market performance).

Resignations

On 12 April 1994, Blakeway Webb announced that Maxmin had 'resigned'. There had, apparently, been a fundamental disagreement between the Chief Executive and the company's long-serving Non-executive Directors over future levels of investment in the brand and human resources. The City was stunned. Two days later Maxmin made his farewell presentation,

Share price from 29/1/85 to 31/1/95, weekly.

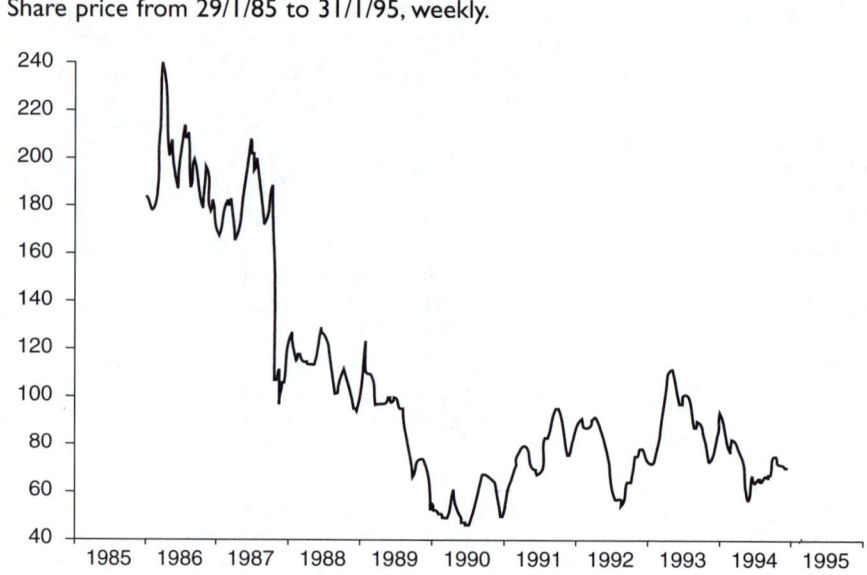

Dividend yield from 29/1/85 to 31/1/95, weekly.

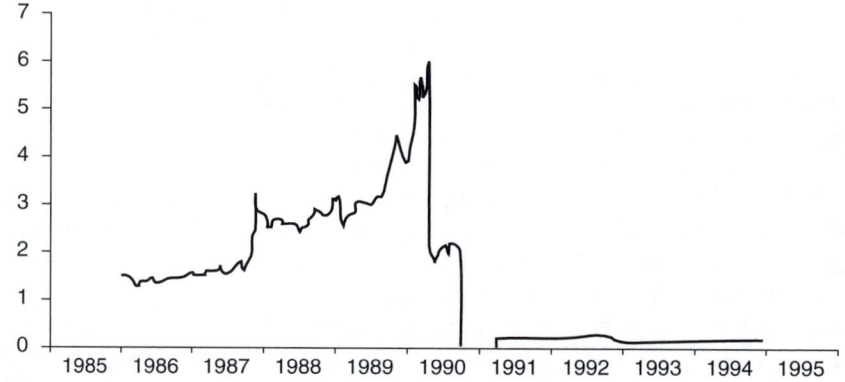

Figure 6.1.1   Laura Ashley Holdings Plc – share price and dividend yield,
1985–94.
Source: Datastream

when he announced a 67 per cent jump in pre-tax profits for the year
1993/94, from £1.8 million to £3 million on a 13 per cent rise in sales.
Analysts and the financial press were not happy (see Figure 6.1.2). They
wanted a fuller explanation of why the company had declined to issue a
regional breakdown of profits, but none was forthcoming.

Blakeway Webb, now elevated to the all-powerful position of Executive
Chairman, informed the City that a new Chief Executive for Laura Ashley
would not be sought until a thorough cost review had been completed,

Jim Maxmin, who was ousted last week as Chief Executive of Laura Ashley, once told me proudly how he had a special 'Bullshit' rubber stamp. This he fearlessly applied to deserving documents. It always seemed an ironic prospect. The documents most crying out for treatment were Mr Maxmin's own mission statements, management slogans, marketing guff and other jargon. The Bullshit stamp could also be applied to virtually every optimistic prediction to come out of the retail group since its flotation in 1985. Time and again Laura Ashley has announced that the worst was over, only to stumble upon another skeleton in the wardrobe ... The pathetic fact remains that shareholders, who paid 135p a share when the group came to the stock market in a massively hyped flotation nine years ago, have been repeatedly disappointed. The shares ended last week at 75.5p.

The man now in charge as executive chairman is Hugh Blakeway Webb. A barrister and former partner with accounting firm Deloitte Haskins & Sells, Mr Blakeway Webb had advised the Ashley family on their tax affairs since 1982 and has been closely associated with the company's management ever since, joining the board in 1990. You might think this would be a very good reason for not putting him in sole charge now. But this is what has happened. Mr Blakeway Webb, who has never run even the smallest of whelk stalls, now heads a business with a turnover of £300m spread across 28 countries. And there are no immediate plans to appoint a new chief executive. It could only happen at Laura Ashley.

Figure 6.1.2
Source: Patrick Hosking, *Independent on Sunday*, 17 April 1994.

early in 1995. He would then have a better idea of the sort of person he wished to appoint. Denise Lincoln resigned soon afterwards, by 'mutual consent': her position was no longer felt to merit a seat on the board. Higginson's resignation followed within a month, together with a declaration from a fourth board member, Alphons Schouten, that he too intended to quit (see Table 6.1.7 for profiles of directors, 1993/94).

The company's future now lay in the hands of Blakeway Webb, Bernard Ashley and the company's other Non-executive Directors. A new Finance Director was quickly appointed, but Denise Lincoln would not be replaced. Her seat on the board was to be taken instead by retail executive Stephen Cotter. As the summer passed, it became clear that cost cutting was once more the order of the day at Laura Ashley. But as the half year end approached, investors were again told that they were unlikely to see a dividend. Profits would be depressed by a £1.2 million severance deal for the departing Maxmin, and a further £200 000 for the removal of Lincoln, together amounting to almost half of last year's pre-tax profits. A spokesman for the company said that 'investors would understand that this was just a one-off charge'.[14]

Table 6.1.7   Laura Ashley – Board of Directors, January 1993

Sir Bernard Ashley, Non-Executive Chairman: Co-founder of the business and husband of the late Laura Ashley. Worked in the City of London from 1947 to 1953, then became the driving force behind the business. Joint Managing Director with John James from 1976 to 1988. (*Stood down as Non-Executive Chairman in favour of Hugh Blakeway Webb, 24 May 1993. He remains a Non-executive Director.*)

Lord Hooson, Non-executive Director: Lord Hooson is a barrister, called to the Bar in 1949, becoming Queen's Counsel in 1960. He was Liberal MP for Montgomery for 17 years before being created a life peer in 1979. He was first appointed to the board as a Non-executive Director in 1985.

Hugh Blakeway Webb, Non-executive Deputy Chairman: Hugh Blakeway Webb is a barrister, called to the Bar in 1975. In 1978 he joined accountants Deloitte Haskins & Sells (later to become Coopers & Lybrand Deloitte), becoming a partner in 1983. He resigned his partnership prior to his appointment to the board in January 1991. He has had a long association with the firm since he first advised the Ashley family on their tax affairs in 1982. (*Succeeded Bernard Ashley as Executive Chairman in May 1993. Became Executive Chairman of the company on the resignation of Jim Maxmin on 12 April 1994.*)

Takuya Okada, Non-executive Director: Representative of Aeon Group and Chairman of Jusco Co. Limited. He is Vice Chairman of the Tokyo Chamber of Commerce & Industry, Special Adviser to the Chairman of the Japanese Chamber of Commerce & Industry, and Vice President of Japan Fashion Association. He became an Honorary Commander of the Order of the British Empire in 1989, and joined the board of Laura Ashley on 5 December 1990.

Motoya Okada, Alternate (Non-executive) Director: Appointed 6 August 1992, as alternate director to Takuya Okada.

Dr Jim Maxmin, Chief Executive: Appointment announced 17 July 1991, moved into post on 16 September 1991. Previously worked for Unilever before becoming Chief Executive of Volvo Concessionaires in the UK, then Director of World Retailing at Thorn EMI. (*Resigned 12 April 1994.*)

Andrew Higginson, Finance Director: Appointed 10 May 1990, formerly Finance Director of Guinness Brewing International. (*Resigned to join Burton Group Plc, 4 July 1994.*)

Denise Lincoln, Human Resources Director: Appointed 3 February 1992, formerly Group Management Development Director at Grand Metropolitan, with experience of working in the US. (*Resigned 7 June 1994.*)

Alphons Schouten, Chief Executive – Continental European Retail: Joined Laura Ashley in 1981 from the Bluebell Corporation, where he was a marketing director. (*Board notified of his imminent resignation 5 July 1994; resigned 1 February 1995.*)

References

1 Bidlake, S. (1991). *Marketing*, 11 April, 10.
2 *Financial Times* (1991). 31 July, 11.
3 Cowe, R. (1991). *Guardian*, 25 April, 15.
4 Mulvagh, J. (1993). *Financial Times*, 3 April, 12.
5 Bevan, J. (1991). *Sunday Telegraph*, 21 July, 46.
6 Bain, S. (1993). Life begins at 40 for Laura Ashley. *Marketing*, 13 May, 18–21.
7 Laura Ashley Holdings Plc (1992). Annual Report, 24.
8 Bowditch, G. (1991). *The Times*, 4 October.
9 Reuter Textline, Regulatory News Service, 27 January 1992.
10 *Ibid.*
11 *Computing* (1992). 22 October, 3.
12 *Financial Times* (1992). 23 January, 44.
13 Ridley, K. (1993). Reuter News Service, 23 September.
14 Gilchrist, S. (1994). *The Times*, 8 June.

Case 6.2   Creating success through relationship marketing at British Airways

This case was prepared by Professor Adrian Payne, Moira Clarke, Andy Coaton and Marcus Hickman as a basis for class discussion rather than to illustrate effective or ineffective handling of an administrative situation.

We would like to thank the following members of the BA staff who allowed us to interview them in preparing this case: Iain Webster, Relationship Marketing; Richard Lucenti, Head of Field Sales UK; Bernard Harrop, BA National Sales Manager; Christopher Allen, Head of Competition and Industry Affairs; Deborah Kotsirea.

© *1996 Cranfield School of Management*

Over the three years of recession from 1990 to 1993, the airline industry collectively lost £10 billion. Even today, many European airlines still receive state subsidies and some US carriers rely on Chapter 11 bankruptcy laws to prevent them from collapsing. Over this time British Airways has returned profits, funded a global investment programme and has been widely accepted to be one of the world's most successful airlines. Yet 10 years earlier, in 1982, BA posted a loss of £545 million and was universally known for its poor service. At that time it was common for passengers to quip, 'Get me there anyhow, but not BA ...'. What has been the transformation and how has it come about? More importantly, how has British Airways maintained the momentum of success?

History

British Airways (BA) can be traced back to the first international scheduled air service run by Aircraft Transport and Travel Limited from August 1919. After a series of mergers in the British airline industry, resulting in the formation of Imperial Airlines and British Airways Limited, these two were joined to establish the nationalized British Overseas Airways Corporation (BOAC) in 1939.

After World War II, the Atlee government created British European Airways (BEA) to take over BOAC's European and domestic routes. Together with Britain's great names in airline manufacture such as Vickers, Hawker Siddeley and Rolls Royce, both airlines made a major contribution to world aviation. BEA effectively created European air transport, while BOAC operated the first international jet aircraft passenger services in the 1950s.

The staff of both airlines were recruited from the military. Fresh from their victory over the German Luftwaffe, employees wore uniforms with their rank clearly displayed and were addressed by title. Many job titles ended in 'officer' and at Speedbird House, BOAC's Headquarters, a Senior Managers' Mess operated for many years.

Like many airlines around the world, BOAC and BEA were proud national flag carriers for the British Government and were heavily regulated and heavily subsidized. The challenges were operational: creating routes, getting aircraft into the air and down again on time and in safety.

Market forces and customer service were alien to the culture of most airlines and a cartel existed where all aspects of air travel, including fares and routes, were negotiated between national government agencies.

Selling and distribution were handled by independent travel agents and, traditionally, agents were regarded as commercial partners, reflecting the power of travel agents to recommend airlines. Carriers spent huge sums on building relationships by various means:

- generous ticket price commissions of 9 per cent and more
- tailored incentive commissions with target for class of seats related to prizes, e.g. free tickets and gift vouchers
- training programmes in the airline 'product', use of technology such as for booking systems, and familiarization trips for new routes or revised services.

1970s

In 1971, the merger of BEA and BOAC created British Airways, although each company remained autonomous, with its own Board and Chief Executive. This reflected and reinforced the cultural differences between the short-haul European route served by BEA and the high flying and international BOAC, where air hostesses were regarded as glamorous and sophisticated compared to their BEA counterparts. Although in 1974 a consolidated financial report was issued, the cultural differences remained throughout the 1970s.

The operational style of BA did not prevent the company from making a profit in every year bar one in the 1970s. The management were very proud that as a civil service agency they were providing such a valuable service and not making any demands on Her Majesty's Treasury.

However, as the 1980s approached, it was becoming clear that the failure to integrate BEA and BOAC properly meant that economies of scale were not being achieved and divisional loyalties were preventing the creation of a common focus.

Soon after, a radical new Conservative government under Margaret Thatcher was elected and in 1979 plans were announced to privatize BA. However, these were almost immediately postponed due to the second oil crisis and a severe recession in the early 1980s. Rising fuel prices, combined with high staff costs and a diverse and ageing fleet (fuel-inefficient Tridents, DC10s and 707s), resulted in losses of over £100 million in 1981 and a stock market flotation was not viable. Margaret Thatcher, however, remained determined and, to ensure that BA was 'put in shape' for privatization, appointed Sir John King, later Lord King, as Chairman in 1981.

Privatization plan: The cost-cutting phase

Lord King was the Prime Minister's kind of businessman. A self-made man from outside the airline industry, he had amassed a personal fortune from Ferrybridge Industries, a ball-bearing company that he founded in 1945. He was later Chairman of the UK giant engineering concern Babcock International, and was knighted in 1979.

King's first priority was to stem BA's losses. The total staff of 52 000 was cut by 10 000 in the first nine months of his appointment, while another 7000 went in 1982. Although staff were offered generous severance payments, tough decisions were taken. At a reception in the US, King was reportedly approached by BA's US property manager, keen to impress the new Chairman. King asked how many staff she needed to manage BA's US property, to which she replied 11. Within a couple of months, she managed the property portfolio by herself. In total, over £150 million was saved through a pay freeze, the closure of six routes, eight on-line stations and two engineering bases. Cargo-only services ceased and parts of the ageing fleet were sold.

BA's management wanted a quick and decisive period of cost cutting and, in the 1982 financial year, posted a massive headline loss of £545 million. Much of this was due to 'creative' accounting, pulling as many costs forward as possible. The effect was to create a 'clean slate' and to add to the sense of crisis among the staff.

Corporate identity and advertising

A new livery was designed to support the claim of being the World's Favourite Airline. A BA official said, 'People have to see a difference as well as experience it if their perception is to truly change'.[1]

Prior to 1983, BA's advertising had involved tactical or promotional campaigns emphasizing product features such as seating, entertainment or scheduling. Price was a key focus and worldwide advertising was controlled by local country managers who were free to create their own copy and message, ensuring adaptation to local conditions but creating variety in BA's image around the globe. The total worldwide advertising budget was £19 million in 1982–83.

Saatchi & Saatchi
Saatchi & Saatchi, the rising star of British advertising agencies which espoused the concept of global brands, won the BA account in 1982 with a

brief to develop a worldwide campaign to improve BA's image. A Central Policy Unit was set up to plan and coordinate the account worldwide and overseas offices of Saatchi & Saatchi had little autonomy from London. The campaign attempted to position BA as the world's leading and most successful airline. 'It had to be simple and single minded, dramatic and break new ground, instantly understood throughout the world, visual rather than verbal, long lasting, likeable, and confident.'

The world's favourite airline
Saatchi created a series of 11 television commercials with the theme the 'World's Favourite Airline'. The lead commercial, 'Manhattan Landing', was broadcast worldwide and lasted for 90 seconds with only 35 words. It depicted the island of Manhattan rotating slowly across the Atlantic to London with the simple statement, 'Every year we fly more people across the Atlantic than the entire population of Manhattan.'

Saatchi's brief was to communicate the message:

British Airways every year carries twenty five million people, of all nationalities, across the world's most extensive international route network and makes every one of them feel good. In other words, despite the airline's great size (traditionally associated with 'cold' or impersonal qualities in advertising lore), British Airways' quality of service is such that it communicates a feeling of warmth and friendliness, and it makes the business of flying twenty five million people not just professionally efficient, but also a lot of fun. In short, it brings a smile to everyone's face.[2]

The new campaign was extensively communicated to BA's country managers. But for some of them there was a concern that the World's Favourite Airline slogan lacked credibility. Also, the tactical budget for which country managers were responsible was to be reduced by nearly 50 per cent. This caused country managers to protest and they succeeded in reinstating some of the tactical budget and in doing so raised BA's total advertising spend to £31 million for the year.

Delivering the world's favourite airline ...

Promoting BA as the world's favourite airline is one thing, delivering was another. King had cut costs, but in an attempt to improve customer service he appointed Sir Colin Marshall as Chief Executive in February 1983. Marshall had strong marketing skills with a background in consumer markets with car rental companies Avis and Hertz, as well as a brief period as CEO of retailer Sears Holdings.

Marshall quickly identified with BA's corporate goal, 'To be the best and most successful airline in the world'. He realized that to achieve this required BA to hold a decisive competitive advantage and Marshall chose to focus on customer service:

> It was ... obvious to me that the organisation was extremely introverted, had really no grasp of what the market place wanted, what the customer wanted ... I believed the most critical thing was for us to address the issue of customer service. In so doing, we recognised the need to create some motivational vehicle with the employees so that we had a better prospect of raising their morale and, in turn, seeing better customer service flow.[3]

Elaborating on this issue 10 years later, Marshall argued:

> In a mass market, the primary influence is reckoned to be price, whereas in a segmented marketplace there are other influential factors, including the quality of service and the caring and comfort. A lot of people maintain these days that price is the be-all and end-all. I don't buy that. A lot of customers are interested in price, of course, but what they are really concerned about is value. They will pay for better service.[4]

Marshall defined the British Airways' offer as, 'we start with the belief that in a service business, what you are selling is an intangible feeling. A passenger is left with an empty ticket stub, and he looks back at what it costs and reflects on the experience'.[5] Success was therefore believed to be critically linked with motivating and giving pride to BA's employees.

'Humanizing' British Airways

Marshall began by insisting that he be called Colin Marshall and not the traditional 'CX' used for the chief executive at BA. Without direct instruction, titles began to be replaced by names. Large committees were axed and ad hoc groups were set up that cut across functions. The hierarchy of the organization was reduced so that customer interface staff had discretionary power. Younger people were promoted and dynamic outsiders were brought into senior roles.

Market research in the mid-1980s revealed alarming gaps between customers' expectations of BA and those of the staff. BA's employees had always concentrated on functional aspects, such as punctuality, speed of answering phones, check-in efficiency, etc. for which strict standards were laid down in detailed manuals. However, the research revealed that these standards only met customers' minimum expectations. What they were really hoping for were behavioural aspects, such as spontaneity, warmth and concern, friendliness and attention to individual needs. What created real positive satisfaction and customer loyalty was the personal experience that customers had with airline staff. These findings:

set the direction for the first of our staff communications and motivation programmes – Putting People First. The aim was to focus attention on customers, the marketplace and later the competition. It started our long term strategy to improve service ... something that would become part of the culture of the organisation.[6]

BA was aware of training programmes at other companies that had failed. Commenting on these failures, the manager responsible, Mike Bruce, said: '[There has been] a failure of programmes which have set out to influence motivation and behaviour. They ignored the issue of values.'[6]

Cultural change – 'putting people first'

BA embarked on a major programme of cultural change that resulted in all staff attending the 'Putting People First' (PPF) training initiative, designed by Time Manager International, who had previously worked with Scandinavian Airlines Systems (SAS). PPF initially focused on customer-interfacing staff, with key themes such as 'Delivering the Promise' and the whole organization working together to achieve customer expectations. Central to the programme was the concept of 'moments of truth' in customer service. Staff were encouraged to own customer problems and be attentive, listening to and understanding their needs.

BA employees were empowered to use their discretion in helping customers. Personal assessment exercises, stress, body language and the psychological basis of interactions with customers were covered. Staff were also trained in the behaviour patterns of teams, working to a common goal, coping with change, role playing common problems and the idea that simple changes in conveying information make a difference in the way it is received. Marshall made a point of concluding two out of every three training days by presenting his vision for the airline.

PPF generated a huge amount of enthusiasm, but unfortunately much of this was dissipated when participants returned to their jobs. This was related to the contradiction that although staff were being called upon to care for customers, they weren't sure that BA's managers cared for them.

To help managers maintain motivation a complementary programme was devised – 'Managing People First' (MPF). A five-day managerial programme, this was created with the themes of urgency, vision, motivation, trust and responsibility. A new management appraisal system was introduced, with measurement against quarterly 'key result areas', set against management behaviours such as evaluating managers by their treatment of subordinates. Mike Bruce commented on the purpose of the exercise:

In BA we believe that if managers make a real effort to keep staff informed of what is happening in the company, treat them as individuals, and listen to their cares and concerns then there is a good chance that this is how they will treat our customers.[7]

'Customer first' teams

'Customer first' teams were developed as a way of putting the new knowledge and enthusiasm into action, with staff volunteers seeking ways of improving customer service at all points of contact. One such improvement involved an enhanced service for children. Children flying alone were traditionally known as 'Unaccompanied Minors' or 'Umms'. Waiting in roped off parts of airport lounges, they often became bored and wandered off. The customer service team developed a new scheme where unaccompanied minors were given their own lounge with toys, snacks and TV which was staffed by play attendants. The problem then became how to keep out children who were flying with other airlines.

The global aspirational advertising campaign the 'World's Favourite Airline' contributed to staff morale, as did the corporate identity and livery change. A weekly newsletter for staff was started, including bulletins about 'customer first' initiatives.

It became impossible for anyone to be unaware of the new creed of customer service at BA, such was the top-down commitment. An on-going series of surveys of BA staff were initiated and they quickly indicated a positive culture change – 'we can get things done'. Other key attitude changes included perceiving BA as a service provider rather than a transport business and a focus on interdepartmental cooperation and increased self-confidence.

However, the bureaucratic and hierarchical structure at BA was clearly incompatible with the new customer-focused strategy. In 1983 the three functional divisions were replaced by eight geographic business units plus cargo, air charter and packaged tours. All the business units reported to Marshall through Jim Harris, the Marketing Director. Jim Harris also controlled central marketing staffs – strategic planning, advertising, market analysis and market research. Thus, marketing and the customer orientation were at the heart of the business.

More crisis?

The heavy financial losses and huge job cuts in 1982 had provided impetus to change among BA's staff and progress was being made. However, 1984 heralded another threat which had to be negotiated if BA was to deliver its claim to be the World's Favourite Airline.

The Civil Aviation Authority (CAA) issued the CAP 5000 'Airline Competition Policy' in 1984 which, in an attempt to improve competition, recommended that the UK government should award smaller airlines, such as British Caledonian, some of BA's routes and landing rights at Manchester and Gatwick airports.

The CAA in the UK has the task of regulating the safety and economics of the civil aviation industry. 'We have a fairly formal relationship with the CAA,' says Christopher Allen, BA's Head of Competition and Industry Affairs. 'We have no reason to believe that the CAA acts in a biased way with respect to other airlines.'[8] However, the CAA's stated policy is for the UK to have a strong multi-airline system, which necessarily means it supports small carriers in their attempts to compete with BA, which holds the dominant market share.

British Caledonian, BA's main UK competitor, had argued to the CAA that a privatized BA would be too powerful a force. In response to CAP 5000, BA mounted a major defence campaign, lobbying government and MPs. Staff sent a petition with over 26 000 names to the Government, MPs and ministers, while bus-loads of employees lobbied Parliament. Regardless of whether this affected the Government, it did galvanize the work force against an external threat. Probably a greater influence was that CAP 5000 conflicted with the Government's desire to privatize BA and the eventual legislation rejected the CAA's recommendation of cutting BA's routes and landing rights.

Freddie Laker and unfair competition

Further competitive trouble was around the corner when in 1985 Freddie Laker, CEO of Laker Airways, alleged that BA conspired with other airlines to create unfair competition to Skytrain, his innovative, low cost transatlantic service. BA settled out of court for an unknown sum.

There was also controversy over British Midland's sister airline, LoganAir. The Chairman, Sir Michael Bishop, claimed that BA Marketing Director, Liam Strong, had approached him about a joint venture to serve the Scottish highlands and islands. Talks went ahead but at the last moment BA pulled out of the deal. Six months later BA established new flights from Manchester and Edinburgh, which were in competition with LoganAir's most profitable routes.

Recruitment problems

The early 1980s had seen major staff reductions but by the end of the 1980s BA was experiencing staff shortages in important skill areas and was having difficulty in attracting the right candidates. BA responded by applying customer service principles to recruitment. They established a Recruitment Marketing Team at Heathrow. Training programmes for managers were developed so that managers would understand recruitment markets, be aware of the importance of equal opportunities and be equipped with selection and assessment skills.

By this time, BA had 2000 pilots who had an average age of 46, while none were under 34. Over 1200 of these would reach the retirement age of 55 by the year 2000 and all would be retired by 2007. BA was clearly going to suffer a shortage without an influx of young pilots. The age profile of

BA's pilots also created other problems. Long-haul flights often required cabin crew to stay away from home for up to a week and so older pilots with families preferred short-haul flights, resulting in shortages on long-haul routes. In November 1985, the company resumed recruitment of pilots and included women for the first time. Reflecting the desire to changed its culture from the 'military' past, BA also recruited civil pilots for the first time.

All of this reflected the seriousness with which BA was attempting to change its culture, something that is always difficult to achieve in large organizations. The success of Putting People First was followed in 1985 by another major company-wide communications project called 'A Day in the Life'. The aim was to increase awareness of the contribution made by each of the airline's major departments and their interdependence. Addressing the issue that BA was still very compartmentalized, the programme had the goal of preventing staffing departments blaming each other and making others scapegoats, instead of taking responsibility.

Bombing Libya

In August 1986, amidst all the internal changes in BA, there was a dramatic external event: the bombing of Libya by American jets, taking off from the UK. Fear of reprisals caused many Americans to cancel flights to Europe. Marshall said at the time, 'The effect ... was extremely serious. At one time, cancellations were outstripping bookings by 27,000 a week, just when our reservations should have been climbing by about 50,000 a week.'

BA's management turned the situation to its advantage by using the incident to add to the sense of crisis that was reducing internal resistance to cultural change. Further, it became an opportunity to present BA as a major global brand to the world, and especially the huge US market. The promotional response was the 'Go for it, America' promotion, a nationwide sweepstake giving away 6000 free seats and a range of valuable prizes, including free use of Concorde for a day with 99 friends, a week at a luxury hotel, rental of a luxury car and a five-year lease on a Chelsea townhouse.

The campaign cost US$9.2 million, but the results redressed the booking shortfall and by the end of the summer bookings were 5 per cent better than in 1985. In the seven weeks before the campaign 8000 bookings were made each week, while after this weekly bookings rose to 17 700. In the long term, BA obtained an additional 1 per cent share of the Atlantic and 3 per cent of worldwide markets.

The crisis at BA which was inspired by the Tripoli bombing in 1986 caused a halt in recruitment in order to save £35 million. However, in December a new programme to train 100 new pilots each year was announced. The package offered was attractive: a Captain earned £43 000 per annum plus overseas allowances which could add another 20 per cent.

All BA staff and immediate family received air flights at 10 per cent of the ticket price. BA received 10 000 applications for the 100 vacancies.

Privatization goes ahead

Once BA had survived the 1986 crisis, King's objective of privatization was becoming increasingly viable. However, there had been occasions when BA was able to use the fear of a failed privatization to influence the Government. For example, South Africa was subject to general trade sanctions, but after a meeting between King and Margaret Thatcher it was reported that 'Lord King, Chairman, is understood to have told Mrs Thatcher that any restrictions [on flights to South Africa] would have a devastating impact on the State airline as well as opening the way for others not supporting an international boycott to pick up business.'[9] It has been reported that this meeting reinforced the Prime Minister's position not to introduce restrictions on South African air travel as part of the general trade sanctions.

BA was finally privatized in February 1987, after British Telecommunications and British Gas. The sell-off coincided with the stock market boom and was 11 times oversubscribed. 95 per cent of BA employees bought shares, creating a stronger commitment to the company. Privatization marked a major milestone in BA's history and turnaround. There was still much to do, with the constraints and protection of government removed, but BA was increasingly being seen as a source of pride. In June that year, Marshall received a knighthood.

Later in 1987 BA launched a drive for cabin crew. A job advertisement for a recruitment officer illustrates BA's approach to staff intake and stated:

> BA have a reputation for excellence that is founded on our concern for people. Our customers have always known this – that is why we are the world's favourite airline.

It went on to state that BA received 2000 applications a week and so 'the care with which they are handled and the quality of our communications with them represents a considerable challenge'.[10]

Investment for growth
BA had successfully negotiated its way into the private sector and this resulted in increased freedom to make, and raise finance for, a series of investments:

● In 1987 BA took a 26 per cent stake in Galileo, an advanced computer reservation system, which it has subsequently purchased completely.

- BA formed a short-lived marketing partnership with United Airlines in 1987 which lasted until BA invested US$300 million in USAir in 1992. This was the first of a three-part investment to give BA a 44 per cent share for US$750 million even though to comply with US regulation, BA is only able to have 21 per cent of voting rights. Although USAir was the fourth largest US carrier, with a hub in Pittsburgh serving secondary airports, it has the highest unit costs of the US carriers and was losing money.
- In 1988, BA acquired British Caledonian after a fierce takeover battle with SAS for control of the British carrier. The BA Charter airline was renamed 'Caledonian' to preserve its heritage, but was divested at the end of 1994.
- BA's attempt to acquire 20 per cent of Sabena World Airlines of Belgium in 1989 with KLM was blocked by the European Commission.

Sustaining improved customer service

After privatization, the company-wide training events continued with, for example, the 'To be the Best' event. This was designed to increase awareness of the highly competitive nature of the airline business, with interactive workshops and presentations to identify the strengths and weaknesses of BA.

This was followed by 'Leading the Service Business', a week-long top management residential course covering strategic issues such as customer service, technology, management style, competitive threats and the post-privatization political and economic environment.

Putting People First had been a major initiative, but the benefits were difficult to sustain. The second wave of company-wide training, planned to involve all BA staff over a three-year period, was 'Winning for Customers' and 'Managing Winners'. The key issue for the programme was customer retention. Learning from the 'Putting People First' programme, these were run simultaneously for staff and managers. The commitment of senior management continued, with Marshall and other directors attending every session.

A key goal for BA had always been to emulate the success of its Scandinavian competitor, SAS, and win business travellers. In 1986 business class accounted for 35 per cent of BA's scheduled tickets. Marketing Director, Michael Batt, who had a background at Mars, had learnt the value of branding and assembled a team with FMCG experience. In the business market he attempted to move the basis of competition away from the price wars of the 1970s, saying, 'it's now a quality elastic rather than a price elastic market'.

Developing the BA brand

In November 1987, Club World and Club Europe were launched as international brands, at a cost of £20 million. Key features of the brands were increased leg room, reclining seats, better food and service. These changes brought BA up to the standards of the competition, but the branding idea was quite original. The £11 million advertising campaign followed the Saatchi & Saatchi model of being aspirational. Michael Batt said of the campaign:

> Most other airlines' advertising tends to be functionally based. When we launched Club World some people couldn't believe we weren't going to mention our new seat in the advertising. Instead, our advertisements are all about business people succeeding.[11]

> What we have done is not remarkable or new, though it happens to be in this particular industry. It has given us increased organisational focus and far greater understanding of our product.[12]

Club World was an immediate success and grew rapidly, opening up markets such as Singapore and Australia. However, competitive response was also quick, with 22 airlines relaunching their business classes between 1988 and 1990. Additional information about the offer of competitors is shown in Table 6.2.1. For information on how BA's services have developed, see Tables 6.2.2 and 6.2.3 and Figure 6.2.1.

Customer perceptions
BA needed to know whether the changes in the company were having an effect where it mattered – with the customers. The perceptions of customers were so important that BA set up a group, Market Place Performance, with its own director to monitor the airline's performance with customers. This group has interviewed over 150 000 passengers each year. Commenting on the importance of customer research, Michael Batt stated:

> To keep us on track we developed from the market research a very detailed set of performance measures which tracked every month how our performance is seen by our customers. Feedback from these measures is communicated to all parts of the company.[13]

Videopoint interactive video booths at Heathrow and Gatwick give BA's customers the opportunity to make immediate comments about the airline. This is used as feedback for staff as to how customers are feeling and is also inputted into subsequent corporate training programmes. The company also runs 'Customer Forums' which allow frequent flyer club members to meet BA's top 50 managers.

Table 6.2.1   Long-haul business class product features of airlines serving the UK, 1993

Airline	Business class name	Special features
Air Canada	Executive Class	Destination videos, CBC News broadcasts on long-haul flights, late check-in, 45-inch seat pitch
American Airlines	Business Class	Executive meal option, 40-inch seat pitch
British Airways	Club World	Arrivals lounge at Heathrow Terminal 4, with showers, breakfast, suit pressing facilities, etc., 'Fastrack' check-in, individual video system
Canadian Airlines International	Canadian Business	45-inch seat pitch, late check-in, 'Fastrack' check-in facilities
Cathay Pacific	Marco Polo	40-inch seat pitch, free limousine service within 60-mile radius of major airports
Continental	Business First	55-inch seat pitch, free limousine service within 50-mile radius of major airports
Japan Air Lines	Executive Class	Free limousine service within 40-mile radius of major airports, 40-inch seat pitch
Malaysia Airlines	Golden Club	Individual video system, 41-inch seat pitch, free limousine service within 40-mile radius of Heathrow airport
Philippine Airlines	Mabuhay	Free first-class rail tickets to Gatwick, 37 to 39-inch seat pitch
Qantas	Business	40-inch seat pitch, free limousine service within 40-mile radius of Heathrow airport
Singapore Airlines	Raffles Class	42-inch seat pitch, free limousine service within 40-mile radius of airport
Trans World Airlines	Ambassador Frequent flyer programme	47-inch seat pitch
Virgin Atlantic	Upper Class	55-inch seat pitch, free first-class rail tickets to Gatwick, in-flight beauty therapist on selected flights

Source: Air & Business Travel News, 6–20 September 1993/Keynote.

In addition, many features are offered as standard by almost all carriers. These include: seat selection at check-in, a late check-in facility, dedicated cabin crews, free newspapers, free drinks, hot towels, and separate 'executive' lounge facilities.

The Gulf War and the 'World's Biggest Offer'

In 1990 the world was entering a major recession that was reducing discretionary consumer spending, especially in the crucial US market. Then in August 1990 Iraq invaded Kuwait, which in February 1991 prompted a

Table 6.2.2 Milestones in BA service and brand development

1986	Services are developed as brands.
1987	Club World and Club Europe launched in November with £8 million ad campaign, total cost more like £20 million. BA challenged Mills of Mills, Smith & Partners ad agency to find a way to fill 9 million empty seats. In 1987, BA signed up as the principal airline supplier and partner and owned 51 per cent of Air Miles UK, launched 1988.
1989	BA's SuperShuttle service relaunched August, cost £11 million. Super Shuttle Executive and Super Savers. Aim of getting 2 per cent more business class reservations. BA's First Class revamped in March.
1990	April Problems of providing enough business class seats, which provide 50 per cent of revenue. Skyflyers launched to cater for unaccompanied minors.
1991	January World Traveller and Euro Traveller new brands introduced. Michel Roux taken on as consultant to kitchens whereas Virgin has Raymond Blanc. BA initiated 'Sprint' package of cuts, including reduced service levels in first class. 'No one used them and it saved us thousands', BA spokesman, December 1991. Air Miles Latitudes scheme, which gives frequent travellers free Air Miles, is launched in March. Air Miles US launched in July. BA not one of the airlines supplying seats to them. June £10 million package of customer service enhancements, called Mission Atlantic, to maintain competitive edge in US. September Club Europe gets £7 million new lounge at Heathrow and new seat selection and improved catering and quick car check-in.
1992	February Executive club frequent traveller programme in USA, which includes a mileage programme towards free flights. By March 4 million consumers were joining Air Miles scheme. October Cargo business rebranded British Airways World Cargo.
1993	January BA offers first-class passengers navy blue pyjamas, duvet and a hot drink on overnight flights. Fast Track launched in April. Dedicated, fast, no queue service from check-in to the departure lounge. Airline launches 'Well Being in the Air', a £100 million programme of enhancements to the airline's premium long-haul brands. Some criticism of Air Miles that it encourages employees to build up longer flights for their own holidays. Planning to increase customer base to 5 million. Can collect Air Miles from between 150 and 200 companies. BA multi-million project to make seats wider for future taller and broader passengers.

Table 6.2.2 *(continued)*

1993	BA offers chance to keep fit, with stretching and relaxation exercises and healthier meals. March BA launched a £8 million campaign to promote Club World. May Chase Manhattan and BA are launching a new credit card which offers many perks in the growing trend for credit card issues to link up with frequent flyer schemes. June Cathay Pacific to link with BA in frequent flyer programme. September BA and Sainsburys linking up again for Buy and Fly discount ticket offer with a promotion that will be the biggest 'freebie' since the Hoover free flight fiasco. September BA has £15 million promotion to double the benefits of frequent flyer for winter, passengers getting double Air Miles for full price tickets purchased. September In-flight telephone calls. October BA launches 'The British Airways Dream Ticket' multi-million pound promotion for business and other premium-class tickets. Dream Ticket is the largest BA give-away since the World's Biggest Offer in 1991.
1994	Club Europe relaunched, in response to the Eurotunnel. January Campaign to celebrate success of the New Club World arrivals lounge. May BA owns Air Miles 100 per cent now. May To build Middle East market, image campaign including Arab in-flight entertainment and food and Arab crew. May New first class, with movable seats and meals anytime you want and new £80 million interactive stereo system and video. June Campaign by Booth Lockett Makin to reposition Air Miles as not just for business customer but also for everybody person. July BA passengers able to play Nintendo video games on long-haul flights. August BA has cut the cost of fares in October and November price war to celebrate its 75th anniversary, sparking a transatlantic price war. September BA £25 million campaign to encourage business travellers in Europe to switch from economy to business class as recession ends.
1995	January World Offer gives 80 destinations for cut price. BA's £85 million advertising account. February BA rated 12 out of a survey of 46 carriers. Zaget surveyed 9394 frequent flyers and 667 travel industry professionals. February Travelex to provide exclusive worldwide currency service for Executive Club members. BA to test headphones which emit silence. Trial shop from an electronic catalogue, make flight and hotel reservations, book car hire, retrieve financial information via BT Racal/Honeywell on BA. In-flight entertainment incorporates credit card swipe reader. In-flight betting introduced.

Table 6.2.3 Passenger preferences, March 1994 (%)

	All	Scheduled	Charter
Short-haul flying			
Good safety record	61	60	64
Good reputation of airline	46	50	55
Cheapest fares	45	57	49
Convenient scheduling/flight times	45	56	58
Comfortable seats	38	43	40
Good food	26	24	27
Previous experience of airline	20	28	30
Special offer/promotion	19	24	22
Good in-flight entertainment	16	18	17
A British airline	16	15	14
First/club class facilities	4	3	4
Frequent flyer scheme	3	4	5
Don't travel by air	7	-	-
None of these	1	*	*
Don't know	2	*	*
Long-haul flying			
Good safety record	57	57	58
Good food	49	54	50
Good in-flight entertainment	49	58	54
Good reputation of airline	42	48	51
Cheapest fares	32	43	34
Convenient scheduling/flight times	32	41	44
Special offer/promotion	18	23	22
Previous experience of airline	14	25	19
A British airline	13	11	13
First/club class facilities	10	7	10
Frequent flyer scheme	4	6	4
Don't travel by air	8	-	*
None of these	1	*	*
Don't know	3	*	1

Source: BMRB/Mintel. * less than 1 per cent.

response from an international force led by the USA. There was widespread expectation that Iraq would fight back with terrorist attacks on the airlines and airports of the participating Western countries. On the first day of the Iraqi invasion, BA Flight 149 was detained during a scheduled stop in Kuwait and, although the passengers were eventually released, the Boeing 747 was destroyed. Air traffic across the Atlantic was decimated and the cost of jet fuel doubled, reducing BA's revenue and raising its costs at the same time.

British Airways' leading role in world air transport is reflected in its performance in the various award schemes for the airline industry. The following summary lists some of the main accolades collected by the company and its people in 1993.

- Business Traveller Airline of the Year Award – for the four consecutive years 1990 to 1993.
- Air Transport World magazine's 'Airline of the Year' award – for the third time.
- A MORI survey conducted by Travel Agents voted British Airways their favourite airline.
- Lord King was named Businessman of the Year for 1992.
- Influential UK magazine Campaign names British Airways 'Advertiser of the Year' for the 'Global campaign' and 'The World's Biggest Offer'.
- 'The World's Biggest Offer' campaign picked up seven awards, including a top award for Best International Campaign, in the annual awards presented by the Institute of Sales Promotion.
- Travel Trade Gazette presented British Airways with its 'European Airline of the Year' award.
- The World Travel Market awarded British Airways with its Global Award.
- British Airways was voted 'Best Holiday Airline in the UK' by readers of the Radio Times and viewers of ITV's 'Wish You Were Here' holiday programme.
- Lord King was presented with a special 'World Globe' award at the Executive Travel magazine Airline of the Year Awards for his outstanding contribution to international aviation during his Chairmanship of British Airways.
- Travel Weekly magazine voted British Airways 'Best European Airline'.
- British Airways Sales Teams received an award for 'The World's Most Direct Accessed Airline' in the Amadeus computer reservations system in 1992, after dominating several key markets in Europe.
- British Airways was named 'Best Transatlantic Airline of the Year' by Ashington-Pickett Airlines and Travel Report.
- British Airways won PR Week's Public Relations Award for the Best Employee Communications Programme.
- British Airways was awarded the Association of South African Travel Agents' 'Airline of the Year' award for the third year running.
- British Airways chefs received a Gold Medal at the prestigious International Exhibition of Culinary Art.
- Sir Colin Marshall received the Institute of Management Gold Medal.
- The ASTA-Smithsonian Environment Award was presented to British Airways by the American Society of Travel Agents for achievements in conservation and preservation.
- Lord King received a Gold Medal from the British Association of Aviation Consultants for outstanding services to British aviation.
- US magazine Financial World named British Airways 'Best European Airline of the Year'.

Figure 6.2.1 Awards won by British Airways.

To compound BA's problems, in March 1991 the USA and the UK signed a bilateral government agreement that allowed increased competition on the North American routes from United Airlines and American Airlines – two of the US's strongest competitors. A number of airlines responded with fare reductions and BA cut US–London fares by 33 per cent and American Airlines launched 'the largest fare promotion in its history'.

BA's management realized that the situation imposed by the Gulf War required a more creative solution. The response was 'The World's Biggest Offer', a promotion modelled on the 'Go for it, America' promotion in 1986. On 21 March 1991 BA announced that on 23 April every seat would be free – 50 000 seats. Those already holding seats on that day would fly free of charge. All the other seats would be raffled. Hotels, car rental, stores, restaurants, theatres were invited to tie in with their own promotions.

The displaced revenue bookings for 23 April amounted to £10 million and £6 million was spent communicating the offer, mostly via public relations methods. All the major US television networks gave it coverage and in the UK Channel 4 ran an hour-long documentary on the promotion which was syndicated around the world. Newspaper advertisements provided entry forms and promotional material was available at travel agents.

To obtain the greatest publicity, 'Up & Away Days' were organized at 62 airports, with reporters, photographers and broadcast teams from 60 countries being carried on flights to create feature coverage of the BA service.

The value of the resulting media coverage was estimated to be worth £60 million. BA believed that the promotion stimulated the whole travel and tourism industry. In March the air travel market was down 28 per cent on the previous March but by May the fall was only 5 per cent.

Sustaining success in the 1990s

BA had performed a remarkable turnaround during the 1980s. Its financial performance and customer image had been transformed. A full analysis is presented in Table 6.2.4 and Figure 6.2.2, including a breakdown of BA's turnover, profitability and other statistics from 1984 to 1995. However, Mike Batt, Director of Marketing, recognized that there remained a lot to do:

> There is a huge range of marketing issues which need to be tackled. In a service business like BA's the sheer size of the organisation makes change difficult to manage. If you want things to change you cannot succeed without the wholehearted support of the entire organisation around you. There has to be an integrator and, in British Airways, the integrating and co-ordination is managed by the marketing department.[14]

Table 6.2.4 British Airways' financial performance, 1984–95

		1984	1985	1986	1987	1988	1989	1990	1991	1992	1993	1994	1995
Group results post privatization													
Turnover	(£m)				3263	3756	4257	4838	4937	5224	5566	6602	7177
Change	(%)				3.6	15.1	13.3	13.6	2.0	5.8	6.5	18.6	8.7
Operating profit	(£m)				173	236	336	384	47	344	310	468	618
Change	(%)				−12.6	36.4	42.4	14.3	−87.8	631.9	−9.9	51.0	32.1
Profit before taxation	(£m)				162	228	268	345	1.30	434	185	280	327
Change	(%)				−16.9	40.7	17.5	28.7	−62.3	233.8	−57.4	51.4	16.8
Profit after taxation	(£m)				148	151	175	245	95	395	178	274	250
Change	(%)				−23.3	2.0	15.9	40.0	−61.2	315.8	−54.9	53.9	−8.8
Capital, reserves and convertible capital bonds	(£m)				605	633	749	1232	1275	1604	1534		
Change	(%)				26.0	4.6	18.3	64.5	3.7	25.5	−4.4		
Earnings per share:													
Basic	(p)				20.5	21.0	24.3	34.1	13.2	54.6	24.1	30.0	26.2
Change	(%)				−23.5	2.4	15.7	40.3	−61.3	313.6	−55.9	24.5	−12.7
Basic adjusted	(p)				20.5	21.0	24.3	34.1	13.2	35.3	24.1	30.0	29.3
Change	(%)				−23.5	2.4	15.7	40.3	−61.3	167.4	−31.7	24.5	−2.3
Fully diluted adjusted	(p)				−	−	23.3	31.2	13.4	31.0	22.3	27.6	24.5
Change	(%)				−	−	−	33.9	−57.1	131.3	−28.1	23.7	−12.7
Dividends per share	(p)				4.116	6.90	7.75	8.85	8.85	10.18	10.60	11.10	12.4
Change	(%)				n/m	67.6	12.3	14.2	0.0	15.0	4.1	4.7	11.7

Group turnover and operating profit by geographical area (£ million)

Turnover:

United Kingdom	238	255	285	331	378	453	539	576	536	623		
Continental Europe	783	897	979	1085	1231	1169	1286	1374	1528	1615	2734*	3015*
The Americas	683	876	1008	982	1175	1374	1619	1615	1645	1709	2029	2185
Africa	167	190	179	185	237	323	356					
Africa, Middle East and Indian subcontinent								590	665	757	900	953
Middle East, Far East and Australasia	600	687	660	662	735	938	1038					
Far East and Australasia								782	850	862	939	1024
	2471	2905	3111	3245	3756	4257	4838	4937	5224	5566		
Discontinued activities	43	38	38	18								
Total	2514	2943	3149	3263	3756	4257	4838	4937	5224	5566	6602	7177

Operating profit/(loss):

Europe	97	81	56	56	36	16	3	(10)	20	30	16	2
The Americas	88	131	85	65	131	181	249	123	1/9	87	140	245
Africa	23	27	14	20	37	49	52					
Africa, Middle East and Indian subcontinent								13	119	134	209	226
Middle East, Far East and Australasia	60	55	46	33	32	90	80					
Far East and Australasia								41	86	59	103	145
	268	294	201	174	236	336	384	167	344	310	468	618
Discontinued activities		(2)	(3)	(1)								
Total	268	292	198	173	236	336	384	167*	344	310	468	618

*UK and Europe combined.
Note: In 1991, exceptional charges of £120 million (comprising costs associated with employee reductions of £93 million and the write-down to estimated realizable value of Tristar aircraft surplus to requirements of £27 million) are not allocable by geographical region.

Table 6.2.4 British Airways' financial performance, 1984–95 *(continued)*

	1984	1985	1986	1987	1988	1989	1990	1991	1992	1993	1994	1995
Group turnover and operating expenditure (£ million)												
Turnover:												
Traffic revenue												
Scheduled services												
Passenger and excess baggage	1905	2254	2376	2481	2858	3445	3941	4057	4281	4533	5388	5838
Freight and mail	208	252	268	265	287	34.3	389	376	396	396	461	518
	2113	2506	2644	2746	3145	3788	4330	4433	4677	4929	5849	6356
Non-scheduled services	112	131	151	141	165	111	101	105	111	153	164	176
	2225	2637	2795	2887	3310	3899	4431	4538	4788	5082	6013	6532
Other revenue	246	268	316	358	446	358	407	399	436	484	589	645
Total continuing activities	2471	2905	3111	3245	3756	4257	4838	4937	5224	5566	6602	7177
Discontinued activities	43	38	38	18								
Total	2514	2943	3149	3263	3756	4257	4838	4937	5224	5566	6602	7177
Operating expenditure:												
Employee costs	521	579	638	722	851	976	1372	1544	1506	1551	1754	1898
Depreciation	112	140	153	185	214	305	292	296	295	312	447	458
Aircraft operating lease costs	4	12	61	86	113	124	179	284	236	224	140	132
Fuel and oil costs	509	610	570	371	394	417	544	598	521	560	666	622
Engineering and other aircraft costs	115	123	176	193	217	244	266	285	262	262	376	424
Landing fees and route charges	172	198	208	233	264	294	330	376	425	497	587	619
Handling charges, catering and other operating costs	195	259	298	313	338	427	564	564	608	714	840	936
Selling costs	238	304	323	350	412	498	569	566	652	732	886	1012

Accommodation, ground equipment												
and currency differences	242	269	349	418	479	507	338	377	375	404	458	458
Total airline	2108	2494	2776	2871	3282	3792	4454	4890	4580	5256	6134	6559
Non-airline	95	117	134	200	238	129						
Total continuing activities	2203	2611	2910	3071	3520	3921	4454	4890	4880	5256	6134	6559
Discontinued activities			41	19								
Total	2246	2651	2951	3090	3520	921	4454	4890	4880	5256	6134	6559

Operating statistics

Scheduled services:

Traffic and capacity:

Revenue passenger km (RPK) (m)	34 206	38 386	41 334	41 356	49 123	57 795	61 915	64 734	65 896	73 996	81 907	87 395
Available seat km (ASK) (m)	53 386	56 031	60 759	61 722	69 970	82 984	86 601	92 399	93 877	104 507	116 974	122 063
Passenger load factor (%)	64.1	68.5	68.0	67.0	70.2	69.6	71.5	70.1	70.2	70.8	70.0	716
Cargo tonne km (CTK) (m)	1122	1292	1356	1444	1793	2249	2400	1463	2510	2691	2991	3349
Total revenue tonne km (RTK) (m)	4244	4810	5155	5267	6345	7636	8290	8641	8778	9730	10 792	11 667
Total available tonne km (ATK) (m)	6699	7275	7956	8141	9427	11 404	12 035	12 929	13 379	14 695	16 240	17 201
Overall load factor (%)	63.4	66.1	64.8	64.7	67.3	67.0	68.9	66.8	65.6	66.2	66.5	678
Passengers carried (000)	14 206	15 951	17 016	17 276	20 169	22 578	23 671	24 243	23 788	25 905	28 656	30 552
Tonnes of cargo carried (000)	229	259	274	291	361	459	498	506	502	532	607	666
Frequently flyer RPKs as % of total RPKs												

Financial:

Passenger revenue per RPK (p)	5.57	5.87	5.75	6.00	5.82	5.96	6.37	6.27	6.50	6.13	6.36	6.36
Passenger revenue per ASK (p)	3.57	4.02	3.91	4.02	4.08	4.15	4.55	4.39	4.56	4.34		
Cargo revenue per CTK (p)	18.54	19.50	19.76	18.35	16.01	15.25	16.21	15.27	15.78	14.72	15.41	15.47

Table 6.2.4 British Airways' financial performance 1984–95 *(continued)*

	1984	1985	1986	1987	1988	1989	1990	1991	1992	1993	1994	1995
Average fuel price (US cents per US gallon)	n/a	n/a	n/a	58.36	63.78	60.22	69.72	89.72	70.94	69.32	63.64	59.79
Operations:												
Unduplicated route km (000)	516	521	555	555	692	677	685	665	584	5	643	743
Punctuality (% within 15 minutes)	84	55	82	81	80	72	72	73	79	81	85	84
Regularity	99.3	99.5	98.8	99.2	99.2	99.0	98.9	98.7	99.2	99.3	99.3	99.5
Total group operations												
Total revenue tonne km (RTK) (m)	4650	5267	5673	5784	6895	8002	8627	8979	9111	10 313	11 467	12 380
Total available tonne km (ATK) (m)	7194	7837	8601	8751	10 083	11 868	12 445	13 351	13 818	15 424	17 233	18 311
Passengers carried (000)	16 241	18 397	19 681	20 041	23 230	24 603	25 238	5587	25 422	28 100	32 749	35 643
Average number of employees	37 247	38 137	40 271	40 759	43 969	50 204	52 054	54 427	50 409	48 960	51 530	53 060
ATKs per employee (000)	193.1	205.5	213.6	214.7	229.3	236.4	239.1	245.3	274.1	315.0	334.4	345.1
Aircraft in service at year end	150	158	158	164	197	211	224	230	230	241	294	283
Aircraft utilization (average hours per aircraft p.a.)	2465	2653	2720	2801	2891	2886	2787	2663	2708	2928	2926	2984
Revenue aircraft km (m)	237	258	276	282	312	364	375	389	390	431	522	556
Revenue flights (000)	185	197	210	217	234	269	274	271	261	268	358	381
Total traffic revenue per RTK (p)	47.85	50.07	49.27	49.91	48.01	48.73	51.36	50.54	52.55	49.28	52.44	52.76
Total traffic revenue per ATK (p)	30.93	33.65	32.50	32.99	32.83	32.85	35.60	33.99	34.65	32.95	34.89	35.67
Net operating expenditure per ATK (p)	27.09	29.92	30.19	31.01	30.49	30.02	35.52	32.14	31.16	30.94	32.18	32.30
Break-even overall load factor (%)	56.6	59.8	61.3	62.1	63.5	61.6	63.3	04.8	61.2	62.8	61.4	61.2

Source: British Airways Plc Fact Book and Report and Accounts. Years ended 31 March.

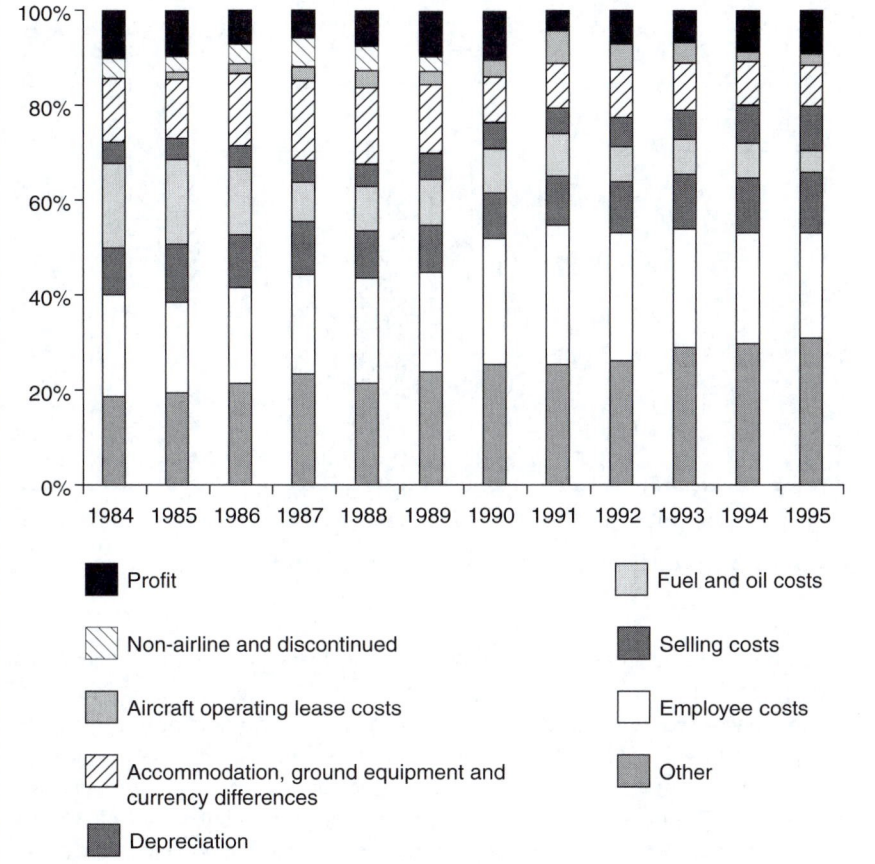

Figure 6.2.2 Breakdown of BA Group turnover.

In June 1990, *The Times* reported a seech made by King at London's Mansion House. He expressed annoyance at the decision to ask the Monopolies and Mergers Commission to investigate the airline's attempt to take a 20 per cent stake in Sabena World Airways. He claimed that he was:

> struggling against not only the efforts of our competitors but the shackles imposed by our own Government ... I have no intention of dying a death by a thousand cuts ... The Secretary of State will finally either permit or block our participation in the Brussels hub. The future will not be determined by the UK Government but our Government may well decide whether we are to be part of that future ...
>
> On June 4 fifty years ago we were called to prepare ourselves to fight on the beaches and on the landing grounds. The tide of commercial conflict has

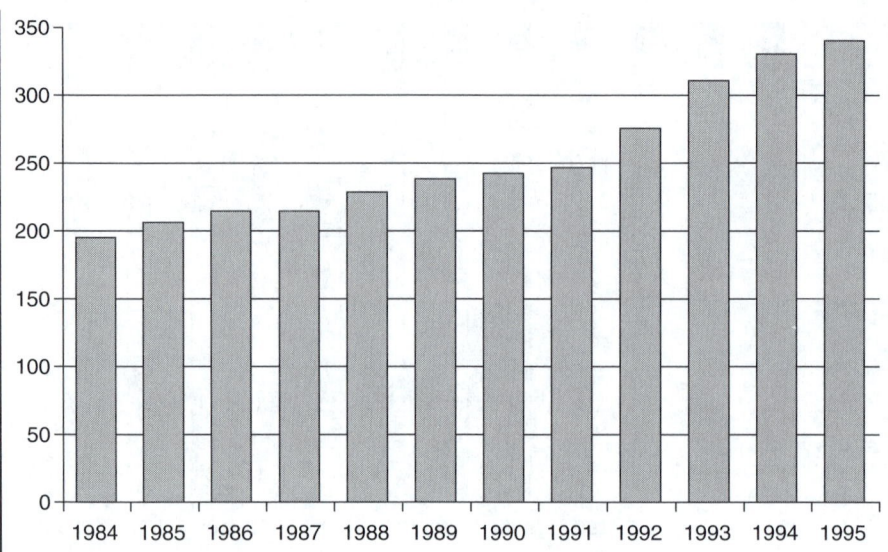

Figure 6.2.3 Productivity – ATKs per employee (000).

rolled over the beaches and the battle for the landing grounds has already been joined. The gateways of today are the airfields deep in the territories which they serve. A commercial battle is raging on the landing grounds of Europe which will determine whether London will remain Europe's principal gateway for long-haul traffic or be displaced by Paris, Frankfurt or Amsterdam. The position of London as a great financial centre and of BA at the hub of a network of international routes are legacies which cannot be reinvented but can surely dissipate by wrong headedness or by negligence. If this country wants to continue to have a major international airline, BA must be allowed to grow and face on reasonably equal terms the emerging air armadas of the 21st century.[15]

When Margaret Thatcher was removed from Government by Conservative MPs in 1990, the BA Chairman lost a friend at the heart of government. That loss, together with the absence of the privatization imperative, meant that BA probably had less influence on the Government. As was mentioned above, in 1991, the two major US carriers United Airlines and American Airlines were granted landing rights at Heathrow. Together with start-up Virgin Atlantic, the arrival of these new carriers has meant that BA's share of business travel began to drop significantly in the 1990s.

At the Annual General Meeting in July 1991, BA unveiled its most traumatic restructuring since flotation. The company also withdrew its annual £40 000 contribution to the Conservative party. King told the meeting that the company had been hard hit by the recession and the Gulf War and stated that BA was also suffering from a series of decisions by the Government that had increased competition on its route network without any compensatory access to the protected US market.

Corporate customers
One of the reasons BA was also losing share of international traffic was a new attitude from the corporate air travel users. Corporate clients began to demand that travel agents assist in reducing travel spend. One method was for the travel agents to pass over a slice of their ticket commission in return for volume deals. As a result, the agents started to lose their role as distributors and in 1991 BA realized that it needed to start working directly with large corporate clients.

The task was to identify corporations 'willing and able' to enter into deals with BA, companies motivated and willing to control and direct travel expenditure throughout their organizations. At the time an AMEX survey showed that about 10 per cent of companies controlled their own travel arrangements.

The primary method employed to increase corporate use was volume discounts. Company Executive Club card agreements with lower thresholds than individual executive club cards were also offered. The other approach was a dialogue with key corporate individuals through direct

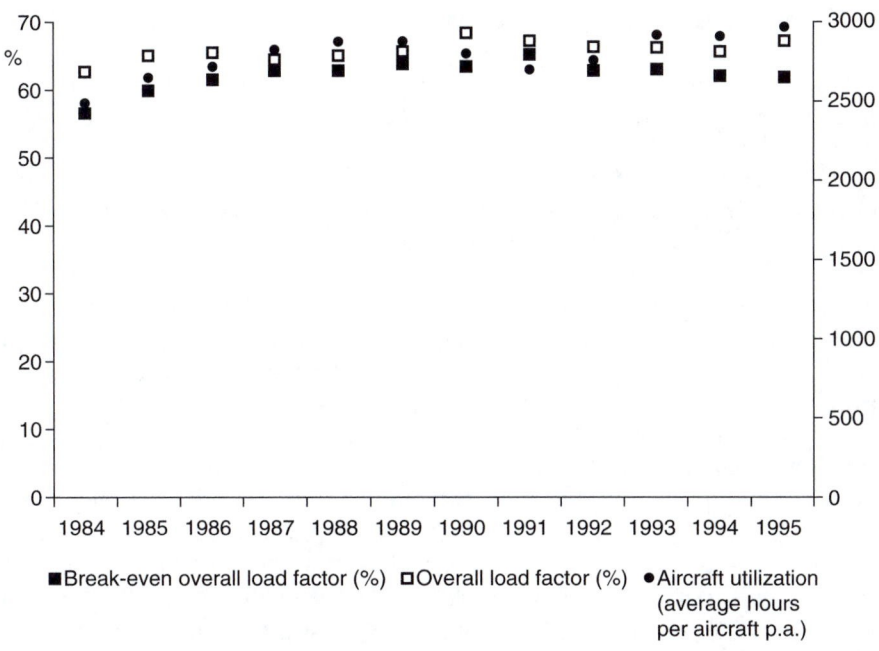

■Break-even overall load factor (%) □Overall load factor (%) ●Aircraft utilization (average hours per aircraft p.a.)

Figure 6.2.4   Efficiency measures.

mail. As travel arrangements are now the responsibility of purchasing managers, it is no longer so easy for airlines to influence through cosy familiarization trips.

The loyalty imperative

With the slower growth in the 1990s, BA focused on building loyalty with its existing customers rather than on winning new customers. Up until 1991, the purpose of the Executive Club was to increase revenue by selling lounge facilities and other features. However, this changed when competition increased at Heathrow with the arrival of American and United Airlines, which introduced customer reward programmes for 'frequent flyers'.

BA's Relationship Marketing Department was set up as a defensive move and BA launched the UK mileage programme – Air Miles Latitudes – which was run in tandem with Executive Club. However, because the existence of both programmes caused confusion, later in the year Executive Club was re-launched with service benefits and mileage rewards combined. By 1996 this was a global 'frequent flyer' programme, with 2 million members.

There are three grades of membership in the Executive Club – Blue, Silver and Gold. 85 per cent of BA's customers are in Blue and progression to Silver and Gold is rewarded by the amount spent with BA. If customers use BA less frequently, they are demoted to a lower grade.

The branding and benefits of Executive Club are centrally controlled, but the mileage programme is regionally driven, with four frequent flyer programmes around the world. This ensured that BA offers the same service in comparison to other airlines and is a tactical sales tool by creating a database for direct mail offers.

BA's prime target market is business travel, which provides 60 per cent of BA's revenue. 85 per cent of this comes through travel agents, while the other 15 per cent is from direct telephone sales and tickets bought at airport check-in.

As was previously mentioned, large corporations have been looking to make savings in their travel budgets which, for many, amount to several millions of pounds a year. Improved information systems and centralized purchasing has given them market power. As a result, there has been a polarization and consolidation in the travel agent market with business travel management companies and smaller leisure-oriented travel agents. For BA, this means that just three travel agents provide the majority of its business revenue: Hogg Robinson, Carlson Wagon Lits and Amex.

These changes have led to BA reconsidering the relationship it has with travel agents. For many, there had always been a suspicion that money was not well spent on travel agents and Richard Lucenti, BA's Head of Field Sales UK and Ireland, reflected this view:

There was always a question mark over how much influence a travel agent has. The suspicion was that they took bookings but rarely recommended an airline, that their influence on ticket sales was marginal.[16]

As travel agents are now being tasked by large customers to minimize their travel bill, while agents' income is commission based, there is a clear conflict of interest. The future may lie with fee-based travel management – the customer gets all the commission from the airline and pays for the service it receives from the travel agent – one major UK telecommunications company already has such an arrangement.

Loyalty to Boeing and Rolls Royce?

Each year BA spends a huge sum on equipment, in particular in purchasing and leasing aircraft. With the demise of the British aircraft manufacturing industry, BA has sourced most of its fleet from the American company Boeing's range of 737, 747, 757, 767 and 777 aeroplanes. With Air Nippon Airways (ANA), BA has one of the biggest orders of 15 new widebody 777s which cost, depending on specification, between US$104 million and US$129 million. Combined with 24 new 747-400s, this order totalled over £4 billion in 1991 and led to the remark, 'King is a Boeing man – he likes its people, its style and products'. BA even assigned four planning directors to work with Boeing on the design of the 777 to ensure that it was 'customer friendly'.

European aeroplane makers, Airbus, which includes British Aerospace, complained in 1991 to the European Commission that BA never chose its A300, A310, A319, A320, A321, A330 or A340 aeroplanes. BA has 10 A320s which it inherited from British Caledonian but has argued that they are around 10 per cent less fuel-efficient than Airbus claims.

The engines for new planes are specified separately and BA chose American General Electric (GE) GE90 engines for the 777 deal, upsetting BA's traditional supplier, Rolls Royce. In the same deal, BA remained loyal to British GEC Avionics flight control systems (£500 million) and Smiths Industries electronic management systems (£380 million).

The decision not to specify Rolls Royce engines was very controversial, with some commentators saying BA wanted to win US customers and government support by choosing GE. Others argued that it was GE's revenge because it missed a £600 million deal with BA in 1986 for political reasons associated with the privatization of BA. The deal also involved GE buying the BA Engine Maintenance facility in South Wales for US$454.8 million and a $200 million discount on the GE 90s. GE agreed to continue to overhaul BA engines at South Wales and at a reportedly lower price than BA was charging itself.

Expanding the global alliance

In 1992 BA took a 49 per cent stake in leading French independent carrier, TAT. The firm previously had a very close relationship with Air France, which had had a 35 per cent holding in TAT until mid-1992. Also, following termination of BA services to Ireland a marketing initiative with Aer Lingus was launched, effectively linking the Irish carrier's services with those of BA at London.

BA was not the only player to be forming alliances. In November 1993 United Airlines and Lufthansa formed a marketing alliance, while KLM, SAS, Swissair and Austrian Airlines launched an abortive attempt to form a pan-European group called Alcazar.

Deregulation and the opening up of international airways continued in 1992 with the UK Government signing a bilateral agreement with Japan. The terms didn't meet with approval at BA. Later that year David Holms, previously at the Department of Trade and Industry, was recruited by BA to be the first Director of Government Affairs, a function previously handled by the Legal Director. The department comprises 24 people, a figure which has not changed over the previous 10 years, and is divided into International Affairs and Competition and Industry Affairs, which covers the CAA, Europe and technical aspects. It communicates with the UK Government to ensure understanding of BA's position when it negotiates international deals. The 1990s has seen the decline in influence of the CAA due to liberalization, a more settled industry and the growing importance of the EU.

Developing the relationship marketing programme

The objectives of BA's relationship marketing programme developed over time and were differentiated from those of other airlines. The scheme became more tightly focused on frequent flying, high yield business adults and meant that BA does not give miles or points on discount fares. In contrast, many other airlines have a mass-market approach to their frequent flyer programmes with, for example, USAir having 15 million members in its scheme.

BA updated its offer in the 1990s and some of the means of increasing loyalty include:

- rewards in the form of free air miles – this originally gave BA a huge advantage but by 1993 had become a hygiene factor as all airlines offered such a scheme
- benefits – 'the more you fly the easier it gets'. BA's members are speeded through airports and by the time customers reach Gold Card level, BA has so much information that it can anticipate their requirements.

BA collects data regarding the flights taken by its club members and this allows BA to communicate offers and information on these locations. BA also knows the history of any contacts they've had with the Executive Club, e.g. if a customer has complained previously, BA's response will acknowledge this and tailor the action taken. However, BA is less good at using this knowledge during journeys as the cabin crew don't have access to the database.

To ensure that the club is meeting the needs of its members BA regularly undertakes market research. This involves speaking directly to several hundred members in each region. The technique used is to ask about a specific flight to determine factors influencing carrier choice. This data is then programmed into an operational research model to determine how much revenue is being driven by each criterion. The main factors are schedule, master brand preference, membership of Executive Club, corporate policy and price. Schedule is pre-eminent on short haul, but on long haul, if there is a daily flight, other factors are more important.

This modelling allows BA's Relationship Marketing Department to determine the additional revenue that the Club's activities are likely to create and by deducting the cost of the programme, the additional profit generated for BA is calculated. By 1996, the Relationship Marketing Department was the largest of the BA marketing departments, with a six-fold increase in staffing since 1991. Details of the marketing department's organization structure for 1995 are given in Figure 6.2.5.

Customer service
BA argues that its success is based on cutting costs and changing the company's culture to provide first class customer service. At the heart of the customer service ethic is the concept of service recovery. It is not just enough to aim to get it right first time, now staff needed to be able to

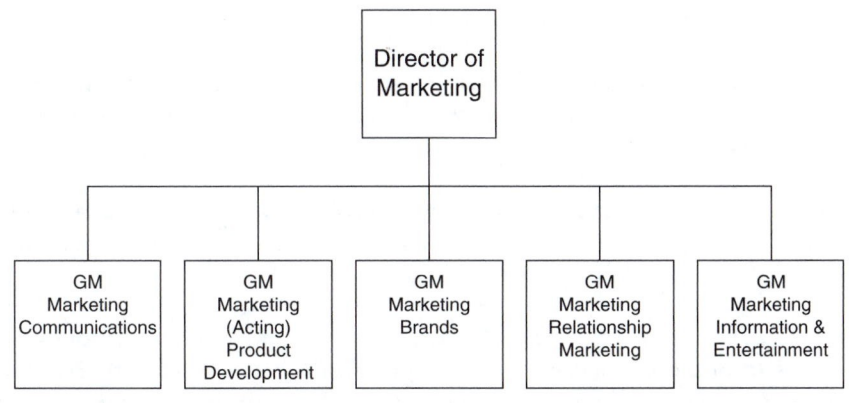

Figure 6.2.5 Marketing department organization structure, 1995.

recover customers when failures occurred. As Group Managing Director, Robert Ayling, said:

> We don't talk about customer complaints, we talk about service recovery. Every single person in my company has been talked to about recovering when things go wrong.

Charles Weiser, BA's Head of Customer Relations, added:

> In a service recovery company like British Airways, our philosophy is to retain the customer, invest in ensuring that we have the proper systems and infrastructure for our employees to do so.

BA invested £4.5 million in a computer imaging system, called the customer analysis and retention system, or CARESS. On average, any one of the 6500 complaint letters received every month used to get a response in 4-5 weeks. CARESS enabled letters to be scanned onto the computer immediately, with replies sent out within three days.

Technology has also helped in providing 'real time' solutions to customers' problems with a freephone careline which enables operators to access 30 BA databases. Charles Weiser says:

> When the customer rings up the entire case history shows up on the screen from telephone contact, to the letters they have sent, to any attachments to the ticket coupons ... it's all in one place.

Staff have also been trained in techniques to deal with unhappy or angry customers, so that relationships can be maintained through problem solving. This recognizes research carried out by BA that showed on average that an unhappy customer told 10 friends or associates of his or her experience. Successful handling of a complaint is the target and BA's Quality and Training Co-ordinator says:

> Fortunately, we are finding out that the actions that we are taking here [in service recovery] are increasing the rate of people who are willing to fly again.

BA believed it was retaining customers in this way and set up a series of 'listening posts' to monitor perceptions and problems. A BA official summarized their attitude to customer relations:

> If you want to build a relationship with your customers, that relationship comes from the individuals. People do not build a relationship with British Airways because BA is a faceless thing. People build a relationship with the

people who work for the company and that is the people they meet on check-in, the people they meet on the aircraft.

Challenges in the 1990s

The poor financial position of USAir was a major concern for BA and in 1995 a large proportion of the investment was written off. Despite the important marketing benefits of the link with USAir its high unit costs resulted in sustained losses – US$459 million in 1994, despite cost savings of US$500 million. BA claimed that it was making £70 million savings each year from the joint marketing of routes and there was a 87 per cent increase in traffic on US codeshare cities in 1993/94. However, during 1994 the US Department of Transport delayed code sharing in an attempt to apply pressure for more landing slots for US airlines at Heathrow and in 1995 BA decided not to invest further funds in USAir.

Richard Branson and Virgin
For at least 10 years BA has been receiving bad publicity over its treatment of small competitors in the UK. The biggest scandal related to Virgin Atlantic, part of Richard Branson's Virgin Group. Richard Branson is a flamboyant, very high profile entrepreneur and a master at self-publicity. Virgin Atlantic is a tiny airline compared to BA, employing 2400 people, with eight aircraft and 1 million passengers a year. However, Virgin successfully brought a libel case against BA and in 1993 BA paid legal costs of £3 million and damages of £610 000. The allegations centred on three activities:

- 'Switch selling campaign' – BA staff were said to have approached Virgin Atlantic's passengers at US airports and tried to persuade them to switch to BA. Also, BA staff 'cold called' Virgin customers and told them that their Virgin flight had been cancelled and offered alternative BA flights. There were also allegations of hacking into Virgin computers.
- 'Operation Barbara' – This was a report by Brian Basham, a PR consultant, at a cost of £40 000, summarizing financial, personal and operational aspects of Richard Branson and his businesses. Parts of it were leaked to the press in order to damage Branson. The report was allegedly sanctioned by Marshall.
- 'Operation Covent Garden' – BA directors were convinced that Virgin was organizing industrial espionage against BA. Ian Johnson Associates were hired by Marshall to investigate security breaches. The court case proved Virgin was not infiltrating BA.

The Virgin libel case in Britain was not an end in itself, but gave an indication of what might happen if the case came to the US courts, where compensation sums are much larger. Consequently, the end of the case marked the start of negotiations between Virgin and BA. Virgin wanted compensation of £9 million and used the situation to apply pressure over landing slot allocations at Heathrow. Although press and media coverage was overwhelmingly unfavourable to BA, ticket sales appeared to be completely unaffected. Other factors were found to be more important in customer loyalty than BA's alleged involvement in 'dirty tricks'.

The Virgin case was followed, in March 1994, by the founder of Air Europe, Harry Goodman, issuing a writ against BA alleging that another 'dirty tricks' campaign had caused the company's failure in March 1991.

Unfair competition in Europe

BA also felt it was a victim and embarked upon a major campaign in tandem with the CAA in 1994 about unfair European competition. The aim was to prevent the French Government 'bailing out' Air France. This provided an example of BA working with the CAA on European and US aviation issues when there is agreement.

Currently BA and Luxair are the only 100 per cent private airlines in Europe and the presence of so many state-owned carriers may be a major challenge for BA in such a deregulated world. But BA has advantages in terms of managing yields from premium business travellers. Marshall believes it to be a serious issue and in 1995 he stated:

> European state aid and subsidy remains a serious threat to open and fair competition. But, it has to be said, we are encouraged by our own government's decision to back our litigation against it in the European courts.[17]

Purchasing strategy

During 1994, Clive Mason, the Director of Purchasing and Supply for BA, spent more than £6 million a day. Mason advocated a broader strategic approach to purchasing which embraced the entire company. He believed in strategic purchasing and extended supplier relationships, which led BA to reduce the numbers of its vendors from 10 000 to 3500. The emphasis was shifted from a price-based to a cost-based selection and now 'Within ten minutes of Heathrow we have nearly 50 000 different components available to us,' says Mason. The average contract length was increased from one year to two or three years. Mason related his philosophy:

> A lot of people talk about the language of partnership. We believe in it. With the right supplier, we can truly talk about partnership. But partnership means very tough. It means that both sides are driving to make the relationship better. It's not soft and cuddly and cosy. It's absolutely saying I'm going to do better, year on year on year.[18]

A global airline

Marshall made clear his vision of a deregulated airline industry in the USA and Europe. He argued that in time global trade talks, including air travel and foreign investment in US carriers, will be allowed freely. Speaking at the Institute of Directors Annual Convention in 1995, he said:

> Our own strategy, against the background of the emerging global market, has been to work towards the objective of becoming the first truly global airline. To achieve it, we took the free market promise – or threat – at its word and we planned to secure presence and extend reach in key market areas through a programme of investment in other, compatible and strategically located airlines.[19]

In addition to USAir and TAT, discussed above, BA has also made a large number of other acquisitions, including Dan-Air; 50 per cent of Deutsche BA in Germany; 25 per cent of Qantas in Australia; and 40 per cent of Air Russia and 12.8 per cent of Air Mauritius. According to merchant bank Warburgs, BA invested more than £600 million during 1994/95 to achieve its global aspirations. The investments recorded a net operational loss of £47 million in 1994, but they are seen as part of a long-term and strategic policy aimed at contributing to the building of international and domestic route structures.

There are also a number of joint marketing initiatives. In July 1993 CityFlyer Express, a small independent from Gatwick, bought the right to use the BA brand. Investment in Brymon Aviation in July 1993 led to BA taking over the debts and operations of Birmingham European Airways. In October 1994, BA formed a marketing agreement with Loganair, Scotland's airline. Loganair aeroplanes were painted in BA livery as part of a franchising agreement on 46 domestic and international routes.

Union influence
As a result, by 1994 BA had a number of subsidiaries. In December recruitment advertisements were run for cabin crew to fly a newly launched Glasgow transatlantic flight run by British Airways Regional. The advertisements offered salaries which were about half the average for shuttle crews in other parts of BA. Union officials were concerned that the move would undermine staff long-term conditions. The issue came to a head when BA ground staff and cabin crew staged a 24-hour strike over working conditions at European Operations Gatwick (EOG), a subsidiary covering Dan-Air. This was resolved when BA management undertook not to set up further EOG type subsidiaries and not to expand EOG arrangements outside the short-haul Gatwick network.

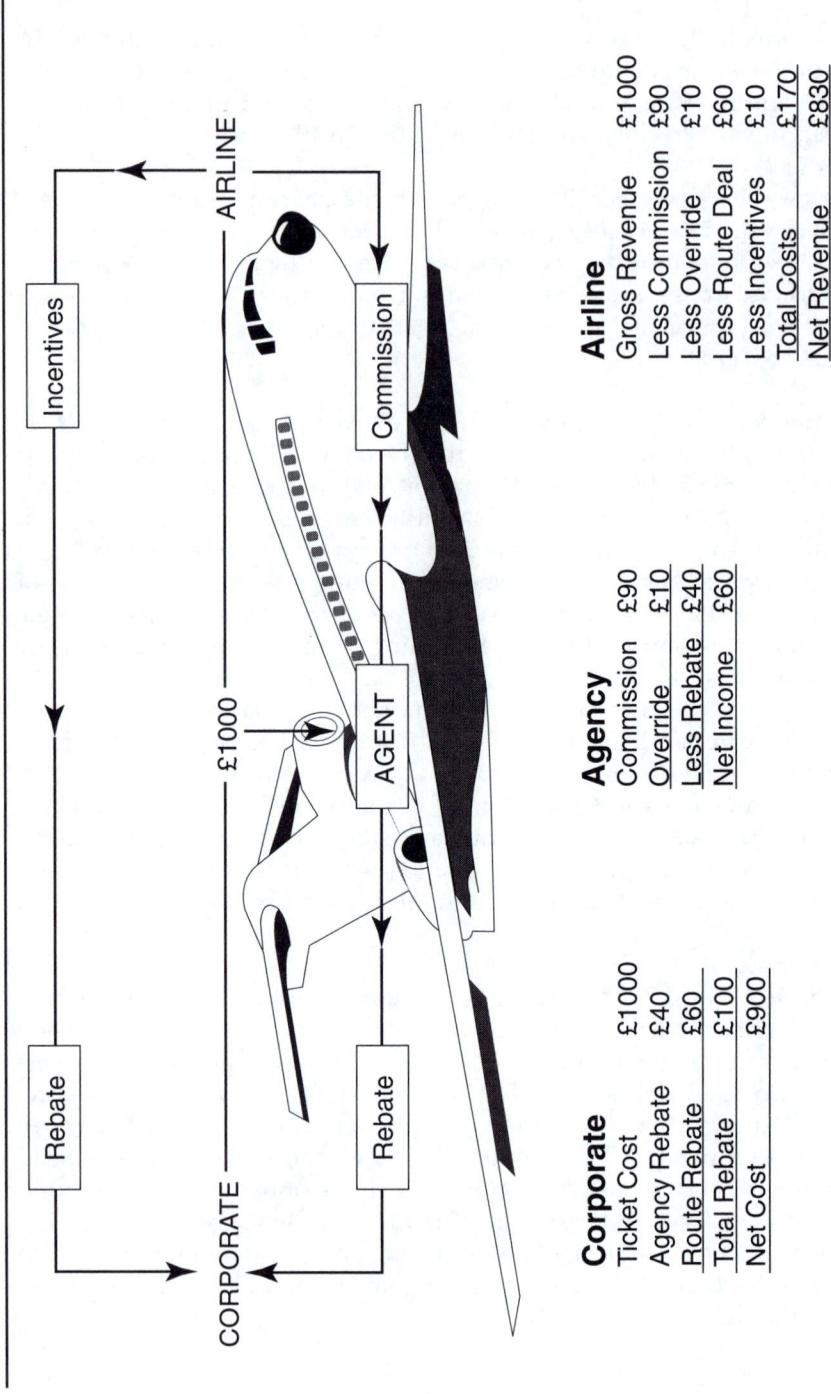

Figure 6.2.6 BA, agent and corporation money flow.

Supplier strategies

Throughout the 1980s and 1990s, the World's Favourite Airline theme has been one of the most enduring for any company. At the same time, the relationship with Saatchi & Saatchi was maintained. With the departure of the founders, the Saatchi brothers, from the agency in December 1994, BA decided to review its agency arrangements. While it was reported that BA's marketing department wanted to stay with the old Saatchi agency, now renamed Cordiant, Marshall and Marketing Director Ford Ennals insisted on loyalty to the Saatchi brothers and moved the account to their new agency, Saatchi M & C.

In June 1995, *The Observer* reported that Airbus had launched an intensive campaign to persuade BA to buy A330s which, they argued would save millions for BA on its European routes. BA indicated that they may change from King's Boeing policy and buy all future aircraft from Airbus. One reason for this was because of the move of Charles Masefield from British Aerospace to Airbus and because Boeing was not committing resources to the development of extra large super Jumbo. By comparison, Airbus seemed more willing to build such a plane with its A3XX.

By 1995 the move to centralized buying of air travel by large customers, discussed above, was gathering pace and BA reorganized its sales teams, regrouping the top corporations and top agents, and according to National Sales Manager, Bernard Harrop:

> We aim to create better dialogue between corporates, agents and BA ... so we can redesign the distribution process to find efficiencies.[20]

There are considerable movements of funds when an air ticket is sold. The traveller pays an agent who passes the ticket price to BA and then receives 9 per cent commission from BA and later any additional overrides. The agent passes up to 3 per cent of the commission on to the corporate client. Finally, BA gives the corporate client a separate rebate. A full explanation of agent and corporation money flow is given in Figure 6.2.6. This complicated flow of payments and information is inefficient and adds to costs and, as a result, the cost of sales to BA has been rising steadily.

In 1995, BA segments corporate customers according to how far they have matured along the travel policy life cycle. (See Figure 6.2.7.)

Each of the national agents has its own sales team assigned to a particular category, with the top 10 independent multiple location agents having a sales team. Local agents are segmented regionally. This organization means that the level of support from BA is business volume dependent. Figure 6.2.8 illustrates the structure of the sales departments.

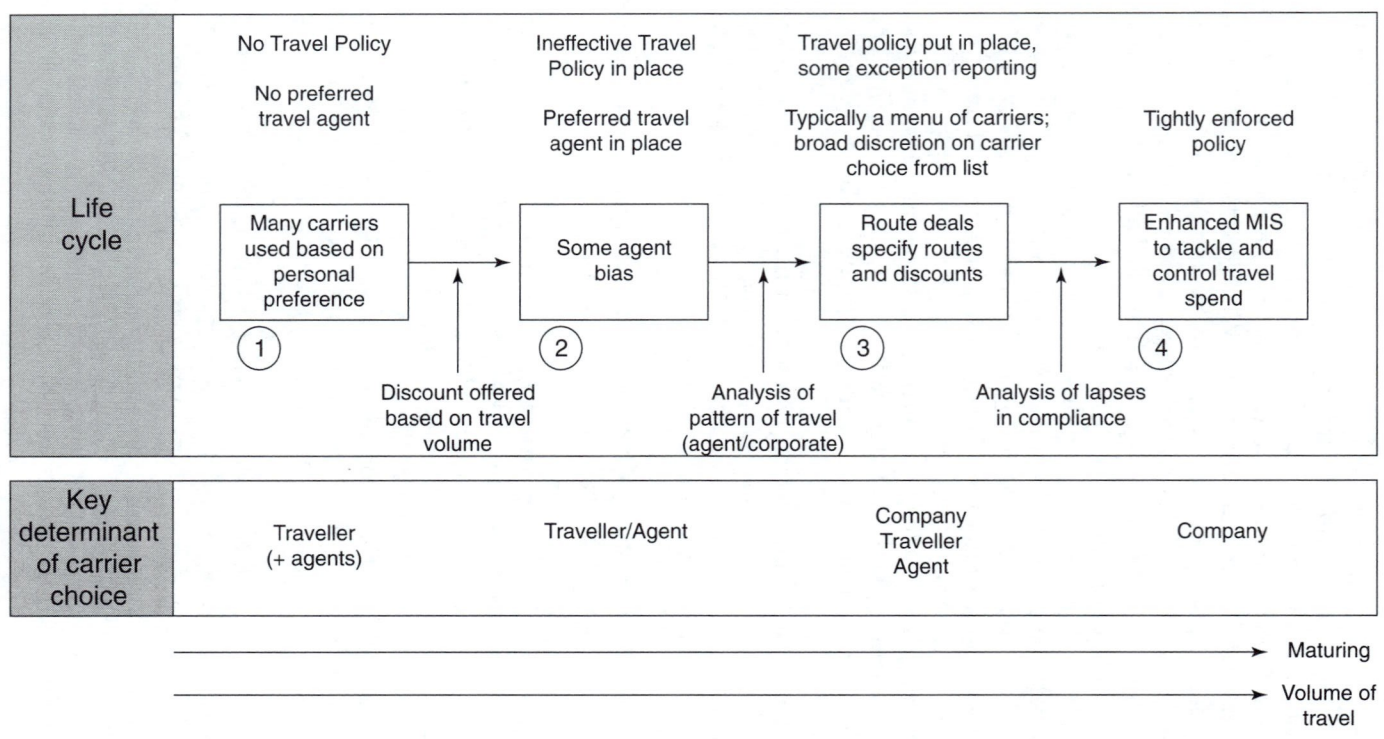

Figure 6.2.7 Key decision maker changes at different points in corporate travel policy life cycle.

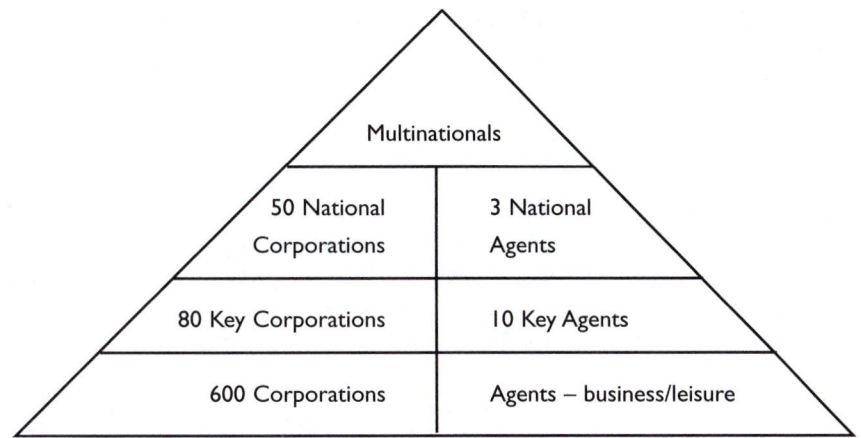

Figure 6.2.8 BA UK sales departmental structure.
Source: Datastream.

The future

BA has come a long way since the 1970s and it is now the world's largest international passenger airline. The company has maintained steady profits while many of its competitors have recorded losses year on year and has emerged as one of the true global players in a rapidly deregulating industry with ever increasing competition. Table 6.2.4 shows a breakdown of BA's turnover, profitability and other statistics from 1984 to 1995. To compare how BA has done against some of its competitors, see Table 6.2.5 and against the industry as a whole, see Tables 6.2.6–6.2.10.

There are still many challenges for BA. Its reputation is not in the clear, with Virgin Atlantic's 'dirty tricks' court action still possible, and there is internal debate over the appropriateness of the strapline the 'World's Favourite Airline'. Should it be changed to reflect the global alliances that now exist for BA? Major corporate buyers are becoming better organized about travel purchase and demanding reductions in their expenditure. BA's alliance partners provide network connections but don't match BA's customer service and return financial losses; and its competition is responding ever more rapidly to BA initiatives.

Getting to the top is one thing, staying there is something else. Visitors to a Heathrow engineering site saw two BA staff fast asleep, feet on their workbenches, at 10 am, in what looked like a scene from a nationalized industry in Britain in the 1960s. In fact, the visit was made in February 1996. Complacency is not an option for leading companies and BA recognizes that it must continually renew itself and sustain the pace of change in order to keep its position in this global industry. Over the next five years, what can BA do to maintain its position as one of the three most profitable airlines in the world?

References

1 Colin Marshall, quoted in a BA internal brief.
2 Bill Muirhead, Saatchi and Saatchi, The Global Image Makers.
3 Hampden-Turner, C. (1994). *Corporate Culture – from vicious to virtuous circles*, The Economist Books, Hutchinson, 86.
4 Colin Marshall, 1985, Institute of Management.
5 *Ibid*.
6 Bruce, M. *Learning in British Airways*.
7 Bruce, M. (1988). Putting customers first – the British Airways experience. *Insight and Outlook*, 22 April.
8 Christopher Allen during interview with the authors, September 1995.
9 *Daily Telegraph*, 8 June 1986.
10 *Marketing Week*, October 1987.
11 Batt, M. (1990). *Marketing Business*, April.
12 Batt, M. (1990). Putting a brand on British Airways. *Marketing Business*, April.
13 Batt, M. (1988). *op. cit.*
14 Batt, M. (1990). Putting a brand on British Airways. *Marketing Business*, April.
15 *The Times*, June 1990.
16 Richard Lucenti, Head of Field Sales UK, during an interview with the authors, September 1995.
17 Colin Marshall, Director, Annual Convention, 1995.
18 *Financial Times*, 10 January 1994.
19 Colin Marshall, Director, Annual Convention, 1995.
20 Bernard Harrop, BA National Sales Manager, during an interview with the authors, September 1995.

Table 6.2.5 Comparative financial performance of selected airlines

	1987	1988	1989	1990	1991	1992	1993	1994	1995
Total trading income									
British Airways (UK)	100	115	130	148	151	160	171	193	220
Singapore Airlines (Singapore)	100	115	131	146	142	156	162	179	
Delta Air Lines (USA)	100	130	152	161	172	204	226	232	229
Lufthansa (Germany)	100	107	118	131	146	156	160		
Air France (France)	100	97	107	131	135	903	874		
Southwestern Airlines (USA)	100	111	130	152	169	217	295	333	
Pre-tax profit									
British Airways	100	141	165	213	80	176	114	186	202
Singapore Airlines	100	135	222	292	236	225	192	177	
Delta Air Lines	100	102	159	105	-112	-176	-146	-148	110
Lufthansa	100	123	142	40	-251	-267	-46		
Air France	100	348	58	-10	81	-1960	-3819		
Southwestern Airlines	100	294	382	257	151	505	893	1030	
Net EPS									
British Airways	100	102	118	166	64	172	117	152	133
Singapore Airlines	100	134	218	259	197	199	182	172	
Delta Air Lines	100	107	159	98	-131	-180	-179	-175	69
Lufthansa	100	148	94	0	0	0	0	0	
Air France	100	619	271	0	109	0	0	0	
Southwestern Airlines	100	293	378	262	150	463	751	872	
Dividends per share									
British Airways	100	168	188	215	215	247	258	281	314
Singapore Airlines	100	120	183	287	330	374	387	392	
Delta Air Lines	100	120	120	170	120	120	70	20	20
Lufthansa	100	114	121	0	0	0	0	0	
Air France	100	110	160	0	30	0	0	0	
Southwestern Airlines	100	102	108	112	115	122	135	138	

Source: Datastream. Index 1987 = 100.

Table 6.2.6   Financial statistics of world international scheduled operations, 1977–92

Year	Operating revenue (US$bn)	% vly	Operating result before interest (US$bn)	Operating result after interest (US$bn)
1977	19.3	16	0.9	0.3
1978	23.5	22	1.0	0.4
1979	28.8	23	0.2	-0.4
1980	35.6	24	-0.9	-1.9
1981	36.1	1	-0.6	-1.9
1982	36.7	2	-0.2	-1.8
1983	37.9	3	1.3	-0.3
1984	39.5	4	2.2	0.8
1985	40.7	3	1.6	0.2
1986	45.2	11	1.2	-0.3
1987	53.8	19	2.5	0.9
1988	60.9	13	3.4	1.6
1989	70.7	16	26	0.3
1990	90.0	13	0.5	-2.7
1991	91.7	1	-0.6	-4.0
1992	103.5	12	-1.0	-4.8

Source: IATA.

Table 6.2.7   Forecasts of growth in major air travel markets, 1994–2013 (%)

	Average annual % change
Asia to Europe	7.2
Intra Asia Pacific	6.8
Trans Pacific	6.8
Intra Latin America	5.6
North America to Latin America	5.5
Europe to Latin America	4.8
North Atlantic	4.4
Intra Europe	4.4
Europe to Africa	4.3
North America	4.0

Source: Boeing Commercial Airplane Group.

Table 6.2.8 BA's share of world scheduled air traffic

Year	World international RPKs (bn)	BA's RPKs (m)	Share of international RPKs (%)	Total world RPKs (bn)	Share of total RPKs (%)
1984	555	34 206	6.16	1277	2.68
1985	590	38 386	6.51	1367	2.81
1986	603	41 334	6.85	1452	2.85
1987	688	41 356	6.01	1589	2.60
1988	761	49 123	6.46	1705	2.88
1989	824	57 795	7.01	1780	3.25
1990	893	61 915	6.93	1894	3.27
1991	860	64 734	7.53	1843	3.51
1992	986	65 896	6.68	1967	3.35

Table 6.2.9 Airline performance ratings –1985 IAPA Survey*

The ten airlines most frequently flown by members (%)

1	British Airways	73.4
2	Lufthansa	53.8
3	Air France	51.8
4	Swissair	51.6
5	KLM-Royal Dutch Airlines	47.7
6	Pan Am	45.1
7	Eastern	37.6
8	Alitalia	32.9
9	TWA	32.2
10	SAS-Scandinavian Airlines	32.0

The six airlines most preferred by IAPA members[†] (%)

International members:			UK members:		
1	Swissair	40.1	1	British Airways	48.2
2	Lufthansa	27.5	2	Swissair	29.0
3	Singapore Airlines	26.4	3	British Caledonian Airways	26.0
4	British Airways	21.8	4	Singapore Airlines	20.7
5	KLM-Royal Dutch Airlines	20.4	5	KLM-Royal Dutch Airlines	18.8
6	Cathay Pacific Airways	16.2	6	Lufthansa	17.7

Table 6.2.9 (continued)

Airlines judged to provide the best in-flight service[†] (%)

International members:		UK members:	
1 Swissair	37.9	1 British Airways	38.4
2 Lufthansa	25.0	2 Swissair	28.3
3 Singapore Airlines	25.0	3 British Caledonian	26.4
4 British Airways	18.9	4 Singapore Airlines	20.2
5 KLM-Royal Dutch Airlines	17.2	5 Cathay Pacific Airways	17.3
6 Cathay Pacific	16.1		
7 SAS-Scandinavian Airlines	15.4		
8 Air France	10.6		
9 Thai Airways International	10.6		
10 British Caledonian	10.5		

Airlines judged to provide the best ground service[†] (%)

International members:		UK members:	
1 Swissair	32.7	1 British Airways	37.7
2 Lufthansa	25.3	2 Swissair	21.0
3 KLM-Royal Dutch Airlines	19.5	3 British Caledonian Airways	20.3
4 British Airways	18.2	4 KLM-Royal Dutch Airlines	19.3
5 Singapore Airlines	17.0	5 Lufthansa	16.2
6 SAS-Scandinavian Airlines	13.4		
7 Cathay Pacific Airways	9.4		
8 British Caledonian Airways	8.0		
9 TWA	7.9		
10 Pan Am	7.2		

Table 6.2.9 *(continued)*

Factors judged to be most important when choosing an airline[**]

	Under 2 hours (%)	(Rank)	2-5 Hours (%)	(Rank)	Over 5 hours (%)	(Rank)
Convenient schedule	67.8	1	59.9	1	44.2	1
Frequency of flights	33.8	2	13.3	11	6.5	11
On-time performance	30.7	3	22.0	4	16.1	10
Past experience	27.8	4	28.4	3	29.5	3
Low fares	18.0	5	19.1	6	22.0	6
Safety record	17.1	6	18.7	7	22.3	5
Attitude of personnel	16.8	7	21.4	5	23.6	4
Quality of in-flight service	15.7	8	29.5	2	38.5	2
Aircraft type	10.4	9	14.2	9	21.1	7
Availability of Business Class	7.8	10	14.1	10	16.2	9
Quality of ground service	6.8	11	5.8	12	4.8	13
National flag carrier	6.6	12	4.9	13	5.3	12
Quality of food and drinks	6.5	13	15.1	8	20.5	8
Frequent Flyers Programme	3.2	14	3.8	14	4.3	14
Others	1.0	15	0.9	15	1.1	15

[*]The International Airline Passengers Association (IAPA) is an independent body representing frequent travellers. The data are from its 1985 survey and based on 19 000 replies.
[†]Members could nominate up to four airlines.
[**]Members could nominate up to three factors.

Table 6.2.10 Airline performance ratings – 1995 business traveller survey

Overall:		Business class travel:	
1	British Airways (1)	1	Virgin Atlantic
2	Singapore Airlines	2	British Airways
3	Swissair (3)	=3	Singapore Airlines
4	Virgin Atlantic (4)	=3	Qantas
5	Cathay Pacific (%)	5	Swissair
6	Qantas (6)	6	Emirates
7	KLM (7)	7	KLM
8	Thai International (9)	8	Air Canada
9	Emirates (–)	9	Thai International
10	American Airlines (8)	10	Cathay Pacific

Short-haul flights:		Long-haul flights:	
1	British Airways (1)	1	Singapore Airlines (1)
2	Swissair (2)	2	British Airways (3)
3	British Midland (3)	3	Virgin Atlantic (2)
4	Air UK (6)	4	Qantas (4)
5	Crossair (–)	5	Cathay Pacific (5)
6	KLM (5)	6	Swissair (7)
7	SAS (4)	7	Thai International (5)
8	Lufthansa (5)	8	KLM (9)
9	Singapore Airlines (7)	9	Malaysia Airlines
10	Thai International (–)	10	Lufthansa (–)

Note: 1994 position in brackets ( ).

Appendix Abbreviations, definitions and specialist terms

ABBREVIATIONS AND DEFINITIONS

AEA:	Association of European Airlines
BAA:	BAA plc and its subsidiary undertakings, the successors to the former British Airports Authority
BOAC:	The former British Overseas Airways Corporation
BEA:	The former British European Airways Corporation
CAA:	Civil Aviation Authority
IATA:	International Air Transport Association
ICAO:	International Civil Aviation Organization
vly	Variation compared with last year

SPECIALIST TERMS

Capacity measurements

Available seat kilometres (ASKs):	The number of seats made available for sale multiplied by the distance flown.
Available tonne kilometres (ATKs):	The number of tonnes of capacity available for the carriage of revenue load (passengers and cargo) multiplied by the distance flown.

Volume measurements

Revenue passenger kilometres (RPKs):	The number of revenue passengers carried multiplied by the distance flown.
Freight tonne kilometres (FTKs):	The number of revenue tonnes of freight carried multiplied by the distance flown.
Cargo tonne kilometres (CTKs):	The number of revenue tonnes of cargo (freight plus mail) carried multiplied by the distance flown.
Revenue tonne kilometres (RTKs):	The revenue load (passengers and cargo) in tonnes multiplied by the distance flown.

Load factors

Passenger load factor:	RPKs expressed as a percentage of ASKs.
Overall load factor:	RTKs expressed as a percentage of ATKs.
Break-even load factor:	The load factor required to equate total traffic revenue with operating costs.

Yield measurements

Revenue per RPK:	Passenger revenue from scheduled operations divided by scheduled RPKs.
Revenue per ASK:	Passenger revenue from scheduled operations divided by scheduled ASKs.
Revenue per RTK:	Revenue from total traffic divided by RTKs.
Revenue per ATK:	Revenue from total traffic divided by ATKs.

Financial definitions

Operating ratio:	Operating revenue divided by operating expenditure.
Scheduled traffic revenue:	Revenue generated from passengers and their excess baggage, and freight and mail on scheduled services.

Index